ENGINEERING THE PANAMA CANAL

A Centennial Retrospective

PROCEEDINGS OF SESSIONS HONORING THE 100TH ANNIVERSARY OF THE PANAMA CANAL AT THE ASCE GLOBAL ENGINEERING CONFERENCE 2014

October 7–11, 2014
Panama City, Panama

SPONSORED BY
The History and Heritage Committee of the American Society of Civil Engineers

EDITED BY
Bernard G. Dennis, Jr.

Published by the American Society of Civil Engineers

Library of Congress Cataloging-in-Publication Data

ASCE Global Engineering Conference (2014 : Panama, Panama)
  Engineering the Panama Canal : a centennial retrospective : proceedings of sessions honoring the 100th anniversary of the Panama Canal at the ASCE Global Engineering Conference 2014, October 7-11, 2014, Panama City, Panama ; sponsored by the History and Heritage Committee of the American Society of Civil Engineers ; edited by Bernard G. Dennis Jr.
     pages cm
  ISBN 978-0-7844-1373-9 (print : alk. paper) 1. Panama Canal (Panama)—Congresses. 2. Channels (Hydraulic engineering)—Congresses. I. Dennis, Bernard G., Jr., editor. II. American Society of Civil Engineers. History and Heritage Committee, sponsoring body. III. Title.
  TC605.A83 2014
  627'.1370972875—dc23
                                                                                                2014025986

Published by American Society of Civil Engineers
1801 Alexander Bell Drive
Reston, Virginia, 20191-4382
www.asce.org/bookstore | ascelibrary.org

Any statements expressed in these materials are those of the individual authors and do not necessarily represent the views of ASCE, which takes no responsibility for any statement made herein. No reference made in this publication to any specific method, product, process, or service constitutes or implies an endorsement, recommendation, or warranty thereof by ASCE. The materials are for general information only and do not represent a standard of ASCE, nor are they intended as a reference in purchase specifications, contracts, regulations, statutes, or any other legal document. ASCE makes no representation or warranty of any kind, whether express or implied, concerning the accuracy, completeness, suitability, or utility of any information, apparatus, product, or process discussed in this publication, and assumes no liability therefor. The information contained in these materials should not be used without first securing competent advice with respect to its suitability for any general or specific application. Anyone utilizing such information assumes all liability arising from such use, including but not limited to infringement of any patent or patents.

ASCE and American Society of Civil Engineers—Registered in U.S. Patent and Trademark Office.

Photocopies and permissions. Permission to photocopy or reproduce material from ASCE publications can be requested by sending an e-mail to permissions@asce.org or by locating a title in ASCE's Civil Engineering Database (http://cedb.asce.org) or ASCE Library (http://ascelibrary.org) and using the "Permissions" link.

Errata: Errata, if any, can be found at http://dx.doi.org/10.1061/ 9780784413739.

Copyright © 2014 by the American Society of Civil Engineers.
All Rights Reserved.
ISBN 978-0-7844-1373-9 (paper)
Manufactured in the United States of America.

Cover credit: The drawing on the front cover is of a 1912 lithograph entitled, "The Approaches to Gatun Lock" by Joseph Pennell (1857–1926). Source: Library of Congress Prints and Photographs Division Washington, DC 20540.

Back cover credits: Historical Panama Canal images property of the Panama Canal Authority (ACP).

Contents

Early Efforts

UK Britain and the Trans-isthmian Dream 1
    Michael Mark Chrimes

The French Attempt to Construct a Canal at Panama 14
    Reuben F. Hull Jr.

Building the Canal

Building the Panama Canal (Men, Machines, and Methods) 27
    Raymond Paul Giroux

Men Responsible

George S. Morison and Philippe Bunau-Varilla: The Indispensible Men of Panama 70
    Francis E. Griggs Jr.

The American Engineers that Built the Panama Canal 112
    J. David Rogers

Remembering Joseph Pennell and the Panama Canal 350
    Augustine J. Fredrich

Akira Aoyama's Achievements on the Panama Canal Project 358
    JSCE International Activities Center USA Group

After Completion

Gatun Dam History and Developments 367
    Luis D. Alfaro, Manuel H. Barrelier, and Maximiliano De Puy

The 1915 Panama-Pacific International Exposition in San Francisco and Panama Canal Model, Conference and Proceedings 384
    Jerry R. Rogers and Luis D. Alfaro

Introduction

The history of a canal at the Isthmus of Panama extends for centuries, one of many outcomes of Christopher Columbus' search for an ocean route from Europe to the Indies. The early chapters of this tale involve some well known characters associated with the discovery of the Americas (Ojeda, Pizarro, Balboa, Cortéz); Spanish colonies and the Spanish Main; and, French, Dutch, and British pirates of the Caribbean (Drake, Morgan, Cook). In 1513, Spanish settlers discovered two oceans separated by a narrow strip of land - the Isthmus of Panama. The idea of a canal to link the two oceans was first proposed by Álvaro de Saavedra Cerón who accompanied Balboa and Cortéz in explorations of Central America and did extensive surveys of the Isthmus between 1517 and 1529.

Multiple locations for a canal were surveyed and mapped, but no actions were taken to initiate construction. Interest was renewed toward the end of the 18th and early 19th centuries, when British interests encouraged surveys of four routes for a canal. Spain decreed that a canal be constructed through the Isthmus and subsequently, the US entered the picture when approached by the Federal Republic of the United Provinces of Central America after it seceded from Spain in 1823. Several efforts to launch contracts for a canal were unsuccessful. However, an offshoot of these efforts resulted in the construction of a railroad linking the oceans which, when completed in 1855, still fell short of the desired canal.

The subject of our attention for this centennial celebration is the next chapter of the saga, which begins in 1876, when Columbia granted the French a concession to build a canal across Panama. It's a multifaceted story of exploration, adventure, political intrigue, engineering and social challenges, financial wheeling and dealing—a story with sufficient twists and turns to hold one's interest and keep the pages turning.

We are fortunate to have a bounty of records and photographs documenting this period of the canal development. As a product of the US government, it was well documented in regular reports to the President and Congress, as well as the nation via continuing newspaper coverage to a population that devoured the epic stories. Books, such as Ira E. Bennett's 1915 *History of the Panama Canal*, David McCullough's 1977 *The Path Between the Seas,* and Julie Greene's 2000 *The Canal Builders*, among others, give wonderful accounts of this era. Other sources include the US Congressional Records; the Library of Congress; and records in the US National Archives, which include many thousands of photographs. Library collections from the University of Florida's George A. Smathers Libraries and the Panama Canal Museum (http://ufdc.ufl.edu/pcm), and the Linda Hall Library in Kansas City, MO provide wonderful access to information and documentation on the birth of the Panama Canal.

To celebrate the 100th anniversary of the opening of the Panama Canal, ASCE's annual conference was held in Panama in October 2014. Recognizing this opportunity, ASCE's History and Heritage Committee worked with ASCE's Conference Planners to include history sessions in the conference. To deliver the history sessions, speakers were asked to submit their papers for publication in conference proceedings. What resulted is this document in which we present superb papers discussing a number of topics, providing conference attendees and lay readers a glimpse of the story of the Panama Canal. These papers document the lives and experiences of the engineers and others who struggled to get the dream accepted, funded, launched, and built, in order to ensure its successful operation over the last 100 years.

The proceedings are organized in the same sequence as the presentations during the four history sessions at the ASCE Conference. The "Early Years" provides two papers focusing on British efforts to study the possibility of a canal across the Isthmus, followed by the French efforts at constructing a canal across Panama. It is interesting to note that, although the French efforts failed, it was not due to the engineering. The adage "timing is everything" can be applied here in the subsequent US efforts, which benefitted from the lessons learned from the French. Their experiences in fighting malaria and yellow fever in Cuba; advances in equipment resulting from the westward expansion and development of the railroad in the US; spoils and experiences gained from the Spanish-American War; and, good old Yankee ingenuity, allowed them to adapt means, methods, and machinery to unique conditions.

The second history session addressed "Building the Canal." The railroad was a key component in the canal construction, and the experience of the construction itself. US engineers with railroad construction, equipment, and operations experience were essential to the successful completion of the canal. The ability to operate track mounted equipment for both excavating and hauling dirt and rock excavated to form the canal channel made the construction possible. Laying out and moving tracks continued throughout the job, a story recounted in this segment. The construction of the locks and gates from movable, rail-mounted cranes is another key element of the construction process.

The third history session examines the "Men Responsible" for moving the canal through the decision process, and those who were responsible for designing and constructing the canal. The first paper provides insight into the many hurdles two individuals, Morrison and Bunau-Varilla, faced in moving the dream forward. It's a lesson on perseverance in championing an idea—one you believe in—despite the obstacles in your path. Following this is a paper on the men who actually worked on constructing the canal. It's a lesson in giga-project development, management, and execution. It shows that the right men in the right situation can achieve what seems impossible. They each had the knowledge and on-the-job training that prepared them for the challenges faced in Panama. A third paper presents an interesting view of the "Wonders of Work"—the art of engineering and construction as seen through the lithographs of a famous American artist of the time, Joseph Pennell. Pennell travelled to Panama to document the fleeting moments of activity at construction sites. It is interesting to note that images of these works were also captured by many amateur photographers—sightseeing visitors to the site armed with another new technology, the Kodak camera.

A bonus paper is included that outlines the experiences of a young engineer from Japan who arrived in Panama fresh out of school to work on the largest construction project in the world. It chronicles his advancement from draftsman to chief engineer on the Panama Canal, experiences that shaped his future engineering career and achievements with waterworks in Japan. This paper was given during the conference in a separate session by the Japan Society of Civil Engineers, highlighting joint cooperation between JSCE and ASCE.

The fourth history session presented two papers that addressed activities "After Completion" of the Panama Canal in August 1914. The first looks at the Gatún Dam, which was the largest earthen dam ever attempted up to that time. It points out many of the challenges that were encountered well before modern geotechnical engineering theory and practices. It also examines the performance of the dam and efforts to analyze and upgrade it to meet current seismic threats. The final paper outlines a unique event that occurred following the completion of the canal—the 1915 World's Fair held in San Francisco, CA. While the exposition showcased San Francisco's reconstruction following the devastating 1906 earthquake, a major exhibit was a working model of the Panama Canal, specifically the Miraflores and Pedro Miguel Locks. That same year, many engineering symposia sponsored by US engineering societies published volumes of their transaction proceedings. One 1916 two-volume set of "Transactions of the International Engineering Congress, September 20-25, 1915 in San Francisco," included 25 papers by key engineers who worked on the canal, including Goethals, McDonald, Seibert, Williamson, Hodges, Mears, and others. Two of the original 1916 proceedings papers are reviewed. It is important to acknowledge the contribution of engineers from the Panama Canal Authority who contributed to these final two papers during a very challenging period in the canal's Third Set of Locks Expansion Program.

For those who attended the ASCE conference, this publication will be a reminder of the excitement and adventures we shared in Panama City in October 2014. For those who could not attend, this volume will introduce some of the excitement, challenges, and adventures that accompanied the building of the Panama Canal. Hopefully, these papers will inspire you to read and explore this era further. It's the history of an ASCE Historic Civil Engineering Landmark—the story of civil engineers on a world-changing project that achieved a century milestone: 1914 to 2014!

I wish to extend my sincere thanks and appreciation to the authors who contributed their time and talents to these proceedings and the ASCE Conference. It was my extreme pleasure and privilege to work with you, allowing me to be the first to review these wonderful stories of the history of the canal and the engineers who made it possible. They are truly an inspiration, and you are to be commended for sharing their lives and achievements for future generations of civil engineers, just as their efforts have enriched our own. Again, thank you.

Readers, enjoy.

Bernard G. Dennis, Jr.
Chairman,
ASCE History & Heritage Committee

Author List

Alfaro, Luis D., 367, 384

Barrelier, Manuel H., 367

Chrimes, Michael Mark, 1

De Puy, Maximiliano, 367

Fredrich, Augustine J., 350

Giroux, Raymond Paul, 27

Griggs, Francis E., 70

Hull, Reuben F., 14

JSCE International Activities Center USA Group, 358

Rogers, J. David, 112
Rogers, Jerry R., 384

UK Britain and the Trans-isthmian Dream
Michael Mark Chrimes MBE BA MLS MCLIP

Director (Engineering Policy and Innovation), The Institution of Civil Engineers, One Great George Street, Westminster, LONDON SW1P 3AA, United Kingdom. tel: +44 (0)20 7665 2250; fax: +44 (0)20 7976 7610; email: mike.chrimes@ice.org.uk

ABSTRACT

From the late eighteenth century until the end of the nineteenth century Britain was unchallenged as the premier mercantile and maritime nation. With enormous financial resources and a growing territorial empire based around naval power it inevitably took an interest in the possibility of a transport link across Central America, as a financial investment and trade route.

At the start of the century Thomas Telford was called upon to advise upon the Darien Canal scheme. Although this came to nothing, leading British engineers and capitalists continued to look at rail and canal schemes. Finally at the end of the century the consortium of Cutbill, Son and De Lungo, and James Perry worked at Culebra on the disastrous French enterprise.

This paper summarises British engineering involvement, drawing on the ICE archives, and consider how many of the ideas were more than 'castles in the air'.

INTRODUCTION

The successful construction of the Panama Canal in the early twentieth century was a triumph of US engineering management, and medical knowhow. The existing Panama Canal was nearly half a century in the making and is generally acknowledged as one of the engineering wonders of the world. Engineers had to contend with both engineering difficulties, notably major landslides, and also disease, which decimated the workforce. That problem had to be addressed before the civil engineering challenges. However, in the pre-history of the scheme to create a transport link across Central America the British played a significant role over a long period of time.

In the sixteenth century the piratical activities of Sir Francis Drake involved the first crossing of the Isthmus by Englishmen. As British trade and the territorial empire grew, more legitimate interests in shortening sailing times and improving access to Asia and the Pacific Coasts of the Americas resulted in a series of schemes by British speculators and engineers to cross Central America. A number of surveys were made, and canal and rail schemes considered. In some cases there was serious outlay of capital, and transport links were completed.

In the second half of the nineteenth century, whilst British territorial imperialization was limited across Latin America, British capital and engineering were to be seen everywhere, generally in tandem. That France and the United States are generally associated with the Panama Canal has obscured the British interest.

PANAMA AND THE INSTITUTION OF CIVIL ENGINEERS

'Atlantic and Pacific Communications' was the subject of 4 papers and meetings at the ICE in the nineteenth century, reflecting the topics significance to the profession. No other potential project attracted such attention. The first was a summary of potential routes by Joseph Glynn (Glynn, 1847). This was notable for the presence of the future Napoleon III, who was an informed advocate of the Nicaragua Canal. The second, by J. A. Lloyd (Lloyd, 1849) was in a supplement to Glynn, giving further details of his surveys. The third (Kelley, 1856) was by the American, Frederick Kelley, and focussed on a sea level canal by the southerly Atlantic route, although summarising other alternatives. That paper was presented at a period of heightened interest, prompted by the Californian Gold Rush and Napoleon III's active sponsorship of a route. There was then a hiatus until the end of the century when J T Ford gave the fourth paper on summarising the De Lesseps scheme and its aftermath (Ford, 1900).

These discussions were complemented by donations and acquisitions by the ICE Library, encompassing government reports, engineering surveys, company prospectuses, and ranging in extent from single sheets of paper to weighty monographs. In addition were the numerous periodical articles. In terms of extent only the Suez Canal can compare. One reason for this, of course, was that so much was speculative.

What follows below is based on these ICE resources.

THOMAS TELFORD AND EARLY SURVEYS

Following the disintegration of Spanish America at the start of the nineteenth century, the newly independent states sought investment in a number of schemes including communications across Central America to link the Atlantic and Pacific. Perhaps the most interesting early scheme was that originating around 1818. On 27 January the British Consul in Panama was approached by the Colombian Government (then Government of New Grenada) with a view to Captain (later Lieutenant-Colonel) John Augustus Lloyd surveying a route across the Isthmus.

Apparently, ICE's President, Thomas Telford was first approached in 1825. In 1827, Lloyd and the Swedish engineer, Captain Falmark, finally began two seasons of surveys on behalf of Simon Bolivar's Government. Nothing happened. Bolivar died in December 1830 and Lloyd was posted to Mauritius. Telford, however, retained a large bundle of drawings, now lost, on the Isthmus of Darien scheme. It is a little known aspect of his career [ICE (1834)].

Lloyd's proposals were published by the Royal Society and Royal Geographical Society, and later by ICE. Of the Chagres river he noted: "The banks are precipitous, of trap and porphyritic formation, worked to the every edges …" A hint of the arduous environment is given in his description of Portobello … "Such is its dreadful insalubrity, that at no period of its history did merchants venture to reside in it … No class of inhabitants can long exist in it." Telford was fortunate to be in Westminster giving his views, rather than on site with Lloyd.

LIEUTENANT COLONEL JOHN AUGUSTUS LLOYD (1800-1854) (Associate of the ICE)

Lloyd can be regarded as the first great British advocate of a Panama Canal (Lloyd, 1830; 1831; 1847). With rudimentary scientific education and some training in mining, engineering, and surveying he secured an introduction to Simon Bolivar, and became a Captain in his Engineering Corps. Despite his letter of authority of 29 November 1827 to survey the 'provincia del Darien del sur' he met with much harassment by local officials, which must have impacted the value of his surveys. He followed the Camino Real, or track across the summit ridge, and made no attempt to identify the lowest crossing point. He did however determine the levels between the Atlantic and Pacific, identifying both canal and rail routes. He drew two potential railroad routes on his map (Figure 1), believing a railway was both feasible and a necessary precursor to a canal. He believed the engineering of a canal to be practical, provided the finance could be found. Once built, he believed a canal would be easier to maintain with an abundant water supply. He also believed a good harbour could be built at Panama.

Lloyd saw the Isthmus as rich in resources, with an abundance of building materials – limestone and timber. He collected many geological and mineral samples which he presented to the British Museum, and Admiralty. On the Pacific side he believed the canal should start at Le Min, and then a canal be cut to Lima. From there the Rivers Chagres and Torindad would be made navigable to a point inland which could be the construction camp.

Lloyd recognised a labour force would be required. He suggested convict labour could be used – escape would be unlikely in a hostile environment – or black labour from the Caribbean or Africa. He also believed the interior would be attractive to European colonists (Lloyd, 1831).

Lloyd was not the only British engineer in Colombia in the 1820s. Both Robert Stephenson and Richard Trevithick were there, and Stephenson took a view that a canal could be built, and that construction materials were available (Lloyd, 1849, 79).

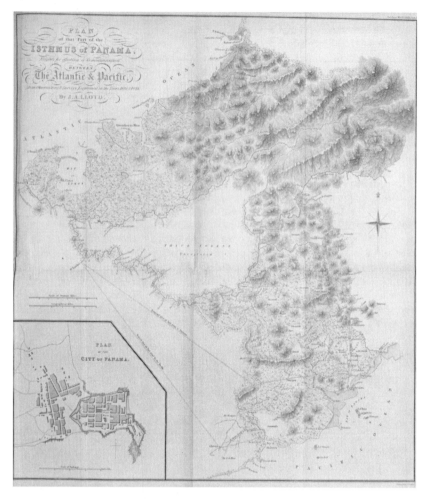

Figure 1. Lloyd's Survey of Panama
Lloyd's map, published by the Royal Society of London (Lloyd (1830)), was the first published map to show potential transport routes across the Isthmus

TEHUANTEPEC ISTHMUS

First surveyed by Craven in 1774, in 1824 Mexico set up the Tehuantepec Commission, and a survey was carried out in 1842-1843. J. J. Williams surveyed the route for a US company in 1851. Political instability and distance precluded progress for more than 30 years.

EVAN HOPKINS' SURVEYS

The Hopkins family were a well-established family of engineers active in South Wales and South West England in the early nineteenth century (Skempton, 2002).

Evan Hopkins (in Lloyd (Lloyd, 1849) 74) was employed in 1847-1848 by the Government of New Granada to survey the Isthmus from Darien to Veraguas. He investigated the geology and topography of the region, and concluded there were two feasible communication routes between the Atlantic and Pacific – in the Isthmus of San Blas, and between Chagres and Panama. His exploration of San Blas and the River Bayano was limited by the hostility of the indigenous population, although he concluded parts of the Bayano could be made navigable. He ruled out an old Spanish made track between Portobello and Panama as impossible to improve. His recommended route (Figure 2), essentially for a railroad, but also for water communication, followed the Chagres river to Gorgona, across the ridge, which he estimated at 260 ft above the Atlantic sea level, to the Rio Grande, following that river to Panama. His view of available building materials and resources was at odds with Lloyd and others.

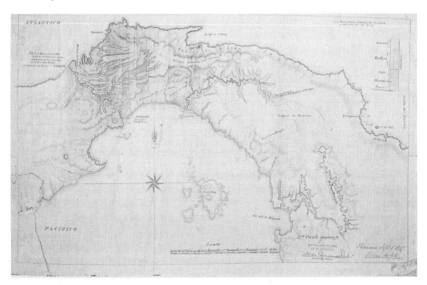

Figure 2. Hopkins 1847-48 Survey
Evan Hopkins map indicates the difficult terrain he encountered, and was arguably the first by a British civil engineer.

PACIFIC STEAM NAVIGATION COMPANY

In the late 1830s British commercial interests promoted a steamship company with connectivity across the Pacific, and in their pamphlets spoke of railroads across the Isthmus. In 1838, the American William Wheelright (Wheelwright, 1838) issued a pamphlet on this proposal.

NICARAGUA CANAL

The appeal of Nicaragua compared to shorter crossings at Panama and elsewhere was the existence of lakes to reduce the need for construction. In 1781 the Spanish engineer, Manuel Galisteo surveyed the levels from the Gulf of Papagayos to Lake Nicaragua.

In 1826, John Baily was invited to survey a crossing through Nicaragua on behalf of an English company. Although nothing was done then in 1837-1838 he surveyed a route for the Government of Central America, following the Rio San Juan from the Atlantic to Lake Nicaragua and thence by a 15 2/3 mile canal to the Pacific at San Juan del Sur. This involved major navigation works on the San Juan, and an enormous volume of muck shifting (Baily, 1844). Harbour facilities were meagre. A further survey of the San Juan River by George Laurance was made in 1840.

An alternative route, first suggested by Gallisteo, was looked at by Dutch engineers in 1830, and the US in 1835. This avoided cutting through the ridge, and instead making use of Lake Managua, the Rio Tipitapa, and joining the Pacific at Realejea.

Further alternative surveys by US-based engineers in the early 1850s all concluded the necessary works were expensive. By mid-century ICE were able to publish Michel Chevalier's map (Figure 3) showing favoured routes (Glynn, 1847).

Under Napoleon III, French interest in the Isthmus increased. In 1859, James Samuel (1824-1874) and Alexander Woodlands Makinson (1822-1886) first went to Panama to verify French surveys on behalf of British business interests. Samuel, by background a railway engineer, in 1863, surveyed the potential route of a ship canal from Greytown, up the River San Juan, across the lakes of Nicaragua and Managua, to Tamarindon on the Pacific. He rejected the proposal as far more expensive than the French had estimated. He went on to become Chief Engineer for the Railway for Veracruz to Mexico City and Puebla (*The Times*, Picayune, 28 August 1863).

NICARAGUAN RAILROAD

A British naval captain, Bedford Pim, promoted the International Atlantic and Pacific Junction Railway through Nicaragua in the early 1860s. An Associate (i.e., non-civil engineer) of the ICE, he visited Nicaragua in 1860 and 1863. A Concession was signed between the Nicaraguan and British Government in 1860, and the civil engineer, John Collinson, among others, surveyed the route (Collinson, 1866; Maury, 1866). Inevitably capital was unavailable.

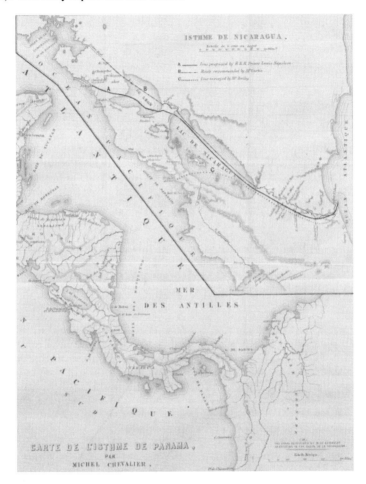

Figure 3. Isthme de Nicaragua
Michel Chevalier's map indicates alternative routes across Nicaragua
(from Glynn, 1847)

(BRITISH) HONDURAS INTER-OCEAN RAILWAY

The American Ephraim George Squires(1821-1886) carried out surveys in 1853 and 1858, for a railway on behalf of an international consortium led by British capitalists.

ATRATO SCHEME (DARIEN): ATLANTIC-PACIFIC CANAL

A southerly route across the Isthmus attracted a great deal of interest in the City of London in the early 1850s. A series of pamphlets were issued by the Atlantic and Pacific Junction Company, promoting a route along the Atrato-Cupica Valleys. Charles Nixon MInstCE and his partner, Lecky, issued an engineer's report, weak on detail. Lionel Gisbourne criticised the route (Nixon and Lecky, 1853).

The route was commented on by the Manx-born American, William Kennish (1799-1862 in 1853. At the end of 1853, the US, France and UK agreed to support further surveys for this company. The route was surveyed by 6 British engineers: Armstrong, Bennett, Bond, Devenish and Forde, led by Lionel Gisbourne, but his critics felt Gisbourne took fright at the challenge of jungle surveys, and his negative views were heavily criticised by Dr Edward Cullen (Cullen, 1856; 1857). He persuaded the contractor William Brady to support the scheme in September 1856. There is considerable doubt over the value of any of these surveys (Michler, 1861).

De LESSEPS SCHEME

There is no intention here to dwell on the history of de Lesseps scheme. Suffice it to say that the successful completion of the Suez Canal gave him credibility both in France and internationally, and meant that his call for an International Conference to discuss an Atlantic-Pacific Ship Canal could attract serious British engineering attention.

(Sir) John Hawkshaw among 6 British engineers, was invited to attend the International Congress in Paris in 1879. The proposal for a sea level canal at that time involved a tunnel through Culebra, which Hawkshaw had rejected as impractical. He had identified the control of the Chagres river as a key to success, and believed the proposal would not achieve that. His son, John Clarke Hawkshaw, had attended that meeting as a young engineer, and believed that the key to the success for the French scheme of the 1890s would be control of the waters of the Chagres.

Soon after construction began, in 1884, John Lewis Felix Target (Cross-Rudkin and Chrimes, 2008) was asked by the Governor of Jamaica to report on the conditions of the Jamaican workforce. This brought him to the notice of the Anglo-Dutch contracting consortium who had been awarded the contract for the Culebra Cut. He acted as their consulting engineer for three years until ill-health obliged him to give up – allegedly the climate inflamed his gout.

CULEBRA CUT AND THE ANGLO-DUTCH CONSORTIUM

The Anglo-Dutch consortium who began work on one of the toughest sections of the Panama Canal comprised a London based grouping of Cutbill, Son and De

Lungo, with T. C. Watson and a Dutch contractor, Van Hullen. Their contract was signed 1883, for 160 million French Francs.

When the American, Lieutenant W. W. Kimball, visited in 1885 his report on the Culebra section suggested a lack of resources (Kimball, (1886)) with the contractors only employing 1,250 men, with a further 100 employed by the company. They were using 12 French steam excavators, 19 locomotives, 277 dump cars, 1 Decauville locomotive and 1,180 Decauville cars, 14.5km of standard contractors railway track, and 17.2km of Decauville track (Figure 4). This was all to excavate an estimated 20M m^3 of material, in a 1,500m cutting up to 140m deep. By March 1886 590,000m^3, was dug, and only about 5% of the cutting had been excavated by 1887 (Boyd,1887).

Figure 4. Decauville railway equipment at Culebra

There is some doubt over when the Anglo-Dutch consortium gave up their contract at Culebra. The work was taken over initially by Artigue Sonderreger, then by the Company Engineer, Philippe Bunau-Varilla, but the financial arrangements once the Anglo-Dutch consortium had gone, were the subject of criticism or, to paraphrase Bunau-Varilla, calumnies against him. Van Hullen, in 1889, argued they should not have lost the contract as one reason they had made slow progress was lack of access to the Panama Railroad.

CUTBILL, SON AND DE LUNGO (McWILLIAM, 2014)

The little-known Cutbill family of contractors and stockbrokers were closely involved with the major British contractors George Wythes (1811-1883) and Thomas Brassey (1805-1870). Thomas Samuel Cutbill (1805-1867) was born on 13 March 1805 in Spitalfields, a son of Thomas Cutbill (1774-1845) and Sarah Soilleux (1785-1862), members of the Huguenot community. Little is known of his early life beyond involvement in the silk trade, but on 12 April 1834 he married Catherine Elizabeth Newton in Shoreditch. They had at least 10 children, most of whom reached adulthood.

The eldest son, Frederick Thomas (1835-1913) became a civil engineer, suggesting that from the 1850s Thomas Samuel could find a civil engineering pupillage for his son through his business connections with civil engineers in the City of London. In the 1861 UK Census he is described as a merchant, but he was a London agent for Thomas Brassey by that time and in the late 1850s he was working on George Wythes' projects in India. He also pursued the claims of the Canadian Grand Trunk Railway Shareholders, including Brassey. He was involved in the Brassey concession for the Mont Cenis Railway over the French Alps part of the British mail system to India. His younger son Walter John Charles Cutbill (1843-1915) was working for his father from 1859, and succeeded him as Secretary to the Mont Cenis Railway. Frederick worked on many international contracts until he returned in the 1890s.

The early connections with Thomas Brassey and George Wythes may have been for Indian railways, but increasingly there was a Latin American focus. At the time of Thomas Cutbill's death in 1867 his business was reconstituted as Cutbill, Son and De Lungo.

Ulysses de Lungo (c.1841-1897) was an Italian born civil engineer, who became a British subject, although dying in Florence where his estate of c.£140, was administered by the engineer Paolo Stacchini. His long-term involvement with the Cutbills was only part of his business activity – he registered a patent for soap manufacture (Patent no. 3633, 1873).

On Brassey's death in 1873, Cutbill, Son and De Lungo carried on some of his contractual obligations as partners in Thomas Brassey and Company, and also remained associated with George Wythes over the following decade until Wythes' death. They were also agents for railway companies such as the Northern of Canada. Although the best-known later contract was as part of the Anglo-Dutch consortium as contractors for Culebra Cut (Figure 5), they were particularly involved in Uruguayan and Venezuelan Railway concessions.

Much of the construction work was carried out with the Perry family. James Perry (b.1812) had been one of Brassey's and Wythes' agents in India, and brought his sons Frederick James (b. 1859) and Lionel (b. 1866) into his contracting business which operated under Frederick's leadership from 1884. It was generally known as James Perry and Co.

Another partner at Panama and elsewhere in Latin America was Thomas Colclough Watson (1822-1890) (Cross-Rudkin and Chrimes, 2008). Watson had been contractor on the North Sea Ship Canal in the Netherlands. Henry Gale (1836-1898) was another engineer closely involved with Cutbill and Perry from the 1870s, notably on its South American contracts. He was based at the London offices from 1888.

Cutbill Son and DeLungo was closed in February 1894 due to the retirement of Cutbill and De Lungo. James Perry and Co. was wound up the following January although the Perry family remained active in a successor business established in 1892-The Railway and Works Company-of which Gale was a Director. This was later involved in the Trans Africa Rail Syndicate. Lionel Perry is known to have visited Panama in 1907.

The Anglo Dutch Co failure at Panama would appear therefore to be one of challenge not lack of experience.

Figure 5. Cutbill's contract at Culebra Cut in 1885

MEXICO: A FOOTNOTE

In 1893, Mexico finally succeeded in building its Atlantic-Pacific route, but it was under-capitalised and lacked the necessary port facilities to rival attendance routes. In 1898, the Great British contractors Weetman Pearson & Son obtained a 50-year concession to rebuild the railway and provide harbour facilities (McWilliam, 2014). The railway contract, completed in 1906, was worth of 2.5 million. The ports cost similar amounts. The concession was bought out by The Mexican government in 1918.

CONCLUSIONS

British interest in Atlantic and Pacific Communication was driven by commercial interests and associated foreign policy. British civil engineers followed these commercial interests, but were also fascinated by one of the world's great engineering challenges. The success of the Americans ultimately reflected a change in the balance of civil engineering leadership from the Telford era to that of Goethals a century later, but the activity of the British, and the Irish navvy should not be forgotten.

REFERENCES

American Railroad Journal (1862) **36**, 618.
"Atlantic and Pacific Junction Company" (1853) *The Atlantic-Pacific Canal*. London: Committee for Surveys.
Baily, J. (1844) "On the Isthmus between the Lake of Grenada and the Pacific". *Jnl. Royal Geographical Society*, **14**, 127-129.
Bigelow, J. (1886) *Panama Canal*. New York: Chamber of Commerce.
Boyd, R. N. (1887) *Notes on the Panama Canal, Civil & Mechanical Engineers Society*.
Collinson, J. (1866) *Descriptive account of ... Pim's project for an International Atlantic and Pacific Junction Railway*. London.
Congrès International d'études du Canal Interocéanique. Paris, 1879.
Cross-Rudkin, P.S.M and Chrimes, M. (2008) *Biographical Dictionary of Civil Engineers of Great Britain and Ireland*. London: ICE.
Cullen, E. (1856) *Over Darien by a ship canal: reports on the mismanagement of the Darien expedition of 1854*. London: Effingham Wilson.
Cullen, E. (1857) *Letters on the Isthmus of Darien Ship Canal*. London: Effingham Wilson.
Davis, C. H. (1861) *Report on Inter-oceanic Canal and railroads*. Washington, D.C.: USGPO.
Falconer, T. (1844) "Survey of the Isthmus of Tehuantepec ... 1842-1843". *Jnl. Royal Geographical Society*, **14**, 306-315.
Ford, J. T. (1900) "The present condition and prospects of the Panama Canal works", *MinProcsICE*, 144, 150-170; Discussion, 171-.

Institution of Civil Engineers (1834) *Telford Bequest*. MSS.
Glynn, J. (1847) *MinProcsICE*, **6**, 399.
Kelley, F. M. (1856) On the junction of the Atlantic and Pacific Oceans and the practicability of a ship canal, without locks by the Valley of the Atlantic. *MinProcsICE*, 15, 376.
Kimball, W. W. (1886) *Special intelligence report on the progress of the work on the Panama Canal during the year 1885*. Washington: USGPO.
Lloyd, J. A. (1830) "Account of levellings carried across the Isthmus of Panama. *PhilTransRoySoc*, London, p.59-68.
Lloyd, J. A. (1831) "Notes respecting the Isthmus of Panama". *Jnl. Royal Geographical Soc*., **1**, 69-101.
Lloyd, J. A. (1849) "On the facilities for a ship canal communication between the Atlantic and Pacific Oceans", *MinProcsICE*, **9**, 58.
McCullough, D. (1977) *The path between the seas*. New York: Simon & Schuster.
McWilliam, R. C. (2014) *Biographical Dictionary of Civil Engineers of Great Britain and Ireland*, *3, 1890-1920*. London: ICE.
Maury, M. F. (1866) *Letter on the physical geography of the Nicaraguan Railway Route*. London.
Micher, N. (1861) "Report of his survey for an inter-oceanic ship canal near the Isthmus of Darien". *US Congress*, **36**, 2. Ex Doc.9.
Nichols, A. B. *Notebook B5*, Linda Hall Library.
Nixon, C. and Lecky, (1853) *Atlantic-Pacific Canal: engineers report*. London: The Committee.
Otis, F. N. (1867) *His story of the Panama railroad*. New York: Harper & Brothers.
Skempton, A. W. (2002) *Biographical Dictionary of Civil Engineers of Great Britain and Ireland*. London: ICE.
The Times-Picayune, 28 August 1863.
Wheelright, W (1838) *Statements and documents relative to the establishment of steam navigation in the Pacific*. London.

The French Attempt to Construct a Canal at Panama

Reuben F. Hull, Jr., PE[1]

[1]Civil Design Engineering Consultants, 1162 Lowell Road, Schenectady, NY 12308; email: ReubenHull47@gmail.com

ABSTRACT

Following the success of the Suez Canal, Compagnie Universelle du Canal Interoceanique incorporated under French law on March 3, 1881, for the purpose of building a canal through Panama, the first attempt at construction of a waterway to cross the Panamanian isthmus. Within eight years, on February 4, 1889, the company was declared bankrupt and dissolved. Five years later, a second French company was created, which continued nominal work, until the United States took control of the project in 1904. The French effort at Panama is regarded as a failure, often relegated to being a footnote in the history of the construction of the canal. This paper chronicles the efforts, the challenges, and the ultimate downfall of the French endeavor at the Panama Canal, calling attention to the achievements and lessons learned from the French experience, which later benefitted the American enterprise at the same location.

INTRODUCTION

In 1854, Ferdinand de Lesseps, former French Consul to Cairo, was invited by Muhammad Said Pasha, Wāli of Egypt and Sudan, to initiate a plan for connecting the Mediterranean Sea and the Red Sea with a canal across Egyptian territory. The proposed canal would save 4,900 nautical miles [8,900 km] for ocean going vessels travelling between Europe to Asia. After four years of planning and engineering, the Suez Canal Company (*Compagnie universelle du canal maritime de Suez*) was founded on December 15, 1858, for the purpose of constructing and operating the Suez Canal. The company was formed with majority control by French investors and led by de Lesseps. Work started on the Mediterranean shore of the future Port Said on April 25, 1859. A decade later, on November 17, 1869, the Suez Canal opened to great fanfare and the event instantly made Ferdinand de Lesseps an international hero.

The elation of the Suez Canal success was short lived. European discord between factions at the time resulted in numerous wars and changes in the balance of power. In 1869, the Spanish parliament offered the throne of Spain to Prince Leopold, nephew of Prussia's King Wilhelm I. Situated between Spain and Prussia, France feared such a powerful alliance. France challenged the candidacy and Leopold resigned. However, Prussian Chancellor Otto Von Bismarck used the opportunity to orchestrate an impression that France was the aggressor and succeeded in provoking the French, who declared war on Prussia on July 15, 1870. By September 2, 1870, Napoleon III surrendered and was taken prisoner with over 100,000 of his soldiers. It was a decisive loss for the French, as the Prussians not only captured the entire French army, but the leader of France as well. Paris surrendered in January 1871, after being under siege for four months. The treaty of Frankfurt was signed on May 10, 1871. France's Second Empire ended and the Third Republic began, but the new French government found itself isolated on the international plane as a unified Germany established itself as the main power in continental Europe.

In the aftermath of the Franco-Prussian War, the defeated and humiliated France lost prestige in eyes of other nations. There was one certain way to restore its glory, by undertaking and completing the most challenging engineering feat in history: build a canal through Central America to link the Atlantic and Pacific Oceans. The leader of that effort would be the hero of Suez: Ferdinand de Lesseps.

FERDINAND de LESSEPS – LE GRAND FRANÇAIS

Ferdinand de Lesseps was born November 19, 1805 into a family of French career-diplomats at Versailles, the city that had been the cradle of the French Revolution. Much of his boyhood was spent in Pisa where his father was the French Consul at Lirna. He attended the Liceo Henri IV and studied at the law school. At age 19, having studied law, he was appointed assistant vice-counsel to his uncle, then the French ambassador to Lisbon. He also served in Tunis later with his father, until 1832 the year of his father's death.

Ferdinand de Lesseps was convinced that his life's work lay in Africa. In 1832, de Lesseps was appointed vice-consul at Alexandria. On the voyage to Alexandria, a passenger died, it was said of cholera. When the ship docked, all the rest were immediately quarantined. While in quarantine, de Lesseps received several books, among which was *Canal des Deux Mers*, a memorandum of the search for the ancient Suez Canal, written by civil engineer Jacques-Marie Le Père, one of the scientific members of Napoleon Bonaparte's late-1700s French expedition. This work struck de Lesseps' imagination, and gave him the idea of the canal that he later constructed across the Egyptian isthmus.

Between 1833 and 1849, de Lesseps served many diplomatic appointments in Egypt and throughout Europe. He happened to be in Egypt during the 1830s when a team of French engineers arrived to survey for a possible Suez Canal route. Egypt's leader Mohammed Ali was not interested, but the idea later greatly attracted his son Said.

In 1849, de Lesseps' diplomatic career culminated with his negotiation for the return of Pope Pius IX to the Vatican. A change in foreign policy led to his being recalled after which de Lesseps retired from the diplomatic service, never again occupying any public office.

In 1854, de Lesseps was granted a concession to undertake the Suez Canal project by Muhammad Said Pasha. With Suez, de Lesseps demonstrated his gift of promotion, succeeding in rousing the patriotism of the French and obtaining subscriptions for more than half of the capital of two hundred million francs needed to form the company. Fifteen years later, on April 18, 1869, the Red Sea joined the Mediterranean Sea. For the opening ceremony, in November 1869, thousands of distinguished guests assembled from all over Europe and the Middle East. The procession of ships through the canal was led by the French imperial yacht with Empress Eugénie on board.

Ferdinand de Lesseps, who surmounted tremendous engineering, diplomatic, and administrative difficulties, was awarded many honors, and was widely proclaimed throughout France. He was honored around the world and celebrated as the greatest living Frenchman. More than twenty thousand people attended a reception in his honor at London's Crystal Palace.

Ferdinand de Lesseps, *Le Grand Français*, had no engineering or technical skills, no financial experience, and only modest administrative ability, but his contacts in Egypt led him to dedicating fifteen years to seeing the Suez project to completion. Above all, de Lesseps was an adventurer. Also, he believed in the notion of progress that by technical and industrial achievement peace and plenty must ensue. By opening the world to the population, his canal was a means to that end, of which there was no nobler cause. Inevitably, he supported the idea of a new Suez between the Atlantic and Pacific in the Americas.

INTEROCEANIC CANAL COMPANY

After the Franco-Prussian War ended Napoleon III's empire, the French government reduced its role in Central America. Humbled before Germany in Europe, France lost every foothold on the American continent. A lack of understanding of the world outside of its borders was considered to be a contributing cause to the nation's weakened position and its loss in the war. The timing of these events coincided with an increased popularity in the subject of Geography. Ferdinand de Lesseps was among those who frequented the Société de Géographie de Paris, (Geographic Society of Paris), the world's oldest geographic society. Also among those who attended Geographic Society meetings was Jules Verne, whose 1872 publication *Around the World in Eighty Days* helped humanize the scale of worldly travel.

In 1872, de Lesseps began his ventures in a canal across the Central American isthmus not at Panama, but at Nicaragua, seeking to obtain a concession for a lock canal. However, the sentiment in Nicaragua favored the United States, so failing to meet his objective, de Lesseps turned his interest to a sea level canal at Panama.

In March 1875, The Geographic Society created a "Committee of Initiative" to seek international cooperation for studies to improve the geographical knowledge of the Central American area for the purpose of building an interoceanic canal. That summer, the Geographic Society sponsored an international congress and it was at this congress that de Lesseps first made public his interest in a canal across the Central American isthmus. The congress coincided with a geographical exhibition at the Louvre, which brought public attention, and de Lesseps was the center of that attention.

Also, in 1875, Britain became the largest shareholder in the Suez Canal Company, when it bought the stock of the new Ottoman governor of Egypt. After the English took over financial control, de Lesseps' influence disappeared from the Suez Canal Company and the French enterprise in Egypt became Britain's strength. This was yet one more blow to the French psyche that needed to be overcome.

Ferdinand de Lesseps had no interest in participating in any commercial development that might follow from the deliberation of the international assembly of experts held under his leadership. It would be an international coup to make the grand connection between the Atlantic and Pacific in the Americas, not only as a business venture, but in terms of national prestige.

The Committee's agent, Anthoine de Gogorza, convinced the Columbian government that he knew a practical route through Panama and in May 1876, established a contract to report on the exploration within eighteen months. When he returned to Paris with the Columbian concession, La Société Civile Internationale du Canal Interocéanique (Interoceanic Canal Company) was formed for the promotion of canal projects through Central America. The speculative venture was headed by Ferdinand de Lesseps.

Exploration of the isthmus was led by French Navy Lieutenant Lucien N. B. Wyse. Explorations began in December 1876, and were completed in April 1877. After the exploration of several routes in the Darien-Atrato regions, de Lesseps rejected all of these plans because they contained the construction of tunnels and locks. On a second isthmian exploratory visit beginning December 6, 1877, two Panamanian routes were explored: the San Blas route and a route from Limon Bay to Panama City, the current Canal route. In selecting the latter, the plan was to construct a sea level canal. The route would closely parallel the Panama Railroad and require a 4.8 mile [7.72 km] long tunnel through the Continental Divide at Culebra.

In 1879, de Lesseps convened Congrès International d'Etudes du Canal Interocéanique (International Congress for Study of an Interoceanic Canal) in Paris to study the digging of a canal across the Central American isthmus. Fourteen proposals for sea level canals at Panama were presented before the congress, and a subcommittee reduced the choices to two, Nicaragua and Panama. The United States government refused to turn over some maps or to participate officially in the conference, but the American delegation's Nicaragua plan was introduced by Cuban-born Aniceto García Menocal, a civil engineer assigned to the Grant surveys in Nicaragua and Panama. Menocal had also evaluated the Panama route and recommended from the beginning that a lock system in Panama, rather than a one-level, sea-level canal, was the only possible way to surmount the Culebra Range.

Baron Godin de Lépinay, the chief engineer for the French Department of Bridges and Highways, was known for his intelligence, as well as his condescending attitude towards those with whom he did not agree. He was the only one among the French delegation with construction experience in the tropics, having worked on a railroad project in South America where most of his workforce died of yellow fever. At the congress, de Lépinay supported the Panama route, but he made a forceful presentation in favor of a lock canal.

Baron de Lépinay predicted that a sea level canal would be a fiasco. The cost, in his opinion, would be exorbitantly more than the financing would be capable of supporting. Also, the mistaken belief at the time was that fevers were caused by rotting vegetation that emanated "bad air," or in French, "mal air," which yielded the term "malaria." The massive excavations for a sea level canal, de Lépinay warned, would expose the bad air, devastating the workforce.

However, although estimated by the Technical Committee as being cheaper than either the Nicaragua route or the sea level Panama route, the de Lépinay design received no further attention. The delegates at the conference resolved that a canal be built in Panama and that it be modeled on the Suez Canal which had organized and opened under de Lesseps' guidance just a decade before. The resolution read *"The congress believes that the excavation of an interoceanic canal at sea-level, so desirable in the interests of commerce and navigation, is feasible; and that, in order to take advantage of the indispensable facilities for access and operation which a channel of this kind must offer above all, this canal should extend from the Gulf of Limon to the Bay of Panama."*

The resolution passed 74-8. The opposition votes included de Lépinay and Alexandre Gustave Eiffel. The predominantly French support did not include any of the five delegates from the French Society of Engineers. Of the seventy four voting in favor, only nineteen were engineers and of those, only one had ever been in Central America.

THE PANAMA CANAL COMPANY

Ferdinand de Lesseps raised two million francs for the project by selling "founders shares" to a syndicate of 270 wealthy and influential colleagues, who would receive their shares once a company was established. Next, he negotiated the buyout of the Société Civile, including their concession and their maps and surveys, for ten million francs (approximately two million dollars), which was almost all profit.

Following the organization of the Compagnie Universelle du Canal Interocéanique de Panama (The Panama Canal Company) on August 17, 1879, de Lesseps was appointed President of the company despite the fact that he had reached the age of 74. Immediately, de Lesseps embarked on a promotional tour of France to raise the starting capital, estimated at 400 million francs, in a stock offering directly from the public.

The Suez Canal had been built with private money and existed as a publicly held corporation. Ferdinand de Lesseps saw the Central American project similarly, and felt its financing via private money was an important and "American" way to pursue the undertaking. Private financing was a strategic move. The French government repeatedly denied any official relationship with the canal enterprises, in deference to the Monroe Doctrine, which declared the Western Hemisphere as the American domain. It was also a sentimental move which de Lesseps spoke about tying the importance of private capital with the notion that it would merely facilitate him as an executor of the American idea.

The initial response for the stock offering was dismal. Engineer Gustav Eiffel and the technical committee knew that Panama's geology was not understood. The French press and financial institutions were hostile, including banks organizing a campaign against the canal venture. Rumors persisted that Panama's climate was a deathtrap and that the Americans would not allow construction to begin. Rumors also included personal attacks on de Lesseps' age, abilities, and mental capacity.

In response, de Lesseps launched a public relations campaign to calm concerns about technical and practical issues of the canal. His grandiosity made him a gifted promoter. A new survey was ordered and an International Technical Commission of well-known engineers went to Panama, accompanied by de Lesseps. He reassured shareholders that progress would quickly be made, insisting that the canal would be open in 1888. It was only though the zeal, skill, experience, and confidence of de Lesseps that within a year the Company had raised enough money to launch the operation.

On New Year's Day 1880, de Lesseps visited Panama to lead a ceremony at the mouth of the Rio Grande, scheduled to become the Pacific entrance to the future canal. The steamer that was transporting the ground-breaking contingent arrived behind schedule and was prevented from landing due to the low tide. The ground-breaking ceremony was improvised ceremonial pickaxe blow in a soil-filled champagne box.

Ferdinand de Lesseps decided that another ceremony should inaugurate the section of the canal that would have the deepest excavation, the cut through the Continental Divide at Culebra. A ceremony was arranged, and on January 10, 1880, dignitaries and guests gathered at Cerro Culebra, which included witnessing the blast from an explosive charge set to break up a basalt formation just below the summit.

PREPARATIONS

As de Lesseps was a trained diplomat and not an engineer, his son Charles took on the task of supervising the daily work. Ferdinand de Lesseps handled the work of promoting and raising money for the project from private subscription. Without scientific or technical understanding, de Lesseps relied upon a naive faith in the nature of emerging technology. He worried little about the problems facing the colossal undertaking, believing that the right people with the right ideas and the right machines would appear at the right time. His boundless confidence and enthusiasm for the project and his consummate faith in technology attracted stockholders.

Meanwhile, the International Technical Commission set about the task of charting the canal route. Survey findings were compiled into a final report by the commission headquarters in Panama City. The International Technical Commission was required to verify all previous surveys, to determine the final canal alignment leading to the preparation of design plans and specifications. Another goal was to convince investors that de Lesseps was not just a promoter for a hastily conceived, half understood, imperfectly planned project.

The few weeks' time allowed for this survey work was too short for an investigation of such importance. The content of the technical commission's report, submitted on February 14, 1880, was scientifically and professionally weak. In approving a sea level canal, the commission reported no significant construction difficulty in cutting the deep channel through the Continental Divide at Culebra Cut and estimated that construction would take approximately eight years.

In February, 1880, de Lesseps arrived in New York to raise money for the project. When he stayed at the Windsor Hotel, its staff flew the French flag in his honor. He met the American Society of Civil Engineers and the Geographic Society while touring the area.

Two years were spent conducting surveys, constructing service buildings and housing, recruiting an enormous workforce, assembling machinery, and establishing communication lines. Armand Réclus, the Agent Général (chief superintendent) Company, led the first French construction group of about 40 engineers and officials, landing at Colon on January 29, 1881.

Ferdinand, de Lesseps contracted Belgian construction firm Couvreux and Hersent, with whom he had worked at Suez. Réclus expected to need a year for preparatory tasks, but Panama's sparse population inhibited labor recruitment, its thick jungles inhibited the ability to accomplish the work. Gaston Blanchet, Couvreux and Hersent's director, accompanied Réclus to the Isthmus. Ten months into the project, Blanchet, the company's driving force died, apparently of malaria.

Work disruption began due to difficulties from New York because the Panama Railroad, which was in American possession, held the transportation monopoly for the Panama isthmus. In lengthy discussions between Paris and New York, an agreement was reached in June 1881. The Panama Canal Company was forced into taking over the railway, whose capital was now held almost solely in the hands of one speculator. Ferdinand de Lesseps was aware that the railroad was important to the work, and control of this vital element was gained by the French in August 1881, but at an inflated cost of over one hundred million francs (twenty million dollars), two and one half times its value and about a third of the Company's resources. Although under the Company's control, the railroad was never organized to serve anywhere near its full potential, especially in moving material from the site of excavation to deposit areas.

CONSTRUCTION

After two years of surveys, work on the canal began in 1882. A banquet on January 20, 1882 in Panama City, marked the official beginning of Culebra Cut excavation. However, little actual digging was accomplished due to a lack of field organization. Engineers continued performing surveys and preliminary work, sending reports to Paris. When excavation began in January 1882, it proceeded rapidly, but after the removal of 40 million cubic yards [30 million cubic meters] of earth, only twenty feet of material had been removed from a ridge over two hundred feet high. Deforestation led to landslides, which in turn led to more excavation. Men were buried in landslides and railroads tracks repeatedly disappeared into the excavated trenches. Couvreux and Hersent decided to withdraw from the project and wrote to de Lesseps requesting cancellation of their contract on December 31, 1882.

Ferdinand de Lesseps decided to assume control of the operation, from Paris, working with a Director in Panama. Jules Dingler was appointed as the new Director General. An engineer of outstanding ability, reputation, and experience, Dingler arrived in Colon on March 1, 1883, accompanied by his family, along with Charles de Lesseps. Dingler concentrated on restoring order to the work and the organization. He instituted a new system of small contracts for which the Company rented out the necessary equipment at low rates. It was an inefficient system, requiring a tremendous paperwork and involving numerous lawsuits in Colombian courts, but work progressed, making use of the available labor force.

As the work force increased, so did illness and death. The first yellow fever death occurred June 1881, soon after beginning of the wet season. A young engineer died on July 25, supposedly of "brain fever." On July 28, international finance authority Henri Bionne died. Yellow fever appeared in epidemic form during the wet season of 1882. The Company established medical services presided over by the Sisters of St. Vincent de Paul, and established hospitals on both the Atlantic and Pacific sides of the isthmus by 1882. However, contractors often allowed workers to die on the job rather than pay for a hospital bed.

With the knowledge of mosquitoes' role in the transmission of yellow fever and malaria not yet discovered, the hospitals unwittingly committed a number of acts that fostered the spread of the diseases. Waterways were constructed around flower and vegetable gardens, and water pans were placed under beds to repel crawling insects. Both insect-fighting methods provided breeding sites for the *Stegomyia fasciata* and *Anopheles* mosquitoes, carriers of yellow fever and malaria. Many patients who came to the hospital for other reasons often fell ill after their arrival and people began to avoid the hospital whenever possible.

Excavation work was progressing in Culebra Cut and was expected to be finished by May 1885. However, there was growing concern about bank stability and the danger of slides. At the Atlantic and Pacific entrances, dredges worked their way inland. Machinery arrived from France, the United States and Belgium but equipment was constantly modified and used in experimental combinations as most was too light and too small. An amassing of discarded, inoperative equipment along the canal line testified to earlier mistakes. As late as July 1885, only about ten percent of the estimated total had been excavated. Ultimately, the unresolved problem of the slides would doom the sea level canal plan to failure.

In January 1884, tragedy struck the Dingler family as his daughter died of yellow fever. A month later his twenty-year-old son, Jules, died of the same disease. Shortly after, his daughter's young fiancé, who had come with the family from France, contracted the disease and died also. In 1885, Dingler's wife died of yellow fever, just about a year after her daughter and son. Dingler stayed on the job until June 1885, when he returned to France, never to return to the isthmus.

The project had a large turnover of labor. The laborers were mostly Jamaicans, men of the Antilles, South American Indians, and Chinese, officered by Frenchmen. At whatever level a man performed work; he was not likely to stay long unless he was well paid. With some 10,000 men employed, work was going well in September of 1883. The maximum force employed by the French at any one time was reached in 1884, with more than 19,000. All the while, the toll in human lives climbed, peaking in 1885. Yellow fever was constant and malaria continued to take more lives than yellow fever. Because the sick avoided the hospitals whenever possible because of its reputation for propagating disease, much of the death toll was never recorded. Along with yellow fever and malaria cases, there were an estimated 27,000 laborers and engineers killed between 1881 and 1889.

PHILLIPE BUNAU-VARILLA

After the departure of Dingler, the new acting Director General was 26-year-old Philippe Bunau-Varilla, a young, capable, and energetic engineer. Under Bunau-Varilla, worker morale improved, and excavation increased along the line. Still, there was inadequate equipment and work organization. In a move toward greater efficiency, Bunau-Varilla reinstituted the old method of large contractors.

A new Director General, Leon Boyer, arrived in January 1886. Soon thereafter, Bunau-Varilla, himself, contracted yellow fever, and greatly weakened, he went back to France to recuperate, after which Bunau-Varilla returned his expertise to the Panama Canal and eventually played a pivotal role in the American effort at Panama.

Boyer informed his superiors his conviction that within current time and cost limits, it would not be possible to construct a sea level canal. He recommended the design proposed by Bunau-Varilla of a temporary lake and lock canal that could later, after it was built and functioning, be gradually deepened to sea level. However, by May, he too was gone, another victim of yellow fever.

There were six bond issues between 1882 and 1888, totaling 781 million francs, but only forty percent of this money went into the excavation. Large salaries were paid to directors. An exorbitant sum of the funds raised for the endeavor was squandered on publicity and commissions to underwriters responsible for placing stocks and bonds. Financial journals in France received payments before each bond issue, presumably to ensure positive publicity for the project. Political leaders were receiving payment at every decision-making point. In February 1885, the eighty years old de Lesseps visited the construction site in Panama. When it was realized how little of the excavation work had been completed, investors became nervous. The Panama stocks plummeted on the Paris Stock Exchange.

In spite of improvements in the field, the Company was in a financial crisis. A lack of progress at Culebra was beginning to concern Parisian officials. Charles de Lesseps proposed to Bunau-Varilla to organize a company to take on the work at Culebra. Bunau-Varilla took over the field supervision of the work himself. As American engineers would do later, he moved into quarters at the Cut so he could watch the progress of the work. About six months later, the French work at Culebra Cut had reached peak activity.

It was becoming increasingly clear to nearly everyone except Ferdinand de Lesseps that, under the circumstances, only a high level lock canal had any chance of succeeding at this point. In October 1887, a report was released by eminent French engineers establishing the possibility of building a high-level lock canal. The plan would allow vessel transits while simultaneously permitting dredging of a channel to sea level sometime in the future. Ferdinand de Lesseps reluctantly relented and he drew up a contract with engineer Gustave Eiffel, for the construction of a sluice canal by 1890.

Bunau-Varilla's concept was to create a series of pools in which floating dredges could be placed; the pools would then be connected by a series of locks. The highest level of such a canal would be 170 feet. Work on the lock canal started on January 15, 1888. Eiffel, builder of the Eiffel Tower in Paris, would construct the canal locks.

Under Bunau-Varilla's company, Artigue, Sonderegger et Cie., work was progressing. Some areas of the canal were nearly complete, the Panama Railroad was rerouted away from the Cut, the first lock was nearly ready to begin installation and preliminary work on a dam had been started. In Culebra Cut, where the average level had been lowered only 3 feet by 1886, was lowered 10 feet in 1887 and 20 feet in 1888, ultimately bringing the level to 235 feet at the time work was stopped.

BANKRUPTCY AND THE NEW PANAMA CANAL COMPANY

In March 1888, several ministers formulated a law allowing the Panama Canal Company to carry out a lucrative lottery loan, with guaranteed funds. Three months later, the Panama Canal Company was authorized by law to set up a government supported lottery loan for 720 million francs, for the completion of the sluice canal according to Eiffel's plans. Despite all de Lesseps' efforts, the attempt at placing these lottery bonds failed.

The company stopped its payments and tried to receive a three-month payment moratorium. Shareholders, at their last meeting in January 1889, decided to dissolve the Company, placing it under legal receivership and Ferdinand de Lesseps resigned from the management. Some aspects of the work persisted for a few months, but by May 15, 1889, all activity on the Isthmus ceased. Liquidation was completed in 1894.

In France, popular pressure on the government led to what was called the "Panama Affair." Three years after the collapse of the Company, the largest corruption scandal of the 19th century shook the French Republic. A large number of ministers were accused of taking bribes from de Lesseps in 1888, for the permit of the lottery issue. This resulted in a corruption process being held in 1892, against Ferdinand de Lesseps and his son Charles who were both indicted for fraud and maladministration. At the same time, 510 members of parliament, including six ministers, were accused of bribery by the Panama Canal Company, in connection with the course of events concerning the permit for the lottery issue.

Advanced age and ill health excused the senior de Lesseps from appearing in court, but both were found guilty and given 5-year prison sentences. However, the penalty was never imposed, as the statute of limitations had run out. Ferdinand de Lesseps' mental state was such that he knew little of the situation and he remained sequestered at home. "Le Grand Français" died at age 89 on December 7, 1894, and was buried in the Le Père Lachaise cemetery in Paris.

Charles was indicted in a second trial for corruption and found guilty of bribery. Time he had already spent in jail during the trials was deducted from his one-year sentence. Then, becoming seriously ill, he served the remainder of his sentence in hospital. Charles lived until 1923, long enough to see the Panama Canal completed, his father's name restored to honor and his own reputation substantially cleared.

With the original Wyse Concession to expire in 1893, Wyse negotiated a 10-year extension. The "new" Panama Canal Company, the Compagnie Nouvelle de Canal de Panama was organized effective October 20, 1894. With insufficient working capital to proceed with significant work, the Compagnie Nouvelle hoped to attract investors who would help them to complete an isthmian canal as a French enterprise. Initially, they had no intention of selling their rights. The new company wanted to make a success of the operation and perhaps repay the losses of the original shareholders.

Sailing from France on December 9, 1894, the first group arrived in Panama to again resume excavation in Culebra Cut. Work on the canal was taken up again and at the beginning of 1895, 500 workers were in employment on the canal. In 1898, the canal employed 3400 workers. Between 1895 and 1898, 3.8 million cubic yards [2.9 million cubic meters] of earth were excavated.

The Comité Technique was formed by the Compagnie Nouvelle to review the studies and finished and ongoing work and devise the best plan for completing the canal. The committee arrived on the Isthmus in February 1896 and presented their plan on November 16, 1898. Many aspects of the plan were similar to the canal that was finally built by the Americans in 1914. It was a lock canal with two high level lakes to lift ships up and over the Continental Divide. Artificial lakes would be formed by damming the Chagres River providing both flood control and electric power.

The directors of the Compagnie Nouvelle soon faced with the reality of their situation following the scandal of the old company. The public had lost faith in the project. There would be no funds forthcoming from a bond issue, and the French government did not provide any support for the project. With half its original capital gone by 1898, the company had few choices, abandon it or sell it.

It was no secret that the United States was interested in an isthmian canal. With the technical commission report and a tentative rights transfer proposal in hand, company officials headed for the United States, where they were received by President William McKinley on December 2, 1899. A deal was five years in the making, but was eventually signed. In 1904, the United States bought the French equipment and excavations, including the Panama Railroad, for 40 million dollars, of which 30 million related to excavations completed, primarily in the Culebra Cut.

CONCLUSION

Many reasons can be cited for the failure of the French to complete the Panama Canal, The principal reason was Ferdinand de Lesseps' stubbornness in insisting on the sea level plan. He had neither a technical education nor financial experience. At Panama, de Lesseps was a victim of his own success at Suez. His own charisma and confidence turned out to be his enemy as people believed in him beyond reason. He preferred to rely on his judgment, which was proven by the success of the construction of the Suez Canal.

A large part of the eventual success on the part of the United States in building a canal at Panama came from avoiding the mistakes of the French. A lack of knowledge of sanitation and tropical diseases contributed greatly to the demise of the French endeavor. The American success at Panama benefitted from twenty years of advances in health, medicine, hygiene, and sanitary engineering as well as advances in engineering and construction.

The American effort also benefitted from the project left behind by the French. Beginning in 1888, the sea level project was finally abandoned for a lock canal with the idea that, after the lock canal was functional, the channel could be deepened gradually to make a sea level canal. When the Americans did take over the canal construction, in 1903, a third of the work had been completed and the materials, buildings, and work the French left behind was of highest quality.

The French experience was without question a financial failure, but in retrospect, the failure did not lie with French engineering. Ferdinand de Lesseps was not an engineer and he did not follow the advice of the specialists. The de Lépinay design in 1879, contained the basic elements ultimately designed into the American design. The French company would use these concepts as a basis for a lock canal that they eventually adopted in 1887, following the disappointment of their sea level attempt. Had his plan been originally approved, France may have prevailed in their canal construction effort and had it been adopted at the beginning, the Panama Canal may have been completed by the French instead of by the United States.

REFERENCES

Beatty, C. (1956). De Lesseps of Suez: The Man and His Times, Harper & Brothers, New York.
Brown, F. (2010). For the Soul of France: Culture Wars in the Age of Dreyfus, Knopf, New York.
Brunn, S. D. (2011). Engineering Earth: The Impacts of Megaengineering Projects, Springer, New York.
DuVal, M. P. (1968). Cadiz to Cathay: the story of the long diplomatic struggle for the Panama Canal, Greenwood Pres, New York.
Holborn, H. (1982). A History of Modern Germany, 1840-1945 (v. 3), Princeton University Press, Princeton, NJ.
McCullough, D. (1977). The Path Between the Seas: The Creation of the Panama Canal 1870-1914, Simon & Schuster, New York.
Schoonover, T. D. (2000). The French in Central America: Culture and Commerce, 1820-1930, Scholarly Resources, Inc., Wilmington, DE.
Panama Canal Authority. The French Canal Construction. www.pancanal.com.

Building the Panama Canal
(Men, Machines, and Methods)

Raymond Paul Giroux[1], M.ASCE

[1]ASCE Chairman Brooklyn Bridge 125th Anniversary, ASCE Chairman Golden Gate Bridge 75th Anniversary, Recipient 2013 ASCE Civil Engineering History and Heritage Award; Ridgefield, WA; paul.giroux@kiewit.com

ABSTRACT

During the early 20th century strategic control of the oceans was essential to all of the great industrial powers of the world. In 1904 United States President Theodore Roosevelt would commit the brains of American engineering and the brawn of America's industrial machine to build a canal of unprecedented scope and challenge. Panama Canal's successful construction was the result of the convergence of extraordinary men, machines, and methods. In the decades preceding Panama Canal's construction, tremendous advancements were realized in every discipline of engineering. In the realm of heavy civil construction, these collective engineering advancements provided the construction technology that made Panama Canal possible. This paper highlights how the right men, the right machines, and the right methods all came together in 1904 to build a project of unprecedented scope and challenges.

INTRODUCTION

In the mid-19th century speedy passage to the burgeoning west coast of North America became increasingly important. Yet, travel by steam ship from New York to San Francisco by the Magellan Route around the tip of South America was an arduous 13,700 mile journey.

Yet, it was known that if travelers were brave enough to face the challenges and dangers of traversing the jungles at the Isthmus of Panama, they could take an 8,400 mile shortcut. Spain's Vasco Balboa had blazed the Camino Real Trail in 1513 after he led a 25 day expedition through the Darien wilderness from the Atlantic Ocean through the jungles to the Pacific Ocean at the Isthmus of Panama. So for the next 350 years travelers would use Spain's Camino Real Trail as it was originally developed, a foot trail.

The trail was established a quarter-century before the industrial revolution; which was to be a period of rapid technological advancement and transition from muscle to machine, from wood to coal, from the horse to the iron horse railroads. While the Industrial Revolution began in Great Britain, within a few decades it had spread to Western Europe and the United States. Coincidental with the industrial revolution was America's western expansion and securing the Panamanian trade route became increasingly important.

In 1846 the Bidlack Treaty was signed, which granted the United States a right-of-way across the Isthmus of Panama. In 1847, a group of New York financiers organized the Panama Railroad Company and further increased the demand for speedy passage across the Isthmus. Construction of the railroad across the isthmus started in May of 1850.

It was extremely difficult to build a railroad through the swamps and high country of Panama. One of its biggest challenges was the construction of the Barbacoas Bridge over the Chagres River in 1854. At the time that this bridge was erected it was reputed to be the longest in the world, consisting of six wrought iron spans each over one hundred feet in length. Finally, after five years of construction effort and some 5,000-10,000 lives lost, the first train crossed the summit at Culebra on January 27, 1855. On a cost per mile basis; when completed the Panama Railroad was the most expensive railroad ever built.

The strategic value of a railroad at the Isthmus of Panama did not go unnoticed back in the United States. Many wondered: if a railroad could be built, could a canal be built? So in January 1870, the Navy Department ordered an expedition to explore the Isthmus to ascertain a point at which a canal could be built from the Atlantic to the Pacific. The expedition set sail on January 22, 1870 as the steam-sloop *Nipsic* slipped out of the Brooklyn Navy Yard, a short distance from where Washington Roebling was working to prepare to sink the Brooklyn caisson of the new Brooklyn Bridge. The Darien expedition would follow the same trail blazed by Balboa in 1513. Between 1870 and 1875 there were a total of seven American expeditions to evaluate a Central American canal. (McCullough, 1977)

During the industrial revolution the promise of emerging science and technologies would capture the imagination of the masses. Yet, during the industrial revolution the science behind many innovations was not always understood. For example when James Watt designed his revolutionary steam engine in 1765, he did so without the science of thermodynamics. So for many living during this era the lines between science and science-fiction were blurry. Many of the great engineering achievements of the day were deemed by many as fanciful and whimsical. The French science-fiction writer Jules Verne was one of the most influential authors of the 19th century. Verne would write about how men could *Journey to the Center of the Earth* (1864), how men could travel *From the Earth of the Moon* (1865), how Captain Nemo could go *20,000 Leagues Under the Sea* (1870), and how Phileas Fogg could travel *Around the World in 80 Days* (1873). To many, it seemed there would be no limits to what man could achieve.

So perhaps with some naivety and plenty of ingenuity engineers would push the limits of emerging technologies. And perhaps no other engineer than Isambard Kingdom Brunel would push the limits of ship design and ocean travel. His revolutionary ship, the Great Eastern, was launched in 1858, and was the biggest ship of its day; 692 feet in length with a displacement of 32,000 tons. The ship's all-iron, double-hulled construction was a first of its kind. It was designed to carry 4,000 passengers around the world without refueling, at a top speed of 14 knots. Brunel's Great Eastern was a virtual time machine on the seas.

Beginning it 1859, the French sought to build their own time machine and broaden their global influence in the Eastern Mediterranean by building a 5,800 mile shortcut between Europe and India at Suez on the Sinai Peninsula. The Suez Canal effort was led by Frenchman Ferdinand de Lesseps. To build the 102 mile long Suez Canal, de Lessups and his engineers began digging with armies of labor. Later de Lesseps would learn to harness the latest technology and deploy some of the most marvelous construction machines of the day. The excavation through the desert sand of the Sinai employed roughly 30,000 men on average during the ten year schedule. Upon completion of the canal in 1869, de Lesseps was a true larger-than-life figure in France. A source of great pride for the French, perhaps many felt de Lesseps could achieve anything.

PANAMA, THE FRENCH ERA

Ten years later, France sought once again to build another great canal, this time at the Isthmus of Panama. In May of 1879, a delegation of 136 men, including de Lesseps, assembled in the rooms of the Geographical Society in Paris. They voted in favor of the creation of a canal, which was to be without locks, like the Suez Canal. Of the 74 that voted in favor of the canal, only 19 were engineers, and only one of those had ever been to Central America. (McCullough, 1977) At the onset it had an estimated cost of $214,000,000 with a planned duration of twelve years.

De Lesseps was appointed President of the Panama Canal Company, despite the fact that he had reached the age of 74. It was on this occasion that the French bestowed upon him the title of "Le Grand Français." After an initial funding tour and a preliminary survey of Panama, Compagnie Universelle du Canal Interoceanique was incorporated.

With France's success at Suez, many young Frenchmen wanted to sign up for duty at Panama to be involved in France's next great adventure. The new venture was exotic, exciting, and it paid well when compared to working in France (Figure 1). Those who would go to Panama were revered by their countrymen. These young engineers and builders were swept away by a heavy mix of patriotism, bravado, and opportunity for adventure in the jungles of Panama.

De Lesseps was a masterful salesman and had little problem raising funds. Initial shares of Compagnie Universelle du Canal Interoceanique were readily sold in 1881 for 500 francs, or about $100. For many this was a year's earnings being entrusted to the "Le Grand Français." (McCullough, 1977) By 1882 the French were ready to begin work and in 1884 French engineer Philippe Bunau-Varilla went to Panama as the chief engineer for de Lesseps. The French plan had many technical challenges with a sea level canal. First and foremost, the variation in tide on the Atlantic side is 2.5 feet as a maximum, and on the Pacific it is 21.1 feet as a maximum. Therefore, ships navigating the canal would be fighting currents with the ebb and flow of the tides. To solve this problem, the French eventually planned a "Tidal Lock" at Miroflores.

Figure 1. This photo of the French Technical Commission was taken in December 1879 on the occasion of their arrival at Limon Bay. De Lesseps is seated in the middle of the second row, fourth from left.

Back in the United States, Americans had a sense of awe and respect for the French. With their gift of the Statue of Liberty to the United States dedicated on October 28, 1886, it must have seemed the French could do anything they put their minds to.

So with a deep sense of national pride the French went to work, deploying many fantastic excavating machines using the latest technology. At the Suez Canal the French were digging sand; however, digging through the Isthmus would require millions of cubic yards of rock excavation. The French machines, given their light design, were not well suited for the harsh rock conditions. With their large number of moving parts, French machines would require extensive maintenance to remain operational.

Unlike Suez where there was no rain, at the Isthmus of Panama on the north side of the continental divide there can be 140 inches of rain per year. Seeking to use the natural course of the Chagres River for much of the French canal, the French quickly learned its perils. With a drainage basin of 1,300 square miles, the Chagres River's normal rainy season flow of 10,000 cubic feet per second could quickly exceed 80,000 cubic feet per second, causing the river to rise over 40 feet in one day in some locations. On one such occasion, the 300 foot wrought-iron bridge at Barbacoas that was built 40 feet above river was wiped out after a three day torrential rain.

The French would find Panama a very dangerous place to work, with one in five workers dying in these early years, mostly from yellow fever and malaria. In addition to the rains, they would face average daily temperatures over 90 degrees Fahrenheit, with 90 percent humidity for much of the year.

There were rats, snakes, and swarms of bugs. On one occasion in December of 1885, Buana-Varilla was in a canoe surveying the aftermath of a devastating flood of the Charges River when he saw the strangest phenomenon. As the flood waters rose, Buana-Varilla noticed a curious change in color of the foliage from green to black in a zone about a yard above the water level. As they moved closer, he realized in horror that the leaves and branches were totally concealed by millions and millions of tarantulas. (Cadbury, 2004)

The dangerous work, yellow fever, and malaria all took their toll on the work force, with the death rate over 200 workers per month. Harper's Weekly posed the question; "Is de Lesseps a canal digger, or a grave digger?" (Cadbury, 2004)

As the deaths mounted, the costs would also soar to over $300,000,000, and public sentiment for the project would wane. However, the French were fighters. At Culebra Cut during 1888 and 1889, Bunau-Varilla and his fellow Frenchmen worked hard to build their sea level canal. Yet, with the reality of their ill-suited equipment, the turbulent Chagres River, yellow fever, and dwindling financial support, it became apparent a locked canal should be evaluated.

In the late 19th century, perhaps no other civil engineer exemplified French engineering prowess more than Gustave Eiffel. A brilliant structural engineer, Eiffel is best known for the world-famous Eiffel Tower, built between July 1887 and March 1889, which was showcased at the 1889 Universal Exposition in Paris. So in November 1887, de Lesseps hired Eiffel to design and build a lock system based upon a lake elevation 161 feet above sea level. This scheme would follow the same line as the original plan, yet it would require a dam be built at Bohio, about fifteen miles upstream of Gatun. The new scheme would also require two huge flights of locks, each with five locks measuring 590 feet by 59 feet (180 meters by 18 meters).

Unfortunately, the abandonment of a sea level canal in 1887 was too late to salvage the French effort. The initial funds raised through the stock sale at $100 per share quickly ran out. For eight years de Lesseps continued to raise funds through seven different stock sales; however, by late 1889, more and more nervous investors turned to the bear market to sell their stock. By December 14, 1889 the bonds were worthless.

The collapse of the Compagnie Universelle du Canal Interoceanique Incorporated rocked the very foundation of the French Government. A generation of wealth was lost in the morass of Panama. In the wake of the failure, investigations and trials would expose huge problems, including mismanagement, broad corruption, inflated progress reporting, and other problems. The reputation of de Lesseps was ruined and he was condemned to five years in prison (although he never served). Even Eiffel was found guilty of misuse of funds, and was sentenced to two years in prison, although later acquitted on appeal. Over 5,000 French people died working on the project. In all, over 25,000 people died during the eight year French effort, mostly from malaria and yellow fever.

In an attempt to salvage something for the stockholders, the French concluded they would have to continue the work. So in 1894 a second canal company; *Compagnie Nouvelle du Canal de Panama* was created with its goal to finish the construction. Yet, the greatest number of people working in the new venture was 3,600, primarily tasked to keep the remaining French assets in saleable condition. In spite of this maintenance effort, the jungles of Panama would slowly swallow the French industrial machine.

PANAMA, THE AMERICAN ERA

While French assets rusted away, the American industrial machine was rapidly building up a head of steam. In the period following the failure of de Lesseps' plan until the United States began the construction of the canal, the era of steel and machinery development had attained full growth in America. (Bennett, 1915) In 1890 the annual American steel production was about 4,200,000 tons, and by 1910 had reached some 22,000,000 tons.

A decade after the French defeat at Panama, Theodore Roosevelt led the Rough Riders to victory in the Spanish-American War of 1898 in Cuba (April 25 to August 12, 1898). During the war, Roosevelt became aware of America's naval strategic weakness, having waited for two months for the Navy's USS Oregon to arrive from San Francisco to Cuba.

With the assassination of President William McKinley in September of 1901, Theodore Roosevelt became president. Roosevelt was determined to make the United States a world power and deemed a canal in Central America essential to the American Navy becoming an effective two ocean power.

Immediately, Roosevelt ordered design studies for a Central American canal in 1901. In an effort to recover some investment for *Compagnie Nouvelle du Canal de Panama* Philippe Bunau-Varilla helped lobby for a canal at Panama in Washington, DC.

To be sure, no written or oral accounts of what had happened could fully provide members of the House and Senate a comprehensive understanding of why France failed in Panama. With some degree of naivety, there was an underlying theme in much of the testimony that industrious, practical, moral men - Americans - might succeed where others had failed. (McCullough, 1977) So after much debate, in 1902 the legislature chose Panama as the site of the proposed canal, and Roosevelt signed the Panama Canal Act into law on June 28, 1902.

Yet the Panamanian territory was under the control of Columbia and in 1903, Columbia decided to not cede the land for the canal to the United States. Acting quickly with the aide of Bunau-Varilla the United States assisted Panama juntas to secede from Columbia. With the signing of the Hay-Bunau-Varilla Treaty November 18, 1903, the United States gained control of the ten mile-wide canal zone across the

Isthmus of Panama. The American plan was to continue where the French had left off and build a sea-level canal. After purchasing the remaining French assets for $40,000,000, the United States began to move forward in 1904. (Cornish, 1909)

THE AMERICAN ERA UNDER WALLACE

Roosevelt named John Wallace Chairman and Chief Engineer of the Isthmian Canal Commission (Figure 2). Expectations about the canal were plainly stated by Roosevelt; "What this nation will insist upon is that results be achieved." To build a sea-level canal, hundreds of millions of cubic yards of earth would have to be excavated. Roosevelt ordered Wallace to "make dirt fly." (McCullough, 1977)

Wallace Biographical Summary:

1852: Born Fall River, Massachusetts September 10
1882: Civil Engineering Degree College of Wooster
1883: Supt. of Construction Iowa Central RR
1886: Bridge Engineer Atchison, Topeka & Santa Fe RR
1889: Resident Engineer, Chicago, Madison, Northern RR
1891: Engineer of Construction Illinois Central RR
1892: Chief Engineer Illinois Central RR
1897: Assistant Second VP Illinois Central RR
1900: General Manager Illinois Central RR
1900: President ASCE
1904 - 1905: Chief Engineer Panama Canal
1921: Dies July 3

Figure 2. John Findley Wallace, Chief Engineer 1904-1905

Having served as the President of the American Society of Civil Engineers (ASCE) in 1900, Wallace was considered one of the most prominent civil engineers of the day. With a background in railroad design, construction, and management, Wallace appeared to be a good choice for the task. Yet from the time of his initial reconnaissance in July 1904, he had been openly skeptical and discouraged. He had only seen "jungle and chaos from one end of the Isthmus to the other." (McCullough, 1977) Hundreds of French machines were lying in huge scrap heaps, slowing rusting, being swallowed by the jungle growth, or sunk in the shallows of the swamps.

At the start of the American effort, crews tried to use some of the antiquated machines left behind by the French. (Figure 3) They soon got a dose of what the French had experienced with flooding, landslides, and terrible living conditions. Wallace realized almost immediately that the Isthmus' harsh terrain would be a serious obstacle to construction. It was a daunting, seemingly impossible task; to dig a channel 50 miles in length and 30 feet below sea level stretching from the Atlantic coast to the Pacific coast.

Figure 3. French machines being used in Culebra Cut in 1904.

Wallace desperately needed time to make a workable plan. However, "in deference to what may be termed the clamor of ignorance, the Commission decided that 'dirt must fly,' and without proper plant — excepting a few modern shovels — and in the utter absence of any intelligent, comprehensive plan, work was begun and carried along in an expensive and unsatisfactory manner." (Sibert & Stevens, 1915)

In addition to the project's technical challenges, Wallace also faced bureaucratic challenges from Isthmian Canal Commission (ICC). The ICC was a seven-member presidential committee established to help avoid the inefficiency and corruption that had plagued the French. Rather than give Wallace sole authority for the work, the ICC was charged with ensuring all decisions were fully vetted and transparent. Thousands of work requests required approval each week from the ICC, some 2,500 miles away. The process was cumbersome and frustratingly slow. On the Isthmus, to hire a single handcart for an hour required six separate vouchers. Carpenters were forbidden to saw boards over ten feet in length without a permit. (McCullough, 1977) Unable to build any significant momentum, work came to a virtual standstill.

Further, the yellow fever and malaria that had plagued the French still remained. The death toll would continue to climb, and fear set into the work force. In the summer of 1905, there were hundreds of deaths due to yellow fever, malaria, pneumonia, chronic diarrhea, and dysentery. Just as the French experienced, daily funeral trains were leaving for cemeteries at Monkey Hill and Ancon Hill.

In an attempt to rid the Isthmus of disease, Colonel W.C. Gorgas, M.D. (Figure 4) was appointed to lead the Department of Sanitation. When he arrived in Panama in June 1904, he found a mosquito paradise. After assessing the conditions, Gorgas wrote, "The experience of our predecessors was ample to convince us that unless we could protect our force against yellow fever and malaria we would be unable to accomplish the work." Even though Gorgas believed mosquitos transmitted disease, Wallace did not support Gorgas' theory, so death, fear, and delay held its grip on the work force. Gorgas pleaded his case saying, "If you get rid of the yellow fever, you get rid of the fear." (McCullough, 1977)

Gorgas Biographical Summary

1854: Born Toulminville, Alabama October 3
1875: University of the South, BA Degree
1879: Bellevue Hospital Medical College, MD
1898: Chief Sanitary Officer Havana
1903: Appointed Colonel
1904: Arrives in Panama
1908: President of American Medical Association
1914: Surgeon General US Army
1920: Dies at age 64

Figure 4. William C. Gorgas, Head of the Department of Sanitation

The chaos, the Commission, and the disease; it was all too much for Wallace. He abruptly resigned in June 1905, and returned to the United States.

THE AMERICAN ERA UNDER STEVENS

Roosevelt quickly replaced Wallace with John Frank Stevens (Figure 5) as the new chief engineer. Many who knew Stevens considered him to be the best construction engineer in the country, if not the world. Stevens was a self-taught railroad engineer who went on to become the chief engineer for the Great Northern Railroad. Roosevelt wrote of Stevens, "a rough and tumble westerner, a big fellow, a man of daring, and good sense, and burly power." Stevens was known as "the hero of Marias Pass" the passage over the Continental Divide, having discovered it in 1889. (McCullough, 1977)

Stevens accepted Roosevelt's offer to run the Panama Canal effort for a salary of $30,000 per year. He was 52 years old. In July of 1905, Stevens and the head of the 2nd Isthmian Canal Commission, Iowa attorney Theodore Shonts, arrived in Panama. Stevens found a mess; $128,000,000 had already been spent, a year had been lost, the Isthmus was in a shambles, and the work force was stricken with fear. (McCullough, 1977) Accommodations on departing ships were at a premium and most of the Americans already in Panama assumed that when Stevens arrived he would tell them to pack up and go home.

Stevens wrote: "I believe I faced about as discouraging a proposition was ever presented to a construction engineer." After studying the management structure, Stevens wrote "I found no organization....no answerable head who could delegate authority....and no cooperation existing between what might be charitably called departments." (McCullough, 1977) Nearly everywhere Stevens looked he found a tendency to postpone action. Finally, he concluded that there were three diseases on the isthmus, "Yellow fever, malaria, and cold feet." (Bennett, 1915)

Stevens would soon learn first-hand the dangers of the Chagres River, which some called "the lion in the path." Like the French, Stevens would soon conclude that it was futile to try to tame the Chagres.

Stevens Biographical Summary:
1853: Born West Gardiner, Maine April 25
1873: Minneapolis City Engineers' Office
1886: Principal Assistant Engineer DSSARR
1889: Locating Engineer for Great Northern RR
1895: Chief Engineer for Great Northern RR
1903: Vice President Rock Island & Pacific RR
1905 - 1907: Chief Engineer Panama Canal
1917: Consults on Trans-Siberian RR
1919: Consults on Chinese Eastern RR
1927: President ASCE
1943: Dies at age 90

Figure 5. John Frank Stevens, Chief Engineer 1905-1907

During the summer of 1905, railroad traffic was nearly at a standstill. Thousands of tons of freight piled high on the docks and would sometimes sit for months and often got lost. (Sibert & Stevens, 1915) Unlike Wallace, Stevens knew that the Panama Railroad was the lifeline of the Isthmus; however, he found the railroad to be a mismanaged mess. On a visit to Culebra Cut, Stevens counted seven trains off of the tracks and every steam shovel idle. The rolling stock and rails were in shambles. Finally, recognizing the futility of "making dirt fly" with no workable plan, Stevens stopped all work in the Cut on August 1, 1905. (McCullough, 1977)

Quickly assessing the challenges of the sea level plan, Stevens would call the plan "an entirely untenable proposition, an impracticable futility." As early as 1879, Ashbel P. Welch, M.ASCE, referred to the possibility of closing the Charges Valley by a dam at Gatun. (Goethals, 1911) After studying the Chagres River, Stevens refined the lake canal plan. Referred to as the "minority plan," Stevens' plan was to construct a locked canal that would harness the water from the Chagres River to create a vast inland lake at elevation +85 feet. In 1905, Major C.E. Gillette,

M.ASCE, presented a plan to build a dam at Gatun to the International Board of Consulting Engineers. (Goethals, 1911) The minority plan for a high-level lake was mostly written by Alfred Nobel.

In spite of the efforts of Stevens and others to promote the minority plan, the Board of Consulting Engineers recommended in a majority report to the ICC, a tide-level canal as practicable and best fulfilling the national requirements, defined by the Spooner Act of 1902. Based on the recommendation of the Board, Roosevelt approved the majority plan. (Cornish, 1909) Unfortunately, even as Stevens worked to restore order on the Isthmus, he was still not sure what he was supposed to build as the so called "battle of the levels" debate to build a sea-level or a locked canal design continued in Washington, DC throughout 1906. (Goethals, 1911)

Stevens told his staff that work would not resume until there was a plan and everything was made ready. Recognizing the importance of things like logistics, living conditions, and sanitation, Stevens would say, "The digging is the least of it all." (McCullough, 1977) Stevens was a man with a plan, and brings to mind the palindrome, "a man, a plan, a canal, Panama."

Stevens ordered the roads of Panama City and Colon to be paved. Over 2,000 new structures on the Isthmus were built including, office buildings, hospitals, hotels, messes, kitchens, shops, storehouses, living quarters, water, and sewage systems. Further, Stevens directed 1,200 French buildings be rebuilt. (Goethals, 1911) He also encouraged women and families to come to the Canal.

Through the summer of 1905 and into 1906, Stevens transformed the old and tired Panamanian Railroad into a modern logistics machine, capable of supporting the construction of the canal. By June 1906, 350 additional miles of heavy-duty track had been built along the Panama Railroad. Double tracks were built to allow for two-way train traffic, and additional track routes allowed more spoil to be transferred around the clock. Massive orders for new rolling stock were placed; 115 heavy-duty locomotives, 2,300 dirt spoils railroad cars, and 101 railroad mounted steam shovels. To keep all of the equipment operating at peak production, Stevens directed the construction of maintenance shops, including the main shop at Gorgona. New terminals, docks, warehouses, and coaling plants were constructed at Colon and Balboa. Stevens also assigned new railroad personnel at all levels of management, with many of the men coming from the United States. Stevens wrote to Secretary of War William Howard Taft that he was determined to prepare well before beginning construction.

Unlike Wallace, Stevens would put his faith in Gorgas to rid the Isthmus of disease. Beginning in August 1905, Stevens threw all of the weight of the engineering department to the aid of Gorgas. By November, 4,000 men were working at the direction of Gorgas. Under his direction lands were cleared, pools and swamps were drained, and fumigation brigades set out to exterminate the mosquito. By December 1905, just as Gorgas had predicted, when the mosquitos were gone, so too was the disease and the fear.

After Roosevelt read Stevens' report in favor of a lock canal, the President became a convert from the sea-level type and ordered work on the Isthmus move forward with a view to constructing a lock canal. (Bennett, 1915) With the support of Roosevelt on February 2, 1906, the ICC overrode the majority decision in favor of building a lock canal. However, it would not be until June 29, 1906, that Stevens had final approval for a locked canal with a lake at elevation +85 feet. Unlike the sea-level plans, which would have been tormented by the Chagres River, the lock canal would harness the waters of the Chagres to power the canal. Further the lock canal reduced the required depth of excavations by 70 feet. (Sibert & Stevens, 1915)

In the locked canal plan, ships would traverse the Isthmus through a double lock at Miraflores, a single lock at Pedro Miguel, cross the continental divide at Culebra Cut, steam through a lake at elevation +85 feet, then return to sea level through a triple lock at Gatun. Yet to create the lake, the largest earthen filled dam in the world would have to be built at Gatun to create the world's largest man-made lake. The proposed 164 square mile Gatun Lake would place the old alignment of the Panama Railroad underwater, in some areas as much as 70 feet.

So the Panama Railroad had to be relocated. In the Gatun Region, half of the realigned track would be washed by the waters of Gatun Lake, so it would have to be at elevation +92 feet, to place it 7 feet above the proposed lake level. Much of the relocated alignment would be through the swamps of the Quebrancha, the Brazos, the Baja, and the Gatun Valleys so huge embankments had to be constructed. Embankment construction began by driving of timber trestles across the swamps and rivers.

Various pile drivers, including steam-powered drop hammers, were used to drive timber pile for trestle bents. To carry heavy loads, railroad designers settled on Cooper E-50 loading for all railroad bridge designs. At the completion of the trestles, spoils from cut excavation operations were side-dumped off of the trestles (Figure 6) to make embankments as large as necessary. During installation of the embankments, settlement in the soft swampy soils was common place.

A typical two stage embankment operation included: installation of a timber trestle for the first deck work; side-dumping first deck embankment to one side of the trestle; installing the second trestle for the second deck work; and, completion of first and second deck fills. Final embankments were at elevation +92 feet. In one three mile stretch through the swamps in the valleys of Quebrancha, Brazos, Baja, and Gatun, crews placed 5,000,000 cubic yards of embankment. Side dumping was accomplished using various sized Western air-powered spoil train cars that tilted their dump bodies 47 degrees during unloading.

During 1904 before Stevens arrived, the railroads only hauled 17,000,000 ton-miles. Stevens and his team would clean up the mess, and through their efforts the rail capacity nearly tripled by 1906 to 42,000,000 ton-miles. By 1907, rail capacity would increase to 150,000,000 ton-miles. (Bennett, 1915)

Figure 6. Side-dump spoil trains dumping into the Chagres River.

During the Wallace reign, he initially had to report to the seven-member ICC. When Stevens started, Roosevelt had reduced the ICC to three members. Yet, Stevens was convinced that decision making needed to be centralized and in Panama, not thousands of miles away.

In November 1906, when he visited Panama, Roosevelt would be the first sitting president to travel outside the US. He would tour the work sites with Stevens and meet the workers. Addressing a crowd he said, "This is one of the great works of the world; it is a greater work than you, yourselves, at the moment realize."

By early 1906, Stevens' plans were far enough along to resume excavation in Culebra Cut. To be close to the work, Stevens ordered the engineering offices to be moved from Panama City to Culebra Cut, where he lived in a modest one-story bungalow. His excavation plan was based upon rail based steam shovels starting at mid-cut and digging a pilot cut. Pilot cuts were excavated through the blasted zone by steam shovels. The bottom of the pilot cuts averaged about 34 feet wide at the bottom and 50 feet at the top, with each cut 8 to 12 feet deep. The following widening cuts were about 26 feet wide and proceeded to the east and west of the cut. When clear, the next pilot cut would start, and so on and so on, with each pilot cut going deeper and deeper.

Stevens worked 12 to 18 hours a day, and made of point of walking the line to inspect the work daily. His ability to marshal men and machines was remarkable. He put into place efficient production methods. Yet in the first three years, the Americans had only managed to remove seven million cubic yards of excavation. By the second half of 1906, Stevens' persistence began brought improved results month by month. In January 1907, crews excavated some 500,000 cubic yards, twice the best monthly effort of the French. In February, 600,000 cubic yards were excavated and Stevens' popularity was at a high. (Rogers & Hasselmann, 2012)

Despite having to comply with the eight-hour work day law and the civil service law imposed by Congress, Stevens had instituted a well-fed, well-housed, well-equipped, well-organized work force. The unskilled labor at Panama was called the Silver Roll. To provide unskilled labor, the ICC set up recruiting agencies in Europe and the Caribbean. Ultimately 12,000 men would come to Panama from Europe, and another 33,000 from British Colonies in the West Indies.

Yet the work force on the Isthmus, like many parts of the world in the early 20th century, was segregated primarily between blacks and whites. Some black Silver Roll workers were paid about $0.10 per hour, while some white European Silver Roll workers earned about $0.20 per hour, due to productivity differences at the time. The so-called Gold Roll workers on the Isthmus performed the high skilled jobs. They were the engineers, steam shovel operators, carpenters, boilermakers, masons, doctors, nurses, averaging $0.50 per hour, plus room and board. Labor transport trains were used to take the men to and from their work stations each day.

Commissioner Shonts believed the best course of action was to contract the work on the canal to private firms. Stevens on the other hand felt that the work would best be managed by the government directly hiring the labor force. In January 1907, private firms submitted bids to Secretary of War Taft; however, upon review none of them met the terms imposed, and they were all rejected. (Bennett, 1915)

Roosevelt would eventually support Stevens' recommendation, and Shonts resigned from the ICC in January 1907 to become head of the Interbourough Rapid Transit Co. in New York. Roosevelt then assigned Stevens as Chairman and Chief Engineer of the ICC. Momentum was building on the Isthmus like never before. Yet, Stevens would come to find that even his "burly power" and tireless work ethic were not enough to get the job done. Stevens was a micro-manager, and as the project progressed, he would be buried under a mountain of details. Stevens also felt there were "enemies in the rear" who sought to undermine his authority. Exhausted, Stevens would resign his post in February of 1907.

THE AMERICAN ERA UNDER GOETHALS

Roosevelt was frustrated with Stevens' departure and felt he was like a commander abandoning his army. In late February 1907, Roosevelt decided to assign a chief engineer who would be compelled under military law to remain on duty. He appointed Major George Washington Goethals (Figure 7) of the US Army Corps of Engineers. Goethals would be paid $15,000 per year, substantially more than what he had made in the Army, but half of what Stevens had been paid.

After assessing Stevens' efforts, Goethals wrote privately, "…I think he has broken down with the responsibilities and an evident desire to look after too many details himself." (McCullough, 1977) Yet upon studying Stevens' work approach for Culebra, Goethals concluded that Stevens' plan was beyond the competence of any Army engineer of the day. So with Stevens' plan and plant, Goethals was poised to finally "make dirt fly."

Goethals was a brilliant engineer, having graduated second in his class from West Point. Now at 48 years old, Goethals rose through the ranks, demonstrating his ability to tackle the toughest projects of the day.

Goethals Biographical Summary:
1858: Born Brooklyn, NY   June 29
1880: Graduates from West Point
1881: Engineer School at Willets Point, NY
1882: Engineering Officer Vancouver, WA
1884: Ohio River Navigation Improvements
1885 – 1889: Teaches engineering at West Point
1889: Tennessee River Navigation Improvements
1891: Promoted to Captain. Muscle Shoals Canal
1898: Spanish – American War
1907 - 1914: Chief Engineer Panama Canal
1914: Promoted to Major General
1917: State Engineer for New Jersey
1919: Retires from active service Consults for NYPA
1928: Dies at age 70

Figure 7. George Washington Goethals, Chief Engineer, 1907-1914

Understanding the futility of reporting to a commission, Goethals sought and received sole authority for the work at the Isthmus. Goethals understood that he would have to divide if he was to conquer, saying "I am commanding the army of Panama. The enemy is Culebra Cut and the locks and the dams." (Bennett, 1915) To divide and conquer, Goethals refined the three division management structure that Stevens put into place. He hand-picked his team, empowered them, supported them, and held them accountable. He delegated responsibility for the Atlantic Division to William Sibert, to the Central Division to David Gaillard, and for the Pacific Division to Sydney Williamson. Also reporting directly to Goethals were the heads of the Departments of Sanitation, Civil Administration, Law, Examination of Accounts, Disbursements, the Quartermaster's and Subsistence Departments, and the Purchasing Department in the US. (Goethals, 1911)

The Atlantic Division under Sibert was a military organization with each of its responsible heads a military man. The Pacific Division under Williamson was strictly a civilian's organization, with not a single army man. The Central Division under Gaillard was made up of a military head with civilian subordinate officials. (Bennett, 1915)

While the division engineers had authority within their area, all would have to work to a common work plan and cofferdam scheme, to build the massive concrete locks and excavate Culebra Cut, in some cases over 70 feet below water level. Major earthen cofferdams would have to be built at Gatun, Gamboa, Pedro Miguel, and Miraflores.

From the beginning of the work under his direction Goethals was a constant presence on the Isthmus. He spent every morning going over the work in the field. When he issued an order he expected it to be obeyed, yet he was careful never to ask the impossible. To encourage cost awareness, Goethals established a unit-cost accounting method. Perhaps to not flaunt his military rank, Goethals wore civilian clothes in Panama. Many called Goethals "the Chief" and as busy as he was, he always maintained an open door policy.

In order to stay close to the work each day Goethals would travel the Isthmus in a day coach on a regular train, but frequently he preferred to walk the line. As he walked the line he talked with everybody he met. By eleven o'clock, the whole field had been surveyed by Goethals' watchful eye. A hundred bits of information had been gathered, and a hundred helpful suggestions had been made. (Bennett, 1915)

In the afternoon he was in his office engaged in administrative duties. It was said that no superintendent knew the details of his own work better than Goethals. No visiting engineer failed to come away with a tribute for the scrupulous attention to detail that characterized the investigations of Goethals. To build the work as efficiently as possible, Goethals established probable unit-cost based on all the facts at hand. Every piece of work, from the breakwater in Limon Bay to the fill in Panama Bay, was gone over, and the estimated cost, both as a whole and for its units, was fixed. (Bennett, 1915)

THE CENTRAL DIVISION, DAVID DU BOSE GAILLARD

Goethals put Major David du Bose Gaillard (Figure 8) in charge of the 32 mile-long Central Division. Graduating from West Point in 1884 and from Willets Point in 1887, Gaillard brought tremendous structure and discipline to the task.

While Gaillard adopted the system that John Stevens had put in place, it would be Gaillard who lead over 12,000 men and master steam powered equipment like no other builder before or after him. With the Stevens' preparatory period complete by March 1907, it was time for Gaillard and his team to get to work.

The biggest challenge for Gaillard would be the nine mile-long Culebra Cut. The French started at Culebra and excavated about 20.5 million cubic yards before stopping in 1889. Gaillard would have to excavate another planned 54 million cubic yards of soil and rock to reach the bottom of the canal, at an elevation of 40 feet with a width of 300 feet.

Conditions were brutal at the bottom of Culebra Cut. Mid-day temperatures were seldom less than 100, more often 120 to 130 degrees Fahrenheit. Culebra Cut would come to be known as "Hell's Gorge." Hell's Gorge indeed as in some areas geothermal activity would cause groundwater to turn to steam. To minimize water from the rains getting into the cut, large diversion channels were built; the Obispo Diversion on the east side, and the Camacho Diversion on the west side.

Gaillard Biographical Summary:
1859: Born Fulton, South Carolina September 4
1884: Graduates from West Point
1887: Graduates Engineer School Willets Point
1887: Commissioned 1st Lieutenant
1895: Commissioned Captain
1898: Commissioned Colonel
1903: Chief of Staff Dept. of the Columbia (Vancouver, WA)
1904: Commissioned Major
1907: Supervisory Engineer Panama Canal
1908 - 1913: Division Engineer, Central Division
1913: Dies at age 54, December 5

Figure 8. David du Bose Gaillard, Division Engineer Central

Culebra Cut is located in the high hills of the Isthmus between mile 31 and mile 40 of the canal. The plan was to excavate uphill to allow for drainage of rainwater through the pilot cuts to dewatering sumps at elevation 32 at each end of the cut that were equipped with pumps capable of pumping 36,000 gallons per minute. With Stevens' railroad based excavation plan, full trains would travel downhill and empty trains uphill on a nominal grade of 1% or less. To power the hundreds of rock drills, three huge compressed air plants were established, for a total output of 37,500 cubic feet per minute. Air and water were distributed along the cut through 181,000 feet of pipes. To allow men access across the cut, a suspension bridge was constructed at Empire.

Just as Goethals had the Canal organized into three divisions, the Central Division would be further divided by into five operational districts, each with its own superintendent. (Bennett, 1915) Every mile of track and piece of equipment in the cut was tracked and planned on a map in Empire. Daily assignments were posted at the roundhouse.

Each month Gaillard was improving the operations in Culebra Cut. By August of 1907, a rainy season record of 1,000,000 cubic yards a month was achieved and President Roosevelt sent to Colonel Goethals and his army a resounding cablegram congratulating them in behalf of the American people for their notable performance. (Bennett, 1915)

About 80% of the required excavation done by the Central Division was some form of rock that had to be blasted before it could be excavated. And to blast, about a half-foot of blast hole drilling was required per cubic yard. (Bennett, 1915) To drill the rock, Americans purchased 725 rock drills between 1904 and 1912, many produced by Ingersoll-Sergeant. Attacking the rock of Culebra (Figure 9), Gaillard organized his rock drills in "batteries" of from 4 to 16 drills, and holes were drilled 15 to 27 feet deep. Holes were spaced 6 to 16 feet apart depending on rock type. (Goethals, 1911)

Peak drilling was in March 1912, when 202 tri-pod drills were in use, each averaging about 43 feet per 9 hour work day. Further, over 200 churn drills were used at various points along the route. They were equipped with boilers and engines and were capable of drilling a 5 inch diameter hole 100 foot deep. At peak, 173 churn drills were in use in Culebra Cut that averaged 52 feet of drill hole per 9 hour work day. All drills combined, at peak, 260,088 feet of drill holes were being drilled per month. (Bennett, 1915)

Figure 9. Typical view in Culebra Cut during excavation.

After blast holes were drilled, DuPont dynamite, with 45 to 60 percent nitroglycerine, was loaded into the holes. At peak, crews were using 700,000 pounds of dynamite per month. Overall, blasting crews would load in about a half pound of dynamite per cubic yard of rock. When the holes in any section were ready for blasting, they were "sprung," that is, four to six sticks of dynamite were lowered to the bottom and detonated thereby forming a chamber for the reception of the main charge. Depending upon the rock conditions, charges varying from 25 to 200 pounds were tamped into the sprung chamber. Drilling and loading the shots was carefully planned throughout the cut so that, shots could be detonated at 11 AM and 5 PM each day. With the warning of the steam shovels whistles, the shots were detonated using a magneto-electric machine or electric current from one of the lighting plants.

The blasting work was dangerous as the nitroglycerin was very unstable. The worst accident occurred on December 12, 1908 when 44,000 pounds of dynamite detonated accidently, leaving 23 men dead and 40 others injured. (Bennett, 1915)

After the rock was shot, excavation crews went right to work. The primary workhorse for excavation was the American-made steam shovel. Seventy-seven Bucyrus shovels and twenty-four Marion shovels were used, ranging in from size from 45-ton to 95-ton machines. These would use a variety of digging buckets, depending on what was being excavated; 2.5 cubic yard buckets for rock, and up to 5 cubic yard buckets for easy digging. Because the steam shovels were only as productive as their operators, shovel operators were among the highest paid workers on the Isthmus earning between $210 and $240 a month. It was grueling work, lowering the dipper, crowding the dipper into the rock face, hoisting the dipper, and swinging the boom. All day long, in the noise and the heat, it was a tough job. In 1905, the average output per shovel per day was 500 cubic yards, and double that by 1912. (Bennett, 1915)

With each cycle of the shovel, the operator would swing his load over a waiting rail car, and the shovel man would trip the locking pin on the bucket flap to dump the load. Back and forth, each cycle would excavate about 3 cubic yards in a one minute cycle. Typical dirt trains consisted of up to 20 flat cars, with each train car carrying about 20 cubic yards.

In order to provide access for the steam shovels and the dirt trains, there was about 100 miles of track within the 9 mile Culebra Cut. As the excavation along a rock face advanced, the railroad tracks would need to be moved laterally for the steam shovels to continue excavation. It would take 600 men to move one mile of track in one day.

There had to be a better way, and sure enough, railroad engineer William Bierd devised a piece of equipment (Figure 10) that could relocate large sections of track and their attached ties intact using large steam-powered cranes without disassembling and rebuilding the track. In one motion, the track shifter raised the track with the ties so as to clear the ground, and by another motion pulled it sidewise. The usual throw is two and a half to three feet, though, if the rails will permit, the track would be thrown as much as nine feet in one throw. (Goethals, 1911) With the track-shifter a dozen men could move a mile of track a day—the work previously done by up to 600 men. At peak, crews were shifting 1,000 miles of track per year. (McCullough, 1977)

With the blow of a whistle, the 11:00 am or 5:00 pm explosives were detonated. Immediately, crews were right back to work. The noise level in the cut was beyond belief. Standing upon Contractor's Hill, one would witness thousands of men, hundreds of rock drills, dozens of steam shovels, and a steady stream of trains moving in and out of the cut. At night, the repair crews and the coal trains came, men by the hundreds to tend to the shovels, which were always being worked to their limit. And though it was official ICC policy that the Sabbath be observed as a day of rest, there was always some vital piece of business in the Cut that could not wait until Monday. So, for seven years the machines of Culebra never slept, not even for an hour. (McCullough, 1977)

Figure 10. William Bierd's track-shifter could move a mile of track a day - the work previously done by up to 600 men.

The spoil trains leaving the cut would travel to 60 different spoil areas, ranging from 1 to 23 miles away from the cut. The largest of the dumps were located at Tabernilla, Gatun Dam, Miraflores, and Balboa. While constructing the Culebra Cut, about 160 loaded dirt trains went out of the cut daily.

When the spoil trains arrived at the dumps, it was essential to empty the trains as quickly as possible to keep the trains moving. Flat cars could haul more as compared to dump cars, yet took longer to unload. To solve this problem, an innovative system was developed by the Lidgerwood Company. The so-called Lidgerwood Rapid Unloader consisted of dirt train flat cars made into a continuous end-to-end platform by placing hinged steel splice plates between flat cars. A skid-mounted plow was positioned at the rear of each spoil train. When a train reached the dump, the plow was drawn towards the front of the train by the steam-powered winch at the front of the train. As the plow moved forward it was kept over the rail cars by side boards on one side of the cars, thereby plowing material off the side of the train to the ground. With the Lidgerwood trains, 300 to 400 cubic yards were unloaded in 7 to 10 minutes. After the Rapid Unloader had plowed the spoils off of the flat cars, another innovation, the "spreader," was used to plow the spoils away from the track. As the Lidgerwood Unloader and the spreader were originally deployed, they failed. Yet, with persistence and ingenuity the men using these tools were able to make the necessary modifications to realize their potential. It was said the spreader underwent some fifty-odd improvements. (Bennett, 1915)

The rains of Panama also created challenges at the spoil areas. In Ira Bennett's 1915 book, *The History of the Panama Canal,* he offered a remarkable description of how the spoil areas might have looked:

"Imagine a dump covering perhaps 1,000 acres and with tracks over its several terraces. Then picture a rainfall twice as heavy as that which occurs in the United States, dashing down and converting this great dump of freshly excavated material into a sea of mud, with the track sinking three or four feet, and shifting to one side or the other. Then watch the trainmen working and toiling to extricate their trains. That is what might have been seen hundreds, if not thousands of times at Panama. But through it all, and in spite of it all, the trains kept running and disposing of the spoil, for when the trains stopped, all other work ceased." (Bennett, 1915)

The Toro Point Breakwater required about 2,500,000 cubic yards of material. Armor stone came from Porto Bello. Work began in August 1910, and was completed by December 1912. At the Pacific entrance, 22,000,000 cubic yards of spoil were unloaded at Balboa (Figure 11), primarily with the use of the Lidgerwood System.

It was said that Colonel Goethals was always cautious in prediction and generous in fulfillment. Today, we might say "under promise, over deliver." Perhaps today's builder might call him a "sandbagger." In 1908, he stated that the high-water mark in the excavation of Culebra Cut had probably been reached. Goethals said, "You see, as we go down deeper the ditch becomes narrower and there is less elbow room for our steam shovels and our dirt trains. There will be a gradual slow-down, and thus the latter half of the work will move forward much more slowly than the first half." (Bennett, 1915)

Yet Goethals was not only interested in total production, but also how efficiently the drill, shovel, and spoil train crews were working. And so, Goethals insisted his people know their production costs to the penny. Many resisted the need to know their daily unit-costs and called the process "kindergarten for accountants." In 1908, Central Division steam shovel excavation total unit cost was $1.01/ cubic yard. (Bennett, 1915) Yet, for Goethals, success was not only measured in production. Goethals took a stand on safety and added sick or injury leave time cost to the responsible superintendent's total costs to encourage safe work practices. Although Goethals had predicted reduced production as the cut went deeper and narrower, yet, dirt would continue to fly.

To foster friendly competition Goethals had production unit-costs published weekly in the Canal Record so each division could compare their costs against other divisions. One of the valuable aids in reducing unit-costs was the canal newspaper. During the French era, de Lesseps also had a canal bulletin, yet much of its space was devoted to the "great promoter" himself, and drew glowing pictures of achievements that only existed in the brains of those who were responsible for the bulletin. (Bennett, 1915)

Figure 11. A Lidgerwood Rapid Unloader is shown unloading a spoil train at Balboa. Overall 22,000,000 cubic yards of spoil material were dumped here.

Conversely, Goethals' weekly paper published everyone's production rates which established transparency and competition. Throughout the Isthmus competition was stiff, especially for the individual shovel crews. Supported by a well-managed railroad, the 95-ton Bucyrus steam shovel No. 213 set the pace for all shovel crews. During one 5 hours stretch, No. 213 had a sustained production of 900 cubic yards per hour! In one month it had a sustained production of 300 cubic yards per hour. (Bennett, 1915) Both of these standards are remarkable even when compared to today's modern machines. In 1908, American excavation peaked at 37,000,000 cubic yards for the year. In March of 1909 sixty-eight steam shovels excavated a record 2,000,000 cubic yards for a single month. (McCullough, 1977) Year by year, more and more dirt would fly (Figure 12) as Gaillard and until 1908 his team would set record after record.

Remarkably between 1909 and 1912, Gaillard and his team cut the average cost of excavation from about $1.00 per cubic yard to about $0.55 per cubic yard. This achievement stands as a monument to Gaillard. He bore cheerful testimony to the value of the system before a committee of Congress, saying that he spent a great deal of time studying his cost sheets and trying to discover from them where economies could be affected. As difficulties multiplied, unit-costs went down. Although slides might come down like avalanches, they could never force up the unit-cost in Culebra Cut. (Bennett, 1915)

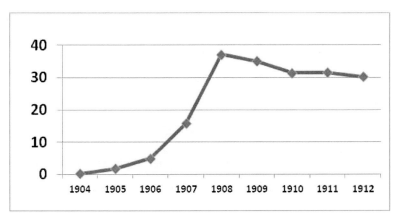

Figure 12. Annual Excavation Totals for the Central Division in millions of cubic yards. (Official Handbook, 1913)

THE BIGGEST DESIGN–BUILD PROJECT OF THE 20TH CENTURY

Because the United States had responsibility for all work at the Isthmus, the Panama Canal would be the largest design-build project of the 20th century. And because the "battle of the levels" wasn't over until 1906, the design of the locks was behind schedule.

During the late 19th century and early 20th century lock design and construction techniques were rapidly changing. During most of the 19th century, natural stone masonry was the material of choice for engineers building their locks. For example, as late as the 1870s, the US Army Corps of Engineers was using stone for the construction of the Muscle Shoals Canal. However, by the 1890s the Corps of Engineers had advanced their designs and construction methods using reinforced concrete; for example, as they constructed the Soo Locks at Sault Ste Marie, at the eastern end of the Upper Peninsula of Michigan.

One of the young engineers learning how to design and build the Soo Locks was Henry Hodges (Figure 13). Graduating from West Point in 1881, Hodges had proven himself as a good engineer and leader. In 1907, Goethals appointed Hodges as the Assistant Chief Engineer. Prior to his arrival in Panama, some limited design work had been accomplished in Washington, DC. Yet, when he arrived in Panama he faced a design scope of an unprecedented scale.

With only two years to complete the preliminary design, Hodges and his Gold Roll design team were busy throughout 1907, 1908, and 1909 preparing the plans. Hodges would divide if he was to conquer.

Hodges Biographical Summary:
1860: Born February 25
1881: Graduates from West Point
1885: Engineer Soo St. Marie Locks
1888 - 1892: Engineering Instructor West Point
1898: Engineer Puerto Rico
1899: River and Harbor Duty
1901: Chief Engineer Cuba
1902: Board of Engineers for Rivers and Harbors
1907: Asst. Chief Engineer Panama Canal
1917: Commanding General 76th Infantry Division
1919: Commanding General 20th Infantry Division
1929: Dies September 24

Figure 13. Harry Foote Hodges, Assistant Chief Engineer

Hodges organized his Gold Role team as follows:

- Masonry and Locks: L.D. Cornish, Designing Engineer.
- Lock Gates: Henry Goldmark, Designing Engineer.
- Operating Machinery: Edw. Schildhauer, Electrical and Mechanical Engineer.
- Emergency Dams: T.B. Monniche, Designing Engineer.
- Spillways: E.C. Sherman, Designing Engineer.

These plans were for the largest locks in the world. Each lock structure was to be designed and built as a pair of side-by-side locks to allow for simultaneous two-way traffic, with each lock chamber 110 feet wide and a thousand feet in length to accommodate the size of ships not yet built. The triple lift locks of Gatun would be nearly 4,000 feet long and would require over 2,000,000 cubic yards of concrete. The double lift locks of Miraflores and the single lift locks of Pedro Miguel would require about 2,400,000 cubic yards of concrete.

To the northeast of Gutun Dam, the Gatun Locks would be constructed. In January of 1906, Chief Engineer John Stevens had fixed the center line of the locks, after approval of the minority lock canal plan. Precision survey was required to build the locks in the right place. To provide dimensional control engineers used transits supplied by Buff & Buff Co. of Boston, among others. To communicate completed designs to the superintendents, plans were duplicated on blue print machines manufactured by Brown & Earle, Inc. of Philadelphia, PA. (Bennett, 1915)

ATLANTIC DIVISION, WILLIAM SIBERT

Goethals assigned responsibility for the nine-mile-long Atlantic Division to William Sibert (Figure 14). Like Goethals, Sibert was a graduate of West Point and brought discipline to the task of building the locks and dam at Gatun.

Sibert Biographical Summary:

1860: Born Gadsden, Alabama October 12
1884: Graduates from West Point
1899: Chief Engineer of 8th Army Corps
1907: Supervisory Engineer Panama Canal
1908 - 1913: Division Engineer, Atlantic Division
1915: Commissioned Brigadier General
1917: Led 1st Infantry Division "the Big Red One"
1918: Commanding General USACE Southeast
1918: Commissioned Major General CWS
1920: Retires from active duty
1935: Dies at age 75

Figure 14. William L. Sibert, Division Engineer Atlantic

Sibert's work at Gatun included building the triple locks, a dam, a spillway, a hydroelectric powerhouse, and the two-mile-long Toro Point Breakwater. Further, the Atlantic Division had about 9,000,000 cubic yards of dry excavation and 40,000,000 cubic yards of dredging.

ATLANTIC DIVISION GATUN DAM

In order to make the locked canal function, the world's largest man-made dam would have to be built at Gatun. When completed the dam impounded the Chagres River to create a 164 square mile lake, 85 feet above sea level, with over 1,000 miles of shoreline. To raise and lower ships from sea level to Gatun Lake, a three lift lock would also have to be constructed. When completed the Gatun locks would be the largest set of locks in the world.

The dam was planned as a hydraulic filled dam; one-and-a-half-mile long, a half-mile wide at the base, and 100 feet wide at crest at elevation 105, making the crest twenty feet higher than the normal surface of the water in the lake. The dam would require 22,000,000 cubic yards of material, and its total weight would be 30,000,000 tons. The mile and a half wide Gatun River Valley is comprised of alluvial deposits, in some places nearly 300 feet deep. (Bennett, 1915)

For many civil engineers during the early 20th century memories of Jamestown flood of 1889 caused by the failure of an earthen dam were still very much on their minds. So, the very idea of an earthen dam being built through the swamps of Gatun brought much criticism from many engineers who felt there would be dangerous amounts of seepage and percolation under the dam. Goethals assigned Caleb M. Saville, one of the foremost earth-dam experts in the world to design the Gatun Dam. (Bennett, 1915) Saville had the dam site area honeycombed with borings, sank test pits to view various foundation strata, and built scaled models of the dam. Saville's testing showed the underlying material to be impervious to water and to have ample strength to uphold the 30,000,000 ton dam structure.

Before the dam construction could begin, the village of Gatun had to be relocated. In late June 1906, workers began clearing the jungle cover at the 600 acre site to make way for the dam and this work was complete by the spring of 1907. In the area of Gatun, the Charges River flowed through four paths to the sea; its old riverbed, the French Canal, an east diversion channel, and a west diversion channel. As the Chagres River was prone to flooding, the sequence of building the dam was critical to its success. The plan was to start work on the eastern end of the dam and the spillway, and then close the west diversion when it was time for the river to be diverted through the spillway.

First, the main channel of the Chagres River and the old French canal were dammed. This diverted the flow of the Chagres into the channel west of Spillway Hill. In April 1907, one steam shovel was transferred to Spillway Hill to start grading for the preliminary track along the axis of the spillway cut. The next step was to cut a channel 300 feet wide through Spillway Hill into which, when completed, the entire flow of the Chagres would be diverted.

About the same time, work began on the west half of the dam, starting by building rail trestles for the dumping of a dam embankment to create "toe" dikes 1,200 feet apart to contain the hydraulic fill core of the dam. By December 1907, crews were ready to begin construction on the Gatun Dam.

When the toes were completed for the east side of the dam, hydraulic fill comprised of silty clays and gravels was pumped into the area between the toes using hydraulic dredges. To prepare for the diversion of the Chagres, temporary culverts were installed across the south end of the spillway channel.

The hydraulic dredges used at Panama were manufactured by Ellicott Machine Co. of Baltimore, Maryland. Ellicott 20-inch hydraulic dredges were equipped with a rotating cutter head, while 1,000 horsepower centrifugal pumps provided powerful suction allowing these dredges to excavate as much as 750 cubic yards per hour in the right soil conditions. The centrifugal pumps on the dredges were manufactured by Morris Machine Works. (Bennett, 1915) About 10 to 15 percent of the dredged material at the end of the 20" discharge line was solids. The fines were allowed to settle at the core, and the excess water was drawn off of the top of the fill.

In advance of the hydraulic fill, dry fill as it was called, was placed on both the upstream and downstream faces to provide containment. An effort was made to keep the dry fill 10 to 15 feet higher than the wet fill. The theory was that the weight of the dry material with the added weight of trains running over it would either compact the underlying hydraulic material or force it to the center of the dam. In 1908, a slip occurred in the south toe of the dam. Upon reevaluation, the Board of Consulting Engineers increased the crest of the dam to elevation +115 feet. (Goethals, 1911)

On the west side of the dam, toes were placed off of rail trestles from both the east and west ends, and converged at the Chagres River. As building of the toe dikes narrowed the channel, the Chagres River fought back with ever increasing velocity. When the stream had been contracted by the dikes to about 80 feet wide and 6 feet deep, the velocity of the water increased so that the dike stone was carried downstream. It was therefore decided to dump car loads of crooked rails into the river above the trestles. The theory was that the rails would form an entanglement that would be able to stop large stones on the upstream side of the trestle. This process was successful and resulted in building the diversion dams.

On the second day after completion of the two diversion dikes, a settlement occurred and the north dam moved leisurely downstream. The sudden release of water caused a slide and only a thin sliver of the upstream diversion dam remained to prevent the Chagres from resuming her old course. Trains were immediately run out onto the trestle, and about 30,000 cubic yards of rock were dumped in to save the work. Water was now diverted through the bypass channel in spillway hill. The Chagres had now left its natural course for the first time.

By the end of 1910, over a million cubic yards were being added to the Gatun Dam each month as it slowly built up in height and strength. By January 1911, the fills were over half way complete. During peak construction of Gatun Dam in 1911, there were 2,000 men working, 100 trains dumping dry fill, and hydraulic dredges pumping in wet fills at Gatun Dam every day.

ATLANTIC DIVISION GATUN LOCKS

Work for the Gatun Locks started in September 1906, and required the excavation of nearly 5,000,000 cubic yards of mostly rock which was performed by steam shovel. Some of the spoils were used at Gutun Dam with the balance going to spoil areas. After three years of excavation and foundation preparation at the Gatun, concrete work was ready to begin.

The amount of concrete required for the Panama Canal was unprecedented: 5,000,000 cubic yards. Averaging one barrel of cement for each cubic yard of concrete placed, 5 million barrels of cement were required for the project. Cement was shipped to Panama from the docks of Jersey City and supplied primarily by Alpha and Atlas Cement companies. Demand for cement in the US grew rapidly during the late 19th and early 20th centuries. Production and quality was greatly enhanced with the introduction of the first long rotary kilns by Thomas A. Edison in 1902, at his Edison Portland Cement Works in New Village, NJ.

Gravel and sand for the Atlantic Division structures came by water to Colon. Gravel was transported 20 miles by barge from the Porto Bello quarry and crushing plant. Sand was transported 40 miles by barge from the quarry at Nombre de Dios. For the Pacific Division structures operations, cement was shipped by train from Colon. Basalt and trap rock gravel were quarried and crushed right at Ancon Hill, and the sand came from Chame Point in the Bay of Panama.

Borings at the Porto Bello quarry site indicated there were 20,000,000 cubic yards of high quality massive andesite available for making concrete aggregate.

Quarried andesite rock was blasted, excavated, and then delivered by rail to the crushing house at Porto Bello. To crush the rock, ICC engineers selected state-of-the-art model No. 21 Allis-Chalmers Gyratory Crushers capable of crushing 5,000 cubic yards of rock in eight hours. (Bennett, 1915)

Cement and aggregates were brought into the old French Canal to the concrete docks. Here, aggregates were unloaded into stockpiles using three cable way cranes. Cement ships were brought under a covered roof for unloading and storage. As the Caribbean Sea was noted for storms, 200,000 cubic yards of gravel and 100,000 cubic yards of sand were stockpiled to eliminate the possibility of shortage.

Delivery of dry mixed concrete was accomplished using an innovative "automatic electric road" on a 24 inch gauge track. There were 42 two-cubic-yard tram cars that traveled in a continuous fashion on this rail system. Each car ran under the cement storehouse where it was charged with two barrels of cement. Without stopping, each tram continued through a tunnel under the sand and stone piles before traveling to the mixer house to discharge its load, only to start the process yet again. At the mixer house there were four 64-cubic-foot cubical-type concrete mixers on each side of the plant, for a total of eight mixers. The typical batch size was 2 cubic yards. At any given time, two mixers on each side were batching concrete so that these could discharge 4 cubic yards simultaneously into the concrete trains. Thirteen 6.5-ton electric-powered mine locomotives then transported each train with two 2-cubic-yard buckets per car to the cableway cranes. The maximum capacity of each mixer was about 70 cubic yards per hour, but would average about 48 cubic yards per hour. The Gatun plant was operated 12 hours per day, and was augmented by an auxiliary plant containing two 2-yard concrete mixers.

One of the typical concrete mixes used was a 3,000 psi mix, referred to as a 1:1:2 mix. (Bennett, 1915) Unlike today's concrete which is weight batched, concrete was batched by volume in this era. The 1:1:2 mix was a volumetric mix consisting of: 1 Part Cement, 1 Part Clean Sand, and 2 Parts Clean 1-1/2" Gravel that was "Crusher Run" from Porto Bello or Ancon.

Typical placing temperature for the concrete was 80 degrees Fahrenheit. To monitor the effects of the heat of hydration in the concrete, thermometers were installed in the concrete. Typical temperature rise would be about 50 degrees to 130 degrees Fahrenheit in the first two weeks after placement. (Bennett, 1915) It is interesting to note that modern cements will hydrate much more quickly, with today's modern mixes generally reaching maximum temperatures in 1 to 3 days. The slow heat rise of the original cement would have lowered the thermal stresses, while the concrete was gaining strength.

Efficient concrete structures operations rely on adequate crane service to ensure that crews remain productive. William Sibert had four Lidgerwood duplex cableway cranes (Figure 15) installed, each with a capacity of 6 tons. These provided eight hooks over the work.

Figure 15. Lidgerwood duplex cableway cranes were positioned to span across the Gutun Locks construction area. Each hook had a capacity of 6 tons. Middle wall steel forms have been erected in the foreground.

The workers at Gatun called the cableway cranes "skyhooks." The cableway cranes were mounted on self-propelled gantry bases that traveled on railroad tracks parallel to centerline. The cableway towers were 85 feet high and spaced 800 feet apart. The cableway transverse speed was about 1,600 feet per minute and a hoisting speed of 400 feet per minute. (Bennett, 1915)

By December of 1909, Sibert's men were rapidly placing invert concrete in the upper locks. With so much repetitive formwork in the Gatun Locks, Sibert elected to use large full section steel forms (Figure 16) to place 36 feet long concrete monoliths extending from the floor to the top of the walls. Each of the 36-foot-long forms held 3,500 cubic yards of concrete.

Running lengthwise through the side and center walls, three 18-foot diameter water culverts had to be constructed within the massive concrete walls to provide water passage from the lake into the lock chambers. From the main longitudinal culverts, there were 14 smaller cross culverts, each with five openings into the floor of the lock chamber. This provided 70 filling holes in the bottom of each lock chamber. When completed, the water passages were all regulated by gates and valves, the largest being the massive 8 ft. x 18 ft. slide gates for the main culverts. Using this design, a single culvert had the ability to fill a lock chamber in 15 minutes, and just 8 minutes using two culverts. (Goethals, 1911)

Figure 16. February 1910 at Gatun. A cableway crane lowers a two cubic yard bucket of concrete into a typical six foot lift of a side-wall monolith. A culvert can be seen in the distant side wall monolith.

The extraordinary size of the culverts and other conduits within the lock concrete necessitated custom formwork. During the construction of the New York subways and other projects, collapsible steel forms had demonstrated their superiority of over wooden forms. The collapsible forms were furnished by the Blaw Steel Construction Company, of Pittsburgh. (Bennett, 1915)

Typical lifts of concrete were 6 feet high, stepping in with each lift. Typically, it would take crews one week to fill each of the 36-foot monoliths.

Concrete was placed using the overhead cableway cranes to deliver concrete into the steel forms using two cubic yard buckets. Average daily placement was about 3,400 cubic yards. The largest amount of concrete laid in any one month at Gatun was 89,401 cubic yards. The average cost of the concrete per yard in place for 1910 was $7.50, including plant charges and division expenses.

As concrete crews worked in the upper locks, excavation crews continued excavating the middle and lower locks, working behind a natural earthen berm left in place to act as a cofferdam to the north of the locks. Lock concrete work continued to the north, and the lower locks were completed by the spring of 1912.

In order to complete the approach wall work, additional excavation was required necessitating that the natural earth berm cofferdam be removed. To prepare for this work, a temporary reinforced concrete dam was built in the locks, and the forebay was watered-up. With water equalized on both sides of the cofferdam, two suction dredges were then used to dredge to about minus 40 feet.

Sibert still had to build the approach wall and the flare sections of the lock. To do this his crews would have to excavate to elevation minus 70 feet, about 30 feet below the reach of the hydraulic suction dredges. To solve this problem, Sibert had a cofferdam berm installed north of the limits of the approach wall, with a hydraulic suction dredge left moored within. Now, the dredge simply pumped out the water, thus lowering itself to about elevation minus 32 to allow excavation to minus 70 feet (Figure 17). Brilliant!

Figure 17. Early 1913. Looking north at Gatun Locks during the construction of the north approach wall. A 20" hydraulic dredge can be seen resting on the bottom of the north approach after excavating the approach area to grade.

When the Gatun locks were in operation, it was essential to regulate the height of Gatun Lake within a tight margin of consistency from elevation 85 to 87 feet. Further, to prevent overtopping of Gatun Dam from a maximum design flow of 175,000 cubic feet per second, a spillway regulating structure had to be built at Gatun Dam. The crest of the spillway was at elevation plus 69 feet. Fourteen massive 45-ton roller "Stoney" gates were installed between the end piers and the thirteen 8.5-foot wide intermediate piers on the spillway crest to level. Overall, the spillway crest formed a 740 foot long arc, and with its 14 gates fully opened had the ability to discharge 145,000 cubic feet per second. Water discharged downstream of the spillway into a concrete lined channel; 960 feet long and 285 feet wide.

To build the spillway, Sibert's crews would place concrete over the top of sluice ways that allowed the Chagres to flow below while the men worked above. Using similar techniques as used elsewhere on the canal, Sibert's crews built the 230,000 cubic yard spillway. Concrete was supplied from the Gatun concrete plant.

PACIFIC DIVISION, SYDNEY B. WILLIAMSON

For the nine-mile-long Pacific section, Goethals assigned Sydney Williamson (Figure 18) to be the Division Engineer. Of all of Goethals' Division Engineers, Williamson was the only civilian. Yet, Williamson was an 1884 graduate of the Virginia Military Institute and was already a proven leader, having worked for Goethals at Muscle Shoals on the Tennessee River and later at Newport, RI. Williamson's scope of work included: building a double lock at Miraflores to raise ships from the Pacific to Miraflores Lake, 55 feet above sea level; and, building a single lift lock at Pedro Miguel to raise ships another 30 feet to the 85 feet level of Gatun Lake. When completed, the Pacific Division locks would require 2,400,000 cubic yards of concrete. Further, to impound the waters of Miraflores Lake, a 1,800,000 cubic yard earthen dam and concrete spillway had to be constructed. Additionally, the Pacific Division was responsible for the three-mile-long Naos Island Breakwater and nine miles of canal and approach channel excavation, totaling some 50,000,000 cubic yards of wet excavation.

Under the command of Williamson, Goethals had assigned an all civilian organization. The Army men of the Atlantic and Central Divisions were not willing to allow the civilians over the mountain to outperform them, and, of course, the civilians on the Pacific side would not think of allowing the Army folk across the Continental Divide to come out ahead of them if that could be avoided; so under the leadership of Sidney B. Williamson, the Atlantic Division toiled as though their lives depended on the job. (Bennett, 1915)

Many of the techniques the Pacific Division employed were similar to those used in the Atlantic and Central Divisions.

However, the adjacent terrain to the locks of Pedro Miguel and Milaflores prohibited the use of the cableway cranes that were used by the Atlantic Division. Instead, self-propelled "Chamber Cranes" were deployed. Further, the adjacent terrain also required the concrete batching to be performed within the limits of the locks. And so, at the Pedro Miguel locks a custom-made double-cantilever crane (Figure 19) was used to transfer sand and gravel from stockpiles into charging hopers on the portable concrete plant. Cement was delivered by rail from Gatun and discharged into the two-cubic-yard cubical mixers. This plant was operated eight hours per day.

Unlike Sibert, who used the large steel forms at Gatun, Williamson used lighter wood cantilevered forms that were raised with each lift of concrete.

Williamson Biographical Summary:

1865: Born April 15 Lexington, Virginia
1884: Graduates from Virginia Military Institute with a degree in Civil Engineering
1886: Engineer St. Paul Duluth Railroad
1887: General Engineer Montgomery, GA
1890: Married Helen Davis
1892: Assist Goethals at Muscle Shoals Canal
1898: Captain 3rd Volunteer Engineers Puerto Rico
1901: Assist Goethals at Newport, RI fortifications
1907 - 1913: Division Engineer, Pacific Division
1939: Dies January 13

Figure 18. Sydney B. Williamson, Division Engineer Pacific

Figure 19. This photo shows one of the portable concrete mixing plants used by the Pacific Division at Pedro Miguel.

Just as Gothals' excavation teams were able to drive down production costs, so too were his concrete structures teams able to improve costs. All structures were able to improve costs about 10% to 30% during the course of the work. At Pedro Miguel, 900,000 cubic yards of concrete was placed at a cost of $5.87 per cubic yard, while at Miraflores 1,500,000 cubic yards of concrete was placed at a cost of $5.34 per cubic yard. The unit-costs of plain concrete fell from $6.67 to $6.03 at Pedro Miguel, and at Miraflores it fell from $8.11 to an average of $5.01 per cubic yard. (Bennett, 1915) Concrete operations were completed at Pedro Miguel during 1911, while the concrete work at Miraflores continued through mid-1913.

Spoil trains leaving Culebra Cut skirted the lock construction at Pedro Miguel and Miraflores as they made their way to the three-mile-long Naos Island Breakwater. The Naos Island Breakwater was one of the most troublesome items of work on the Isthmus. The breakwater was built by dumping spoil material off of a work trestle. As spoils were dumped from the trestle it sank down and shifted to the side, at some places as much as 300 feet from the spot where it was placed. (Bennett, 1915)

LOCK GATES AND CONTROLS

To raise and lower ships from Gatun Lake, three sets of locks were required at each end of the cana,l with each lock requiring lock gates at each end of the lock chambers. In mid-1911, McClintic and Marshall, one of the only private contractors at Panama, started erecting the largest lock gates in the world (Figure 20). All told, 46 pairs of lock gates were installed requiring some 60,000 tons of steel in their construction. (McCullough, 1977) The gates ranged in height from 47 to 82 feet, and were 7 feet thick. Each gate had two independent leaves 65 feet wide, which varied in weight from 390 tons to 730 tons, depending upon the height of the leaf. Each of these gate leaves was hung on huge 18-ton pintle hinges anchored to the concrete walls of the lock.

The Panama Canal has been referred to as a "monument to the electrical art." (McCullough, 1977) All told, the operation of the locks would run on no less than 1,500 electric motors, with about half of them supplied by General Electric of Schenectady, NY. General Electric supplied motors, relays, switches, wiring, and generating equipment. Much credit for the ultimate electric motors and other technology used at Panama goes to Charles Steinmetz, often call "the patron saint of the General Electric motor business." Steinmetz had over 200 patents including the world's first three phase alternating current generator, and a method for long distance distribution of alternating current.

The massive steel gates were so perfectly balanced a small 40 horsepower electric motor had no trouble operating the huge gates. To operate the locks, a state-of-the-art control board was designed and built at each lock to control water filling and releasing valves and the miter gates. Each lock control board was designed to indicate to its operator the actual position of the level of the water and of the lock machinery at any instant. (Bennett, 1915) To ensure safe lockage, the control boards were also designed with interlocking mechanisms to ensure operators performed each lockage step in the correct sequence.

Figure 20. The canal's massive locks were built by McClintic and Marshall one of the only private contractors at Panama.

To ensure the safety of the locks while in operation, an emergency chain fender system was designed and installed. Further, an emergency gate system was designed and installed which could be stored out of the way of the lock chambers, but upon demand swing over the gate chambers and lower dam plates to seal the lock.

After Gatun Lake had filled, it not only harnessed the waters of the Chagres River to power the locks, it created a nice head differential to generate hydroelectric power. Downstream of the Gatun Spillway, Sibert's crews built a six mega-watt powerhouse equipped with three vertical shaft Francis turbines, enough power to operate all of the electrical appurtenances of the entire lock system. To distribute electrical power, a double 44,000-volt transmission line was built across the Isthmus, connecting Cristobal and Balboa. Four 44,000/2,200/240-volt substations, thirty-six 2,200/240-volt transmission stations, and three 2,200/220/110-volt transformer stations were built for the index and control boards at the three locks.

CUCARACHA SLIDE

Back at Culebra Cut, Gaillard and his men faced the constant threat of the steep excavated slopes failing and burying men and machine.

Through much of Culebra Cut, decomposed volcanic sedimentary rocks such as shales, siltstones, and agglommerates containing clays, were laying on top of rock. As the rains saturated these clayey soils, its density would increase, the internal pore pressure would increase, tension cracking could occur, and zone of shear was created. And, as the pore pressure increased, the shear strain along the shear zone would soften the soils until the soil could take more shear stress, thus allowing the soils to slide. One of the most problematic and peskiest slides was called the Cucaracha Slide or Cockroach Slide. In 1910, Cucaracha let go twice.

Yet as Gaillard's crews excavated deeper, the slides persisted with the soil seeking a natural angle of repose. Eventually the cut would widen and widen to a width of 1,800 feet.

Throughout 1913, the slides continued and Cucaracha would start flowing like an earthen glacier. (Figure 21) It was a war with mother-nature as Gaillard's machines struggled to keep pace and not get buried. A total of 200 miles of railroad track was covered up, destroyed, or dislocated in a single year by these slides. (Bennett, 1915)

Figure 21. During 1912 and 1913 landslides caused significant additional excavation in Culebra Cut.

Gaillard's team pressed on, and finally on May 20, 1913, steam shovels No. 230 and No. 222 met head-on, completing the last pilot cut at elevation +40 feet. Even though the pilot cut of the canal had reached the bottom, millions of cubic yards of known excavation remained, and the slides still showed no signs of stopping.

It was more than the mere digging of a ditch that Goethals encountered when 75 acres of the town of Culebra broke away and moved foot by foot into the canal, carrying hotels and club houses with it, until these buildings were removed. It was fight, fight, fight, now with dynamite, now with steam shovels, now with hydraulic excavators, and now with dredges. (Bennett, 1915)

Gaillard had worked tirelessly leading the Central Division since arriving in Panama six years earlier. Unfortunately, by late summer of 1913, Gaillard's health began to suffer from what everyone concluded was nervous exhaustion. Reluctantly Gaillard returned to the United States for medical treatment.

DIVISIONS UNITED

Finally in the summer and fall of 1913 it was time to unite the work of the Atlantic, Central, and Pacific Divisions to make a continuous 50-mile waterway.

At the approaches to the canal locks on the Atlantic and Pacific, wet excavation operations had been underway to ensure ships had adequate channel depth as they traversed the canal. Over shadowed by the deeds at Culebra, the subaqueous excavation, called wet excavation, required at the approaches, proved to be the biggest dredging operation the world had ever known.

The Atlantic Division wet excavation was about 40,000,000 cubic yards, and was primarily soft coral. To blast the coral, an old ship hull was fitted out with a boiler and eight drills, four to a side spaced 15 feet apart. Each day eight holes were drilled, loaded with explosives, and detonated. The wet excavation proved to be much more difficult for the Pacific Division, where crews had to perform about 50,000,000 cubic yards of wet excavation in hard basalt. To complicate matters, the crews would have to deal with large tidal fluctuations. To negate the effects of the tide, an early form of a jack-up barge was used. The barge was 112 feet by 37 feet, and was equipped with four 24-inch timber spuds to lift the barge enough to provide a stable platform for its six drills during operation.

After blasting, various dredges were used, including seven massive ladder dredges. These were one of the few pieces of French equipment used during their work. They were manufactured in Scotland and could dredge over 1,000 cubic yards per hour in the right conditions. The ladder and buckets arrangement weighed up to 240 tons, with 2.5 cubic yard buckets used for soft material and 1.5 cubic yard buckets used for hard material.

During 1913, as Gaillard was fighting his seemingly losing battle at Culebra, Sibert and his men were completing work on the Atlantic Division and making final preparations to start filling Gatun Lake. Throughout the first half of 1913, Sibert's team worked to complete the construction of Gatun Dam and Spillway. It had taken seven years to perform this work. Finally, on June 27, 1913, the spillway sluice gates were closed to fill the biggest man-made lake in the world, 85 feet above sea level.

During the summer of 1913, final work was being completed on all of the lock structures on the canal (Figure 22), and Gatun Lake continued to fill to full height at elevation +85. To test the Gatun locks, the tugboat *Gatun* was the first vessel to be lifted from the Atlantic Ocean to the level of Gatun Lake, on September 26, 1913.

Figure 22. July 1913 - Gatun Locks nearing completion. At the north end of the locks hydraulic dredges can be seen on both sides of the approach wall in the process of removing the sea berm cofferdam.

Even as Gatun Lake filled, the Central Division crews continued to fight the slides with steam shovels within Culebra Cut, working behind the protection of the Gamboa Dike. Desperate for a solution to the slides, some felt that watering-up the cut might help stabilize the side slopes from further sliding. And by late summer 1913, Goethals had decided that further attempts to dry excavate the Cucaracha Shale would prove fruitless, and ordered that the canal excavation be flooded north of the Continental Divide. The last dry excavation by a steam shovel was on September 10, 1913. To prepare for flooding Culebra Cut, all of the equipment and rails within had to be removed.

On October 1, 1913, as Gaillard lay in a hospital bed in Baltimore, Culebra Cut was flooded. With the water level equalized on both sides of the dike, on October 10, 1913, President Woodrow Wilson pressed a button in Washington, that was relayed to Panama, blowing out the center of the dike to complete the flooding of the Cut and join it to Gatun Lake.

Now with water in Culebra Cut, the Atlantic Division dredge fleet was mobilized into the Cut to attack the ongoing landslides. Goethals would use every type of dredge at his disposal, including the ladder dredges, the hydraulic dredges, and the largest dipper dredges ever built, the Bucyrus 15-cubic-yard dipper dredge equipped with a 62 foot boom.

THE FINAL PUSH

Even as the water of Gatun Lake filled the Culebra Cut, the slides choked off water from reaching the Pedro Miguel Locks. Without the waters of Gatun Lake reaching the locks at Pedro Miguel they could not be tested, nor could the Pacific Division dredges be mobilized to help excavate the slides of Culebra Cut. In early October, Goethals' men would literally battle in the trenches at the Cucaracha Slide with their picks and shovels in order to allow water to reach Pedro Miguel (Figure 23).

Figure 23. Men and machines would continue their battle through late 1913 and well into 1914. Left: This photo from October 11th, 1913 shows men digging through a slide to allow water to reach Pedro Miguel.

When water finally reached Pedro Miguel, the locks were used to bring in dredges from the Pacific Fleet. Men and machines would continue their battle through Culebra Cut. Between November 1913 and February 1914, Goethals increased the number of dredges working the toe of the Cucaracha Slide by five-fold, attacking the slide from both ends. (Figure 24)

Figure 24. This photo from February 1914 shows a ladder dredge and a fleet of suction dredges working at Cucaracha Slide.

Ultimately, almost half of the 232 million cubic yards excavated between 1907-1914 was removed using floating dredges.

French 1879-1904:	78,000,000 cubic yards
US 1904-1907:	14,000,000 cubic yards
US 1907-1914:	232,000,000 cubic yards
Total:	324,000,000 cubic yards

Finally by mid-summer 1914, the dredges were able to establish a navigable width through Culebra Cut, and on August 15, 1914, with the passage of the SS Ancon, the Panama Canal was officially opened.

EPILOGUE

America had succeeded where France had failed. Yet, without the failure of the French, America might have also failed. The success of Goethals had been influenced by the failures of Stevens. Stevens' success had been influenced by the failures of Wallace. And Wallace's success had been influenced by the failures of the French.

Americans would celebrate the opening of the canal. On the Panama Canal's official seal it boasts "The land divided, the world united." Yet Germany's declaration of war against Russia and France in early August 1914 (WWI), overshadowed what otherwise would have been a global celebration. The American effort had taken ten years and cost $352,000,000. (McCullough, 1977) From the start of the French effort in 1879, through the end of the American effort, 35 years had elapsed and some $639,000,000 had been spent.

However, the human cost was high: over 500 deaths per mile of canal during the French effort, and perhaps another 100 deaths per mile during the American effort. And there were environmental consequences. The Panama Canal has been referred to as "the greatest liberty man has ever taken with nature."

America proved its industrial might, with the dawn of the communication age; the first North American transcontinental telephone call between Thomas A. Watson in San Francisco and Alexander Graham Bell in New York City, just a year away.

Theodore Roosevelt's visionary leadership for the construction of the Panama Canal officially ended when he left office on March 4, 1909. Yet, upon completion of the canal Goethals would say, "The real builder of the Panama Canal was Theodore Roosevelt." It could not have been any more Roosevelt's triumph, Goethals wrote, "if he had personally lifted every shovelful of earth in its construction." (McCullough, 1977) Acknowledging the significance of Roosevelt's role in the canal, employees who completed two or more years of service in the canal zone received the Roosevelt Medal. After all, it was Roosevelt who acted swiftly and had the moral courage to change his own mind; to side against the majority decision of his board of consulting engineers who wanted to build a sea level canal. (Bennett, 1915)

After leaving Panama, John Stevens went on to help build more railroads, including the Trans-Siberian Railroad and the Chinese Railroad. In 1927, Stevens was elected as the president of the American Society of Civil Engineers. The competence with which Stevens applied his technical knowledge, management ability, and concern for the people who built the canal, was attested to by his successor, Col. George W. Goethals. Years later Goethals was quoted as saying,

"Stevens . . . was one of the greatest engineers who ever lived, and the Panama Canal is his greatest monument."

When Gaillard left Panama, it was not stress that brought Gaillard down, but a brain tumor that took his life on December 5, 1913. It was said of Gaillard and Culebra, "No man who fell, sword in hand, under the flag, died for his country more gallantly than David du Bose Gaillard, the conqueror of Culebra. He gave himself without stint while he lived, and he laid his life on the altar - "the last full measure of devotion." (Chicago Tribune, 1913)

Major Gen. Edgar Jadwin, Chief of Engineers, said of Goethals, "There was something about General Goethals that you find it hard to describe," adding "He was just about the hardest worker I have ever known. He carried to successful conclusion the greatest of tasks without much apparent effort. He was not a society man, just a man who loved family, friends and his work." (The New York Times, 1928)

Above all else, Goethals was a remarkable leader. In his 1911 speech to the National Geographic Society, he commented, "Generally speaking, employees are selected on account of their special fitness for the work in hand, and are then unhampered in their methods of securing definite results, thus bringing out to its fullest extent individual effort and brain power. As a consequence each man has a personal interest in the work and seems imbued with the idea that the success of the enterprise depends on him. The spirit of enthusiasm and of loyalty among the canal workers strikes forcibly everyone who visits the Isthmus, and convinces the doubting that the canal will be built." Simply, Goethals knew how to recognize an individual's talent and he empowered his people to use their skills and creativity to build the work.

General Goethals retired from active service in 1919, and went on to be a consultant for the New York Port Authority. He worked on several planning efforts including, the Holland Tunnel, the George Washington Bridge, and a new truss bridge at Staten Island. On his death in 1928, the Staten Island Bridge was named the Goethals Bridge when it opened in 1929. Now obsolete, the old steel truss will be replaced with a state-of-the-art cable stayed bridge that will deservedly still be called the Goethals Bridge.

In 1999, the United States turned over control of the canal to the Panamanian government. Since opening in 1914, annual traffic rose from about 1,000 ships to nearly 15,000 in 2012, hauling 333 million tons of freight.

The American Society of Civil Engineers recognized the Panama Canal as an International Civil Engineering Landmark in 1984, and named the Panama Canal one of the seven wonders of the modern world. When Theodore Roosevelt visited Panama in 1906, he said, "This is one of the great works of the world. It is a greater work than you yourselves at the moment realize." Now, one hundred years since its opening, it is clear Building Panama Canal was a greater work than even Theodore Roosevelt could realize.

REFERENCES

McCullough, David. (1977). "The Path Between the Seas." Simon and Schuster.
Sibert, W.L. and Stevens, J.F. (1915) "The Construction of the Panama Canal." D. Appleton and Company
Bennett, Ira E. (1915). "The History of the Panama Canal." Historical Publishing Company.
Haskin, Frederic. (1913). "The Panama Canal." Doubleday, Page & Company
Cornish, Vaughan. (1909). "The Panama Canal and its Makers." T. Fisher Unwin, London.
Goethals, George W. (1911). "The Panama Canal." National Geographic Society.
Goethals, George W. (1915). "The Panama Canal, an Engineering Treatise, V1&V2."
Cadbury, Deborah. (2004). "Dreams of Iron and Steel." Harper Collins.
Parker, Matthew. (2007). "Panama Fever." Doubleday
Rogers & Hasselmann. (2012) "The Americans Succeed in Constructing a Canal Across Panama" ASCE
The Secretary of the Isthmian Canal Commission (1913). Official Handbook of the Panama Canal
Chicago Tribune. (December 7, 1913) Gaillard Story
The New York Times. (January 22, 1928) Goethals Obituary
Photos: Earnest Hallen, The National Archives

George S. Morison and Philippe Bunau-Varilla: The Indispensible Men of Panama

Francis E. Griggs, Jr. Dist. M. ASCE

ABSTRACT

Without the efforts of Morison and Bunau-Varilla it can be justifiably stated that the world would be celebrating the 100th anniversary of the opening of the Nicaragua Canal in 2014. This paper gives a status report of engineering boards and political leaders who were in support of a Nicaragua Canal up to March 3, 1899 when the Isthmian Canal Commission was formed. It describes how these two men, through their actions, influenced the decision made on June 28, 1902 by the United States Congress and President Theodore Roosevelt to build a canal at Panama.

AN ISTHMIAN CANAL, A BRIEF HISTORY TO 1899

The idea of a canal across the Isthmus of Central America dates back to the 16th Century. Christopher Columbus on his fourth voyage searching for a northwest passage to Asia sailed along the northern shore of present day Panama in 1502. In fact, he anchored near the mouth of the Chagres River on Christmas day of that year. Nunez Balboa would, in 1513, become the first European to cross the Isthmus and view what he called the South Sea. Pedrarias and Pizarro would later establish a trail across the Isthmus between Porto Bello and Old Panama City. Later a trail called the El Camino Real was used by going by canoe up the Chagres River from present day Colon, named after Columbus, thence by foot over the Cordileras and down to the Pacific Ocean near present day Panama City. In 1520 Magellan sailed around the tip of South America into the Pacific Ocean on his voyage around the world. He determined the vast width of the Pacific and sparked interest in a canal across the narrow isthmus as a better way to the west. Charles V showed some interest in a canal and his successor Phillip II looked at a canal at Nicaragua through Lake Nicaragua at the summit, but nothing came of it. The revolutions in Mexico to the north resulted in its independence in 1812. Colombia to the south, under Simon Bolivar, became independent of Spain in 1819.

Baron Alexander von Humboldt in his tour of south and central America in the early 19th century prepared maps showing six possible routes for a canal between the oceans including many across the Isthmus as shown in Figure 1. They were from northwest to southeast the Tehuantepec Route in Mexico, the Nicaragua Route through Costa Rica and Nicaragua, the Panama Route, the San Blas Route, the Darien Route and the Atrato River Route. These routes were studied over and over again for the rest of the century by the United States, France and England.

Figure 1. Various Routes Proposed

As early as 1835 the United States showed an interest in a treaty with Central American governments to allow American Companies to build a canal through its territory. President Jackson sent Charles Biddle to look at possible routes in Nicaragua, Guatemala and Panama. He obtained a private concession from Colombia in 1836 to build a canal or railroad across the Isthmus. As early as 1839 John Stephens recommended a canal at Nicaragua to President Van Buren, but nothing came of it. In 1843 M. Garella, a Frenchman, prepared a survey for a canal at Panama generally following the El Camino Real that involved many locks and a summit tunnel. In 1847 the Bidlack Treaty was signed with Colombia for "a transit privilege" across Panama and it was approved by the Senate in June 1848. William Aspinwall won the contract from Congress to provide mail and passenger steamships from the West Coast to Panama City. Another firm won the concession to run from the East Coast of the United States to Colon. Aspinwall formed the Panama Railroad Company to connect these two steamship lines. With the discovery of gold in California in 1849 the need for a quick transit across the Isthmus was imperative.

Aspinwall sent a survey party to Panama to lay out a route for the railroad in 1849 under the supervision of Col. George Hughes with a young Edward W. Serrell as an assistant. They returned with a proposed route in 1850. Between 1850 and 1855 the line was pushed across the Isthmus with workers fighting yellow fever, cholera and malaria as well as the Chagres River and the jungle. The 47-mile line was completed on January 27, 1855 and became our first transcontinental railroad. With it American engineers and entrepreneurs connected the oceans with rails, but there was still a desire to build a canal so ships could pass through without transferring goods and passengers from ships to rail and back to ships again.

A canal got another boost on March 19, 1866 shortly after the end of the Civil War, when a Senate Resolution ordered "the Secretary of the Navy furnish through a report of the Superintendent of the Naval Observatory, the summit levels and distances by survey of the various proposed lines for interoceanic canals and railroads between the waters of the Atlantic and Pacific oceans; as, also their relative merits as practicable lines for the construction of a ship canal, and especially as relates to Honduras, Tehuantepec, Nicaragua, Panama and Atrato lines; and whether in the opinion of the superintendent, the isthmus of Darien has been satisfactorily explored and if so furnish in detail, charts plans, line so levels, and all information connected therewith, and upon what authority they are based. (Davis 1867, 4) Rear Admiral Charles H. Davis made his report on July 11, 1866. In it he reviewed all the literature and plans that had been prepared for 19 different routes, eight of which were in Nicaragua, four in Panama, and five across the Darien. Most of his report consisted of 14 maps of various routes compiled by many sources which he reproduced, many of which were published for the first time, including a map of the Panama Railroad. He wrote, "the lower Isthmus had not as yet been satisfactorily explored, and that there did not exist in the libraries of the world, the means of determining even approximately the most practicable route for a ship canal across the Isthmus." He concluded, however, "It is to the Isthmus of Darien that we are first to look for the solution of the great problem of an interoceanic canal...The interoceanic canal, in width, depth, in supply of water, in good anchorages and secure harbours at both ends, and in absolute freedom from obstruction by lifting locks or otherwise, must possess, as nearly as possible, the character of a strait." (Davis 1867, 22) In other words his recommendation was for a sea level canal at the Darien.

President Ulysses S. Grant, who crossed the isthmus in 1852 on his way to the west coast as a young Army officer, was the next president to take an interest in a canal. In his 1869 Address to Congress he stated the policy of the United States was an American Canal, under American Control for use by all nations of the world. Congress approved survey expeditions to Central America under the supervision of Capt. Daniel Ammen who reported to the Secretary of the Navy. Captain Robert W. Shufeldt surveyed the Tehuantepec route, Commander Thomas O. Selfridge the Darien route; Commander Chester Hatfield the Nicaragua route, Commander Edward P. Lull and Chief Civil Engineer Aniceto G. Menocal the Panama Route. Their surveys were far and away the best that had been prepared up to that time. On March 13, 1872 President Grant appointed an Interoceanic Canal Commission to evaluate the surveys, findings and recommendations. The Commission consisted of General A. A. Humphreys, Benjamin Pierce, and Captain Ammen. They decided they needed more information, and a new committee consisting of Major Walter McFarland, Captain William Hener and Professor Henry Mitchell were tasked with studying the Darien and Nicaragua Routes. In addition, Lull and Menocal were sent back to Nicaragua and Panama, and Lt. Fredrick Collins was sent back to the Darien.

With this additional survey information a report was prepared by the Commission in 1876 that concluded, "The Commission appointed by you to consider the subject of communication by canal between the Atlantic and Pacific Oceans across the Isthmus connecting North and South America, have the honor after a long, careful and minute study of the several surveys of the various routes, unanimously to report: That the route known as the Nicaragua Route, beginning at Greytown on the Atlantic, running by canal to San Juan River, up this river to the mouth of the San Carlos River, thence across to Lake Nicaragua, across this lake, and then through the valley of the Rio Bel Medio and the Rio Grande rivers to the port of Brito on the Pacific, possesses both for construction and maintenance of a canal greater advantages, and offers fewer difficulties from engineering, commercial and economic points of view, than any of the other routes, shown to be practicable by surveys, sufficiently in detail to enable a judgment to be formed of their relative merits." (Humphreys 1876) Daniel Ammen also presented a paper to the *American Geographical Society* on November 12, 1878 promoting a Nicaragua Canal and stating why it was superior to the Panama route. It was copied in a later report in 1880 by Ammen entitled, *The American Inter-Oceanic Ship Canal Question*.

The report was not given to congress until 1879 when it was decided to send representatives to Ferdinand de Lesseps international meeting in Paris on a proposed Isthmian Canal. From that time on until 1902 Nicaragua was called the "American Route." Ammen, Menocal and Selfridge were selected to attend this conference that was called to determine the best route to connect the oceans. They were all assigned to the Technical Committee. Selfridge presented the maps and plans for the Atrato River Route, Ammen presented maps and plans for a Nicaragua Canal and Menocal for a Panama Canal with locks. Their plans were the most detailed and complete of any presented at the conference, especially when compared to the French plan for a sea level canal at Panama using the concession of Lucien Napoleon Bonaparte Wyse. A total of 14 routes were presented for consideration of the Committee between May 14-29 but the fix was in and when a vote was taken the sea level canal at Panama was approved by a vote of 74 to 8 with 16 abstaining and 38 absent. Ammen and Menocal, the official representatives of the United States abstained as they believed only trained engineers should participate in the vote. Many French engineers, including Godin de Lepinay and Gustav Eiffel, voted against the French plan.

The story behind the French effort between 1879 and 1903 is beyond the scope of this paper. Suffice to say the initial French effort failed for the reasons many stated at the Paris Conference. Some reasons were a lack of control of yellow fever, malaria, the Chagres River, insufficient financial backing, and what was a lack of the right kind of equipment to excavate a major cut on the route at Culebra. A change was made in the plan in 1888 to build a temporary lock canal. Once the canal was complete, it would over time be converted into a sea level canal. On February 4, 1889, after a nine-year effort, the company went into liquidation. A New Panama Canal Company was formed in 1894 to perform enough work to maintain the concession from Colombia but most observers knew the purpose of the company was to find a buyer for their assets, plans, and concession.

So while it became obvious that de Lesseps was going to fail, and after its final collapse, interest in an American Canal at Nicaragua was awakened in Congress and by American businessmen. The Nicaragua Canal Association was chartered by Congress in 1887 but little work was done. The Maritime Canal Company of Nicaragua was then chartered by Congress on February 20, 1889. It assumed the concession to build a canal across Nicaragua, but it too suspended operations for lack of funds, after appealing to Congress to subscribe to their securities in return for seats on the Board. This plea was approved in the Senate but was never acted upon in the House. The Company suspended its efforts in the financial panic of 1893. The government and President Grover Cleveland got involved once again and on April 25, 1895 appointed a Nicaragua Canal Board chaired by Col. William Ludlow, with a budget of $50,000, "for the purpose of ascertaining the feasibility, permanence, and cost of construction and completion of the Nicaragua Canal" begun by the Maritime Canal Company. He was joined on the committee by M. T. Endicott (US Navy) and Alfred Noble a leading private civil engineer. They were to report back to the President on or before November 1, 1895. They spent five weeks in Nicaragua, and Ludlow later testified to Congress "we went over every foot of the route of the canal, we made as thorough and as exhaustive an investigation of the physics and natural conditions as we could in a journey of that sort. We are the only engineers who have ever been over the entire route, and in making that statement I do not make any exceptions." (Ludlow 1896) Their report, in addition to descriptions of all parts of the canal, had 22 conclusions. Conclusions 12-22 under a heading of General are of most interest. They wrote,

> 12. All locks should have a width of not less than 80 feet if the navigation be intended to provide for the passage of war vessels and for future developments in the case of commercial vessels. The U. S. S. Iowa and others of the same class have a beam of 72 feet 3 inches, and still others are now contemplated with a beam of 75 feet.
>
> 13. All streams affecting the canal route should be gauged to ascertain their regimen, and in particular the regimen of the lake and the San Juan and San Carlos rivers should be carefully determined.
>
> 11. Rainfall observations should be made at several consecutive points over the entire route from Greytown to Brito.
>
> 15. The construction of the eastern division on the line proposed by the company is feasible, but in view of the risks involved in the maintenance of the numerous dams and embankments the final

adoption of any route is inexpedient until all alternative routes shall have been fully investigated and compared.

16. Full explorations for alternative routes in the eastern division should include the so-called low-level line on the left bank of the San Juan to the San Juanillo and thence to Greytown; and, more particularly, full investigation should be made as to the practicability of extending the canalization of the river to the vicinity of the Serapiqui by means of comparatively low dams.

17. The cost of work, particularly in the eastern division, will be increased in consequence of the heavy rainfall, but from observations made, its destructive effect on completed work will be much less than that of frost in the United States.

18. The climate of Nicaragua is mild, equable, and humid, and while the sanitary conditions are not unfavorable as compared with portions of the United States, the value of physical labor is much less. It is probably to be expected that the opening of earth excavations will be attended, as elsewhere, by the development of malarial diseases, but not of a specially malignant type.

19. The volcanic and seismic forces in Nicaragua are manifestly declining, and there seems little reason to apprehend disturbance of so serious a character as to imperil the stability of canal construction.

20. The official estimate by the company of $66,400,880 is insufficient for the work. In several important cases the quantities must be greatly increased, and in numerous cases the unit prices do not make proper allowance for *the* difference in cost of work between the United States and Nicaragua.

21. The provisional estimate by the Board is $133,172,893. It should be understood that the existing data are inadequate as a basis for estimating the cost of many of the structures; some portions of the work may cost more, others less, but in the judgment of the Board the entire project can be executed for about the total amount of its estimate.

22. For obtaining the necessary data for the formation of a final project, eighteen months time, covering two dry seasons and an expenditure of $350,000, will be required. (Ludlow 1896)

The report consisted of 103 pages of text/tables and 11 appendixes along with 34 plates and was very thorough considering the time frame in which they had to work. The Maritime Canal Company took exception to their findings.

Their last conclusion lead President William McKinley to appoint another Nicaragua Canal Commission (first Walker Commission) under Rear Adm. John G. Walker, with Col. Peter Hains and Prof. Lewis Haupt on July 29, 1897 funded under the Sundry Civil Appropriation Act of June 4, 1897. They reviewed the Ludlow

Report and all of the plans of the Maritime Company between December 1897-February 1899, and submitted their report in March 1899. They reported under the heading of Feasibility:

> Under this division of the subject, the Commission would respectfully submit that it has failed to find any competent authority that denies the feasibility of constructing a canal across Nicaragua.
>
> The feasibility of the canal is conceded for the following reasons:
>
> 1. There are at this date sufficient precedents for ship canals capable of passing the largest vessels, so that any question of the navigation of such a channel is eliminated.
>
> 2. The ability to construct and operate locks of the requisite dimensions is sufficiently established by existing structures on the Manchester and Keil canals, at Davis Island on the Ohio, and at the St. Mary's canal, Michigan.
>
> 3. The possibility of constructing the necessary dams, weirs, sluices and embankments, which shall be sufficiently stable and impermeable to control the water required for navigation, as well as to regulate the floods, is within the resources of the engineering profession and is fully demonstrated by the many hundreds of miles of embankments, levees and dams, both at home and abroad. There is no reason to doubt the ability to build them out of the native rocks and earth and to give them the required strength and tightness to retain or to discharge the water with safety.
>
> 4. There is no question as to the adequacy of the supply of water for all purposes at all seasons nor as to its control in times of flood.
>
> 5. Neither is there any doubt with reference to the ability to secure good supporting ground for the trunk of the canal nor suitable sites for locks and dams.
>
> 6. The harbor question is only a matter of money and it is believed that good, capacious and safe artificial harbors can be created at a reasonable cost. In brief, this Commission sees no reason to doubt the entire feasibility of the project, but it realizes the necessity of exercising due care in the preparation of the specifications and in the conduct of the work, that the details of construction be thoroughly inspected and properly executed under competent supervision. (Walker 1899)

They estimated a cost of $118,113,790 for the route selected, and also noted the area "is practically exempt from any seismic influences of sufficient force to cause destruction or danger to any part of the canal route or suspension of its traffic." (Walker 1899, 13) This estimate is $15,000,000 less than the Ludlow estimate and with its complete faith in the feasibility of a Nicaragua Canal, Figure 2, indicated the possibility of action in its construction.

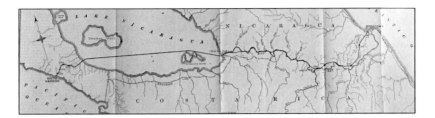

Figure 2. Nicaragua, The American Route 1899

This didn't happen as some people believed the United States should not commit itself to building the Nicaragua Canal without looking at other possibilities once again. The main difference was, of course, that the French had excavated a great deal of soil and rock at Panama and it was known that the New Panama Canal Company was interested in selling its rights, with the permission of Colombia, to another firm or country.

While these studies were being conducted the House and Senate of the United States were considering laws promoting the construction of a canal at Nicaragua. In the Senate John Tyler Morgan, Chairman of the Committee on Interoceanic Canals, a democrat and William Peters Hepburn, Chairman of the House Committee on Interstate and Foreign Commerce, and a republican, were the two main players. The fact that they were from different parties at times prevented cooperation on their goals to build a Nicaragua Canal. Morgan believed the United States should work with the Maritime Company by buying their stock and controlling its construction by membership on its Board of Direction. Hepburn believed the canal should be built by the United States government. In discussing the Morgan Bill Hepburn stated, "The Senate bill, for which your committee recommended a substitute, proposes to amend the charter of the Maritime Canal Co., and then reorganizes the company by appointment of a majority of the board of directors by the President of the United States, and then uses that corporation as its agent for constructing and operating the canal. This corporation is created by the United States, it is a creature of the Government. After creating it the Government proposes by the Senate bill to inject itself into the corporation, and thus masquerading it proposes to do a work that it is in every way capable of doing in its own proper person." On January 21, 1899, for instance, the Senate passed Morgan's bill by a vote of 48-6 and on February 13, 1899 in the House Hepburn still did not agree the canal should be built by a private concern. With these previous overwhelming votes in favor of Nicaragua it seemed as if they could have worked out their differences. This did not happen and some historians believe it was because each wanted his name associated with the bill that was sent to the president. The main problem however was their firmly held views on who should build the canal, a private company or the United States Government.

A 16-member International Commission of Engineers (Comité) had also been formed to report to the New Panama Canal Company on what should be done at Panama. They made their report on November 16, 1898 presenting a plan for a lock canal at Panama and indicated it could be accomplished at a cost of $96,836,100. Another complication for Morgan was that Nicaragua revoked the concession to the Maritime Company to take effect on October 9, 1899. McKinley also received a letter from the New Panama Canal Company President in which he listed the advantages of Panama over Nicaragua. McKinley then, according to Bunau-Varilla, upon the recommendation of Speaker of the House Reed, and Joseph Cannon a leader in the Republican party, both of whom he (Bunau-Varilla) converted to the Panama cause, against the Nicaragua forces in Congress both within and without his party decided to support and act to appoint a new Commission to look at all possible routes once again. On March 3, 1899, an act was attached to the Rivers and Harbors Bill that stated, "That the President of the United States of America be, and he is hereby, authorized and empowered to make full and complete investigation of the Isthmus of Panama with a view to the construction of a canal by the United States across the same to connect the Atlantic and Pacific oceans; that the President is authorized to make investigation of any and all practicable routes for a canal across said Isthmus of Panama, and particularly to investigate the two routes known respectively as the Nicaraguan route and the Panama route, with a view to determining the most practicable and feasible route for such canal, together with the proximate and probable cost of constructing a canal at each of two or more of said routes; and the President is further authorized to investigate and ascertain what rights, privileges, and franchises, if any, may be held and owned by any corporations, associations, or individuals, and what work, if any, has been done by such corporations, associations, or individuals in the construction of a canal at either or any of said routes, and particularly at the so-called Nicaraguan and Panama routes, respectively and likewise to ascertain the cost of purchasing all of the rights, privileges, and franchises held and owned by any such corporations, associations, and individuals in any and all of such routes, particularly the said Nicaraguan route and the said Panama route; and likewise to ascertain the probable or proximate cost of constructing a suitable harbor at each of the termini of said canal, with the probable annual cost of maintenance of said harbors, respectively; and generally the President is authorized to make such full and complete investigation as to determine the most feasible and practicable route across said Isthmus for a canal, together with the cost of constructing the same and placing the same under the control, management, and ownership of the United States. (Isthmian Canal Commission Act, Section 3, 1899)

Summarizing the above, prior to March 3, 1899, the Congress, the President and several Boards of Engineers were virtually unanimous in support of a canal in Nicaragua, and this act only opened the window a small degree for a possible Panama route. In addition, the nation's newspapers, magazines, chambers of commerce, etc. were urging construction of the Nicaragua Canal. The only people in opposition were the nations railroads, but their opposition was to any canal across the Isthmus, as it would take away their transcontinental business. In addition, the New Panama Canal Company, the Commission of Engineers hired by the New Panama Canal Company, and a French Engineer by the name of Philippe Bunau-Varilla were, with the exception of William Nelson Cromwell, the only voices in support of Panama. This situation would change radically over the next four years.

GEORGE S. MORISON - THE FIRST INDISPENSIBLE MAN, AND THE SECOND WALKER COMMISSION

Morison, Figure 3, was born in New Bedford, Massachusetts on December 19,1842 the son of a Unitarian minister. After attending Philips Exeter Academy he went to Harvard University graduating in 1863. He then obtained a Law Degree from Harvard graduating in 1866 but after practicing law for a short time decided he wanted to be a Civil Engineer. He was given an introduction to Octave Chanute who was building the first bridge across the Missouri River at Kansas City and arrived there in October 1867 to begin his new career. After building that bridge he followed Chanute to work on the Leavenworth, Lawrence and Galveston Railroad and then to the Erie Railroad in 1873. After replacing the famed wooden Portage Bridge that burned near Rochester, New York with an iron bridge in 86 days, he went into business with George Field and as a consultant to Baring Brothers who were heavy investors in American Bridges. From 1880-1898 he was chief engineer on many bridges across the Missouri, Snake, Ohio, Columbia, Mississippi, St. Johns Rivers. On several of these bridges he worked in partnership with his classmate from Philips, Elmer Corthell. He was a leader in the change from wrought iron to steel, and his Memphis Cantilever Bridge over the Mississippi River was the longest

Figure 3. George S. Morison

cantilever bridge in the country. He designed a 300-meter tower for the Chicago World's Fair, but it was not funded. The financial panic of 1893 slowed bridgework.

Morison was then selected by Grover Cleveland to a Board of Engineers to review the feasibility of a bridge across the Hudson River. As a part of this service he designed (never built) a 3,000 ft. span suspension bridge. This was followed by membership on a Board to improve the docks of New York City. Cleveland then appointed him, along with William H. Burr and Alfred Noble, to locate a deep-water harbor in Southern California. This followed by a design competition for the Rock Creek Bridge in Washington, D. C. that he won over several other well-known engineers. His multiple span reinforced concrete bridge started construction in 1898 but was not finished until 1907. He also submitted a proposal for the Arlington Bridge across the Potomac River but lost that competition to William H. Burr. He was elected President of the American Society of Civil Engineers in 1895. In March 1899 New York Governor Theodore Roosevelt appointed him to a committee to study the canal system of the state and recommend future policy. He recommended an upgrade to the Erie Canal but suggested the route be revised so that the 60 miles between Newark and the Rome level, the low portion of the canal, be rerouted so the entire canal could be "fed from Lake Erie, thus giving a continuous current in the direction of heaviest traffic." (Morison 1899, 137) He also had a suggestion "to enlarge the portion of the canal west of Lockport, to the dimensions of a ship canal for lake vessels, so that lake steamers could reach that point, and there, by mechanical appliances transfer their cargoes to barges or canal boats at the lower level thus avoiding the 57 feet of lockage at Lockport." (Morison 1899, 138)

Having been active of the profession for over 40 years and known to Presidents Cleveland and McKinley, and soon to be President Roosevelt, he along with Burr and Noble, his close colleagues, was selected to the Second Walker Commission even though he did not have much canal experience. Walker broke his Commission members, Figure 4, up into subcommittees to investigate the following.

- Nicaragua route: Mr. Noble, Mr. Burr, Colonel Hains.
- Panama route: Mr. Burr, Mr. Morison, Lieutenant-Colonel Ernst.
- Other possible routes: Mr. Morison, Mr. Noble, Colonel Hains.
- The industrial, commercial, and military value of an interoceanic canal: Mr. Johnson, Mr. Haupt, Mr. Pasco.
- The rights, privileges, and franchises: Mr. Pasco, Lieutenant-Colonel Ernst, Mr. Johnson.
- The first named men were the chairmen of their respective subcommittees.

In 1895 in his presidential address to ASCE Morison stated, "The second, of more importance to Americans than any water project started since the conception of the Erie Canal, is the Nicaragua Canal, the completion of which, besides opening a new general waterway will make the Atlantic coasting fleet available for the coasting service of our whole country. A board of three engineers, all members of this Society, is now in the Isthmus engaged in an investigation which we all hope will justify the early construction of this great work." (Morison 1895, 480)

Figure 4. Second Walker Commission
Left to Right
First row: George S. Morison, Samuel Pasco, Admiral John G. Walker, Alfred Noble, Prof. Emory Johnson
Back Row: Lewis Haupt, William H. Burr, Lt. Col. Oswald Ernst, Col. Peter C. Hains

After studying all existing reports and plans the Commission was invited to Paris to review the plans, etc. of the New Panama Canal Company. They sailed to France on August 9 and returned on September 29, 1899. While there they reviewed the records and plans of the company as well as the report of the Panel of Engineers (Comité) that was formed to recommend a revised lock plan for Panama. What is important in this paper is that Morison and Philippe Bunau-Varilla, who, though a stockholder in the company, was not an insider with the company, met for the first time. Morison doesn't mention these meetings but Bunau–Varilla described them as follows:

> My plan of attack was soon decided. As I had refused to go to Washington the preceding winter, to attack openly the Nicaragua Canal, I resolved not to come in contact at the beginning with any of the six determined partisans of Nicaragua. I decided to plead the cause of Panama before those whose minds were free from any preliminary bias.

Chance served me well. The Commission was split up into various sub-commissions. The members who were delegated for the study of the Panama Canal route were precisely the three members I desired to meet MM. Morison, Burr, and Ernst.

On the 9th of August the Commission began its labours by leaving for Europe with the intention of studying the maritime canals of Manchester and Kiel and the archives of the Panama Canal in Paris.

A fortunate circumstance placed me, immediately after their arrival in Paris, in relations with MM. Morison, Burr, and Ernst. The last-named was a friend of Mr. Bigelow and brought a letter of introduction. With the two other members I was soon on excellent terms. A common friend, Mr. Frank Pavey, a prominent member of the New York Bar and former Senator of that State, was the connecting link.

Our conferences were long and frequent. I was gradually able to impress their minds with facts showing at the same time the inferiority of Nicaragua and the admirable superiority of Panama.

I had a strong foundation for my theories. It was formed by the two books I had published seven years before. "Here," said I, "is what I published during the most violent moral storm which has agitated France for many years. If in these books, written for the defence of Panama, you find today a solitary fact or a single incorrect figure you may throw them away. If you find nothing erroneous, adopt the conclusions, which are written therein, because they are the expression of the eternal truth."

I was soon able to convince myself that these demonstrations had been fruitful seed fallen on generous ground.

When the Commission left Paris I was certain the scales had fallen from the eyes of at least three of its members. Among these gentlemen was a man characterised by a great energy of conviction. It was George Morison... He was certainly, when he came to Paris, the most prominent of American engineers. To his professional eminence he added an absolute independence of mind. No consideration whatever could have brought him to temper the expression of his wide and precise views in order to truckle to any influential men or body of men.

The immense injustice of which the Panama enterprise had been a victim, was perhaps what most prompted him to take its side when he was convinced of it. He was its first partisan in the Commission. Nearly simultaneously he was supported by Professor Burr and Colonel Ernst. From that moment the Nicaragua party ceased to be omnipotent. At the contact of the ideas of this group of strong men, and under the radiation of scientific truth, the Nicaragua block gradually melted, the members of the Commission passing one by one to the opposite side.

It was not without prolonged study and serious examination that this transformation was effected, a transformation which greatly honoured, alike those who first saw the Truth as those who gradually turned away from Error. It was a slow but steady fight between Truth and Prejudice. (Bunau-Varilla 1920, 166)

After meeting with Bunau-Varilla and the engineers of the New Panama Canal Company Morison and the rest of the Commission, minus Haupt and Johnson, visited the Kiel Canal in Germany, the North Sea Canal in Holland and the Manchester Ship Canal in England as these were the major ship canals in Europe. They returned to the United States on September 29, 1899.

After some more study and discussion the Commission traveled to the Isthmus. They traveled to Nicaragua on February 16 staying there until February 27, 1900. Panama was next from March 3-March 16. Morison then went to the Darien on March 16 staying there until April 7, 1900. Back in the States they prepared a preliminary report to the President. Walker wrote, "It has been found impracticable to complete the work of the Isthmian Canal Commission and prepare a full report upon its investigation it was required to make in time for its presentation before congress convenes in December…It has therefore been deemed best to report the progress that has been made and the conclusions which have been reached, and thus avoid the delay that a postponement might cause in the inauguration of the work which the legislation, under which the Commission is acting, was designed to promote." They determined that both the Nicaragua and Panama routes were feasible and concluded:

> I. The estimated cost of building the Nicaragua Canal is about $58,000,000 more than that of completing the Panama Canal, leaving out the cost of acquiring the latter property. This measures the difference in the magnitude of the obstacles to be overcome in the actual construction of the two canals, and covers all physical considerations such as the greater or less height of dams, the greater or less depth of cuts, the presence or absence of natural harbors, the presence or absence of a railroad, the exemption from or liability to disease, and the amount of work remaining to be done.
>
> The New Panama Canal Company has shown no disposition to sell its property to the United States. Should that company be able and willing to sell, there is reason to believe that the price would not be

such as would make the total cost to the United States less than that of the Nicaragua Canal. [They wrote the New Panama Canal Company was, from some sources, asking for $100,000,000 for their rights, etc.]

II. The Panama Canal, after completion, would be shorter, have fewer locks and less curvature than the Nicaragua Canal. The measure of these advantages is the time required for a vessel to pass through, which is estimated for an average ship at twelve hours for Panama and thirty-three hours for Nicaragua.

On the other hand, the distance from San Francisco to New York is 377 miles, to New Orleans 579 miles, and to Liverpool 386 miles greater via Panama than via Nicaragua. The time required to pass over these distances being greater than the difference in the time of transit through the canals, the Nicaragua line, after completion, would be somewhat the more advantageous of the two to the United States, notwithstanding the greater cost of maintaining the longer canal.

III. The Government of Colombia, in which lies the Panama Canal, has granted an exclusive concession, which still has many years to run. It is not free to grant the necessary rights to the United States, except upon condition that an agreement be reached with the New Panama Canal Company. The Commission believes that such agreement is impracticable. So far as can be ascertained, the company is not willing to sell its franchise, but it will allow the United States to become the owner of part of its stock. The Commission considers such an arrangement inadmissible.

The Governments of Nicaragua and Costa Rica, on the other hand, are untrammeled by concessions and are free to grant to the United States such privileges as may be mutually agreed upon.

CONCLUSION.

In view of all the facts, and particularly in view of all the difficulties of obtaining the necessary rights, privileges, and franchises on the Panama route, and assuming that Nicaragua and Costa Rica recognize the value of the canal to themselves and are prepared to grant concessions on terms which are reasonable and acceptable to the United States, the Commission is of the opinion that "the most practicable and feasible route" for an Isthmian Canal to be "under the control, management, and ownership of the United States" is that known as the Nicaragua route.(Walker 1900, 43)

At this time there appeared to be no disagreements on the Commission about the opinions expressed in the report even though Morison was in favor of Panama at that time. On January 9, 1900 the House of Representatives, not waiting for the final report of the Commission, passed the Hepburn bill calling for a Nicaragua Canal, Figure 5, by a vote of 308-2. Since the concession for the Maritime Company had

Figure 5. Nicaragua "The American Route" 1900, Isthmian Canal Commission

elapsed, Morgan was ready to support Hepburn's bill. Morgan held hearings on the bill and invited the entire Canal Commission to attend and testify. On May 11, 1900 the hearings on House of Representatives Bill 2538, *To Provide for the construction of a canal connecting the waters of the Atlantic and Pacific Oceans*, was held by the Committee on Interoceanic Canals. Walker and all members agreed that a canal at either Panama or Nicaragua was feasible but they were not in a position to make a recommendation until all of the information from their field parties was in hand and considered. Morgan at times became incensed when none of them would state a preference and demanded to know what additional information they could possibly need. Morgan asked Morison:

> The Chairman. Mr. Morison, you are a civil engineer?
>
> Mr. Morison. Yes, sir.
>
> The Chairman. You have not been on either of these previous commissions?
>
> Mr. Morison. No, sir.
>
> The Chairman. So that we have first impressions so far as you are concerned of both the Panama route and the Nicaragua route?
>
> Mr. Morison. You have.
>
> The Chairman. In your examination of the Nicaragua route, commencing at Greytown Harbor and going to Brito, including the harbors as well as the line of the canal, what is your conclusion as to the feasibility and practicability of that route at a reasonable cost?
>
> Mr. Morison. Everything is feasible in construction to an engineer, provided he has sufficient time and money. The Nicaragua Canal can be built, given time and money, and maintained as a serviceable canal.
>
> The Chairman. Have you reached a conclusion that you are willing to state in regard to what the probable cost will be?
>
> Mr. Morison. No, sir; I have not. It is going to be an expensive piece

of work. That is as definitely as it has shaped itself in my mind...

The Chairman. I think there are no further questions.

Mr. Morison. Perhaps I might give a little reference to some other questions which have been asked previously.

The Chairman. Certainly.

Mr. Morison. I would refer particularly to the report of the Comité Technique. They considered three plans. They reduced them down to two plans, the plan with a summit level at the higher elevation, which you call the three-level scheme, and the plan with the level of Lake Bohio for the summit level. Those two plans differed somewhat more than in merely leaving out the third level, or the plan with the reduced lockage. The level of Lake Bohio was somewhat raised to reduce the amount of work in the Culebra cut; the result being that, although they dispensed with two locks on the Atlantic side, they dispensed with but one on the Pacific side. Those were the plans considered by the Comité Technique. They decided in favor of the plan with the higher level, on the ground that that was the only one which could be done within the time that their concession had to run. They perhaps did not put it quite as definitely as that, but that in that way they could open the canal at the earliest possible day, and at the time at which they could get the locks and dam done at Bohio. That was their reason. But if you read the report in full, you will see their preference seems to be to avoid the summit level. If they avoided that summit level, they would have avoided three locks, and they would have avoided the conduit between the Alhajuela dam and the summit level, simply using Lake Alhajuela as a storage reservoir. The cost of that conduit and the cost of the locks which they omitted about balanced the additional cost of excavation, and made the two canals almost the same price. In fact, they were so nearly the same cost that a variation in the unit prices might have brought them together.

They always contemplated using the upper lake above the upper dam at Alhajuela for two purposes, as a storage for water during the dry season and as a regulator to diminish the effect of the floods during the wet season. They had figured, in all their calculations, on spillways having a capacity in the aggregate of 1,250 cubic meters per second. If they were increased to 2,000 cubic meters per second, the Alhajuela dam would not have been necessary as a regulator, but our commission have considered that though the French plan was the one that was set before us, it was not necessarily the one that we thought the best, and if we could find means of improving the French plan at Panama, we considered it our duty to recommend such improvements. (Morison 1902a, 87)

Morison determined, regarding the Darien Route, "the most complete plan involves a tunnel at least seven miles long while not necessarily impracticable, such a tunnel would be very objectionable and would render the line inferior either to the Panama or Nicaragua Route. (Walker 1901a, 18)

The Walker Commission worked on its final report, and it was done in August 1901. Morison now had become a vocal supporter of the Panama Route and informed Walker he could not support the proposed final report as he disagreed with its conclusion and would not sign it. Apparently he, Burr and Ernst supported the Panama Route, but Burr and Ernst changed their position in favor of Nicaragua. He attempted to modify the report but was unsuccessful so he wrote what he called a Minority Report stating his reasons for disagreeing. He later wrote of this disagreement to President Roosevelt who succeeded McKinley in September 1901 after the latter's assassination. Morison wrote on December 10, 1901:

> The recent misleading statements in the public press touching my work in connection with the final report of the Isthmian Canal Commission have made me think it best to write to you explaining my actual position in this matter. I am led to do so largely by the kind assurance which you have given me that you know my record and value my opinion.
>
> The Isthmian Canal Commission completed a report which was then considered final in August 1901. I was unwilling to accept the conclusions in this report. I voted against its acceptance and declined to sign it. I then prepared with much care a statement of the reasons why I declined to sign the report of the Commission, which I signed and sent to the president of the Commission in triplicate so that it could be attached to the report of the Commission following the signatures of the other members and requested that this be done.
>
> In the preparation of this minority report I felt that while I should have been willing to waive some of my exceptions if an agreement could be reached on the others, when once forced to disagree with the final conclusion, it was right to state all points of difference. The grounds on which I took exception were five; the first related to the percentage allowed for contingencies on the Panama and Nicaragua estimates; the second related to the plan of the Bohio dam and the harbor requirement at Panama; the third related to dangers of delay; the fourth to international complications and the fifth to the assumption that the Panama Canal Company would not sell its property or put itself in a position to enable the United States to secure the necessary rights for the construction of a canal on the Panama route.
>
> On the 12th of November the Commission was called together again. A correspondence which had taken place between the president of the Commission and M. Hutin president of the New Panama Canal

Company was exhibited and a modified draft for the final chapter of the report was submitted to the Commission. The Commission continued in session four days and after much discussion and various concessions the final chapter was reduced to such form that the report was signed by all the members of the Commission including myself. In signing it I waived some points of difference in the interests of unanimity. M. Hutin did not appear before the whole Commission but the one thing which more than any other led me to act as I did, was the correspondence with M. Hutin, which was so much like the procrastinating dickering methods of the Oriental trader that the only course was to say no.

On the 21st of November the final chapter of the report, as signed six days before, was printed in the New York Journal and in the Chicago American. On the following day my minority report of last August was printed in the same two papers. This was followed by a dispatch from Washington published in various papers stating that the so called minority report had not been prepared for that purpose, but simply as a test paper and did not represent my views at the present time. This paper was prepared as a minority report and for no other purpose.

On Wednesday the 4th inst. you transmitted to Congress the report of the Commission. On the following morning various Washington dispatches stated this fact and also said that a minority report, from which quotations were made, favoring the Panama route was submitted by me. The correspondents evidently use copies of the New York Journal and the Chicago American instead of ascertaining the actual conditions at the Capitol. The evening papers in their editorials referred to such minority report and I have since received various letters and other communications congratulating me on an independence which I did not exercise. Another denial of the purport of the minority report appeared on the 6th and finally the New York Herald on the 9th gave an account of this whole matter which, though not entirely correct, was approximately correct.

I have not changed my views since I signed the minority report in August. I still feel that the Panama route is very much better than the Nicaragua route and I sincerely hope that matters may yet take a shape which will permit our Government to complete the unfinished work at Panama rather that to build the Nicaragua canal.

I believe that the difference in cost of the work to be done in the two canals will be much greater than the Commission's report states although I have waived this consideration in signing the report. If an allowance of 20 per cent is required for contingencies and other uncertain elements at Panama, an allowance of 30 per cent on the

Nicaragua prices would not be to high. A large portion of these contingent expenses occur at the beginning of the work and have already been met at Panama. The provision for contingencies, maps, drawings, records and omissions in the $40,000,000 estimated as the value of the French property at Panama, is nearly 45 per cent on the net estimated value of the actual excavation. A similar provision would add something like $20,000,000 to the Nicaragua estimate.

My second and third objections are partly covered by brief reference in the concluding chapter as finally adopted, which were not there in August.

[He then got into international complications giving his vision of an expanded Mexico north of the canal and South America to become like Mexico south of the canal]

When my minority report was written and signed, no proposition had been made by the representatives of the French Company to dispose of its property and the report from which I dissented contained the statement that "the Commission has been forced to the conclusion that the New Panama Canal Company has no disposition to sell its property or to put itself in position to enable the United States to secure the necessary rights for the construction of a canal on the Panama route should it desire to select that location." I then said that if a proposition had been made by the French interests this Commission would have had no power to accept it and nothing could be expected from negotiation until both parties had power to act. All this has been changed; the president of the New Panama Canal Company has positively declared that he is able to convey all the canal properties on the Isthmus to the United States by a perfectly clear title. Although the authority to complete the canal with the ample rights which are required must be derived from a new treaty with the Republic of Colombia such conveyance would remove all obstacles in the way of this treaty. But although the representative of the French Company may be able to convey by a clear title, the price which he has named is exorbitant, coming as it does after a series of procrastinating delays and unsatisfactory correspondence, that it can only be rejected as unworthy of consideration. The French Company can find no customer but the United States Government. If this Government sees fit to build the canal at Nicaragua the French investment at the Isthmus will be entirely lost. With the United States it is not the case of an Isthmian canal only, but the position with which the nation must hold among other nations of the world. If it pays an exorbitant price now, it may be called upon to do it in many other cases. Despite the advantages which the Panama route has over the Nicaragua route in cost of construction, in cost of operation and in convenience when done, the United States Government cannot afford to put itself in the position which the acceptance of the terms now proposed in behalf of the French Company would involve.

For this reason I thought it best to accept the report of the Commission. Perhaps I did wrong. I have not changed my views as the physical and political merits of the two routes and if the French would reduce their price to a reasonable sum, say something like one-half what they now ask, I should very strongly recommend its acceptance and the construction of the Panama canal. I sincerely hope that this may yet be the result. (Morison 1901b)

What had happened was that a Commission stenographer had leaked the last chapter of the final report and Morison's minority report to the New York Herald and the Chicago American both W. R. Hearst newspapers. They both published the final report on November 21 and Morison's minority report on November 22, 1901. This gave the supporters of Nicaragua heartburn as the Panama route was well reviewed and it appeared Nicaragua was chosen on the basis of cost and not utility. It, and Morison's minority report gave, for the first time, hope to the Panama people.

Morison stated in his minority report written in June 1901, in part:

I accept the location for the Nicaragua canal as one on which I can suggest no improvement. I consider that the estimate does not make enough provision for unknown conditions and contingencies. The cost of the work on both the Nicaragua and Panama routes had been estimated at the same unit prices, with the addition of the same percentage to cover 'engineering, police sanitation and general contingencies.'

The excavation of the Panama canal had been opened for nearly it entire length and the character the material to be removed can be examined in position.

On the Nicaragua route the character of the material has been determined by borings which, though unusually complete, do not give the definite information that is visible at Panama. At Panama there are fine harbors at both ends, fully adequate for all demands during the construction and connected by a railroad in high condition; the country is settled and many of the necessary accommodations for a large working force are there. Before the eastern section of the Nicaragua canal can be begun a harbor must be created at Greytown, convenient lines of transportation which does not exist must be provided, as must also the means of housing and caring for large population; nearly all of which must be imported.

The preliminary engineering has been done at Panama and the General contingencies have been reduced to a minimum. If 20 per cent. is added to the estimate to cover 'engineering, police, sanitation and general contingencies' at Panama at least 25 per cent and probably 30 per cent should be added at Nicaragua. The former figure raising the estimate for the Nicaragua route to $201,000,000. (Morison 1901a)

He then presented some modifications to the Panama plans that would reduce the cost of that route to $134,000,000 or $67,000,000 less than Nicaragua. He followed this with a discussion of possible delays in a canal over 112 miles long from accidental interruptions which he thought would offset the time advantage of Nicaragua from the east cost of the United States to the West Coast. He continued with a discussion of the franchises at both locations and concluded, the Panama route, "offers advantages in cost of construction, in cost of operation and in convenience when done while it is less likely to lend to local international complications."(Morison 1901a)

The final report of the Commission, dated November 16, 1901, was sent to the President but not sent to Congress until December 4 so the first news of the report was in the leaked version on November 22. The plans, charts, maps etc. were not published until even later. The conclusion was similar to the Preliminary Report and stated:

> There are certain physical advantages such as a shorter canal line, a more complete knowledge of the country through which it passes, and lower cost of maintenance and operation in favor of the Panama route, but the price fixed by the Panama Canal Company for a sale of its property and franchises is so unreasonable that its acceptance can not be recommended by this Commission...

> After considering all the facts developed by the investigations made by the Commission, and the actual situation as it now stands, and having in view the terms offered by the New Panama Canal Company, this Commission is of the opinion that 'the most practicable and feasible route' for an Isthmian canal to be 'under the control, management, and ownership of the United States,' is that known as the Nicaragua Route. (Walker 1901)

Walker withheld the final report for months waiting for the French to give the Commission a final price for their rights, plans, equipment, etc. The only number that had been given was $109,141,500. The use of the words, "as it now stands" implies that if the French were more reasonable and would accept the $40,000,000 that the Commission believed was a fair price, the decision could be changed.

The report had been modified by Walker and the Commission to partially accommodate Morison's second and third points, but Morison stuck to his point on contingencies and international complications in his letter to Roosevelt. As to cost he wrote Roosevelt. "I believe the difference in cost of the work to be done in the two canals will be much greater than the Commission's report states, although I have waived this consideration in signing the report." In other words he signed the final report of the Commission as presented to the President in order to make it unanimous and to promote construction of the canal. The proponents of Panama jumped on Morison's earlier dissent to promote their cause and Bunau-Varilla later wrote, "Most happily, on November 21 and 22, the veil was accidently lifted."(Bunau-Varilla 1920, 207) Newspapers around the country published parts of Morison's minority report and public opinion started to change on the issue of Nicaragua or Panama. This got the attention of the New Panama Canal Company that changed Presidents and on January 4, 1902 the new President of the canal company, Marius Bô, agreed to sell their rights to the United States for $40,000,000 the amount stated in the Commission's report. President Roosevelt, in light of this offer, called each member of the commission, less Haupt, to his office to discuss this new information. Each member reportedly agreed that the only thing keeping them from choosing the Panama route in the first place was the amount the French had demanded. Walker called the Commission together on January 16 to discuss the situation. On January 17 the *Evening Star* (Washington) ran a story with the headline- THE ISTHMIAN CANAL, Sessions of the Commission Expected Today, WILL SURELY REACH A VOTE, Likely That There Will Be Two Reports Presented, (Evening Star 1902, 1) On January 18, on the motion of George Morison, the Commission unanimously changed their vote in favor of Panama with Lewis Haupt agreeing with great reservation. A supplemental report of the commission stated:

> The advantaged of the two canal routes have been restated according to the findings of the former report. There has been no change in the views of the Commission with reference to any of these conclusions then reached, but the new proposition submitted by the New Panama Canal Company makes a reduction of nearly $70,000,000 in the cost of a canal across the Isthmus of Panama, according to the estimates contained in the former report, and with this reduction a canal can be there constructed for more than $5,500,000 less than through Nicaragua. The unreasonable sum asked for, the property and rights of the New Panama Canal Company when the Commission reached its former conclusion overbalanced the advantages of that route, but now that the estimates by the two routes have been nearly equalized the Commission can form its judgment by weighing the advantages of each and determining which is the more practicable and feasible.
>
> There is, however, one important matter which cannot enter into its determination, but which may in the end control the action of the United States. Reference is made to the disposition of the governments whose territory is necessary for the construction and operation of an isthmian canal. It must be assumed by the Commission that Colombia will exercise the same fairness and liberality if the Panama route is determined upon that have been

expected of Nicaragua and Costa Rica should the Nicaragua route be preferred.

After considering the changed conditions that now exist and all the facts and circumstances upon which its present judgment must be based, the Commission is of the opinion that "the most practicable and feasible route" for an isthmian canal, to be "under the control, management, and ownership of the United States," is that known as the Panama route. (Walker 1902)

With this report the work of the Commission was done, but they still had to be available to testify before Morgan's Senate Committee. Morison testified on February 12, 1902 for about four hours under questioning by Senators Harris, Morgan, Hanna, Kittridge and Platt. All of them, except Hanna and Kittridge, were supporters of the Nicaragua Route and therefore critical of the Commission's recommendation. Some of his testimony is as follows:

> Senator Harris. The Panama line was not regarded by the Commission in their first report as the one which they desired to recommend, and apparently-I should like to know if that is correct - the change in their opinion grew entirely out of the reduction in the price at which the property of the Panama Canal Company could he obtained.
>
> Mr. Morison. Well, I can speak only for myself in that respect. I never should have signed any report recommending the Nicaragua route in preference to the Panama route except on the ground that I felt that the United States could not afford to be held up by a French organization.
>
> Senator Harris. The reduction in the price asked for to $40,000,000 made it come within what you thought was a legitimate and proper price?
>
> Mr. Morison. I think that is a perfectly proper price. I think our Government could have better afforded to pay twice that price than to have built the Nicaraguan Canal, if that had been the whole question; but the United States Government, as I look on it, has many other things to do than to build an interoceanic canal; and if it allows itself to be imposed upon through an unreasonable price for one piece of property, it may be for some others, and that feeling was what settled my decision. I felt that these French people had put themselves in a position in which we could only treat them as you would treat an oriental trader-tell them that we could not have anything more to do with them if that was the way they talked...
>
> Senator Hanna. What I wanted to ask you, as, to the choice or practicability of routes for canal purposes, the getting of the, concessions was a necessity in either case; and from an engineering standpoint, from the standpoint of the physical conditions, you were in favor of the Panama route, as I understand you?

Mr. Morison. I am, and always have been so since I have seen the two routes...

Mr. Morison. If the last reports are correct, we can get rid of yellow fever by killing the mosquitoes. The swamps will produce fevers that are worse than yellow fever.

Senator Hanna. Then, summing up the proposition of the Panama Canal route, there are to your mind no physical difficulties that are insurmountable or approaching that, and no engineering difficulties but what can be easily overcome?

Mr. Morison. Well, which it is perfectly practical to overcome.

Senator Hanna. That is better. In view of those facts, do you consider the price at which this property is offered to us a reasonable price?

Mr. Morison. Yes, I do.

Senator Hanna. Do you consider it a low price?

Mr. Morison. I consider it a very fair price. That price represents, as near as we could estimate it, what would have to be expended now, if nothing had been done in the Isthmus, to put the Isthmus in such condition that the cost of the completing of the canal would be what it is now.

Senator Hanna. When that proposition was made by the new company I understand that it did not embrace a great deal of machinery, in the way of locomotives and other machinery there, that has since been added to the scrap pile which we buy.

Mr. Morison. Yes. I would not give anything for any of the machinery down there.

Senator Hanna. Because you could get better?

Mr. Morison. Because it is twenty years old. You cannot afford to use it. The locomotives are too small: they have not more than half the power they ought to have, and there are a great many machines there that nobody has ever discovered the use of...

Mr. Morison. The greatest advantage of the Nicaragua Canal-I think I may say that it is the only advantage it has - is that its west end is 500 miles nearer San Francisco than the west end of the Panama Canal. That means that a slow steamer, the class that usually carry freight, running night and day through the Nicaragua Canal, would get to San Francisco in one day less time than she would if she went through the Panama Canal. If, however, she did not run day and night, but tied up during the night, running only during the day, she would get to San Francisco in the same time by both routes. (Morison 1902a)

 He then read into the record a paper he had written on the savings in shipping time for shipping around the world. He concluded, on balance, "The only conclusion that I can draw from all this is that there is practically no advantage in one route over the other." In other words there would not be, as some claimed, significant savings in time if the Nicaragua route was built.

Morison described the Commission's Canal as follows:

> When this is completed, a ship entering the Panama canal will navigate the canal for 17 miles at the level of the sea. Then by a flight of two locks, it passes up into this lake, called "Lake Bohio;" it will not pass through the entire length of the lake, but after passing through 13 miles it will turn off towards the south and go through the Culebra cut on the original French location.
>
> This cut is about 8 miles long; it is heavy work for about 7 of those 8 miles; but it is exceptionally heavy only for one. The cut will be about 280 feet deep, and the work remaining to take out of it is about 40,000,000 cubic yards. That has been one of the bugbears of the Panama Canal.
>
> Passing through this 8 miles of cut a ship would pass down a double flight of locks [Pedro Miguel] to a lower level, go one mile on this lower level, pass through another lock [Miraflores] of varying lift according to the tide, and then in about six miles more would reach the line of the shore, and after four miles more in an excavated channel would reach deep water in the old anchorage in Panama Bay. The total length of this canal is about 49 miles from deep water to deep water. (Morison 1902c) [Note, as built one lock at Pedro Miguel and two at Miraflores)

Even after hearing all the testimony of the Commission members, and knowing President Roosevelt's desire, Morgan's committee voted 7-4 to forward the Hepburn Bill in favor of Nicaragua, to the entire senate for approval on March 10, 1902. Senator Mark Hanna wrote a lengthy minority report in favor of Panama. Some believe the report was written by Morison or Bunau-Varilla or even William Nelson Cromwell. Senator John Spooner, probably at the urging or President Roosevelt or Mark Hanna proposed the Spooner Amendment which supported the Panama route. The Senate did not take up the recommendation until June 4-19, 1902 with many senators weighing in on both sides of the issue. Morgan used all his powers of persuasion, but he lost on the main issue which was a substitution of the Hepburn act by the Spooner Act by a vote of 42-34. Other votes on amendments to place a time limit on the time the President had to reach an agreement with Columbia failed by smaller margins. After all the amendments were acted upon a final vote was taken and it passed by a vote of 67-6 on June 19. The House took up the senate Law between June 20 and June 26 but the House refused to vote on the Spooner amendment and the two chambers had to go into a conference committee to resolve their differences. The senate conferees would not budge and the House conferees finally agreed to support the Spooner amendment, and Panama. The entire house voted on the amendment and approved it by a vote of 260-8. The President signed it on June 28, 1902. The Spooner act stated in part:

> An Act To provide for the construction of a canal connecting the waters of the Atlantic and Pacific oceans.
>
> Be it enacted, . . . That the President of the United States is hereby authorized to acquire, for and on behalf of the United States, at a cost not exceeding forty millions of dollars, the rights, privileges, franchises, concessions, grants of land, right of way, unfinished work, plants, and other property, real, personal, and mixed, of every name and nature, owned by the New Panama Canal Company, of France, on the Isthmus of Panama, and all its maps, plans, drawings, records on the Isthmus of Panama and in Paris, including all the capital stock, not less, however, than sixty-eight thousand eight hundred and sixty-three shares of the Panama Railroad Company, owned by or held for the use of said canal company, provided a satisfactory title to all of said property can be obtained.
>
> SEC. 2. That the President is hereby authorized to acquire from the Republic of Colombia, for and on behalf of the United States, upon such terms as he may deem reasonable, perpetual control of a strip of land, the territory of the Republic of Colombia, not less than six miles in width, extending from the Caribbean Sea to the Pacific Ocean, and the right to use and dispose of the waters thereon, and to excavate, construct, and to perpetually maintain, operate, and protect thereon a canal, of such depth and capacity as will afford convenient passage of ships of the greatest tonnage and draft now in use, from the Caribbean Sea to the Pacific Ocean, which control shall include the right to perpetually maintain and operate the Panama Railroad, if the ownership thereof, or a controlling interest therein, shall have been acquired by the United States, and also jurisdiction over said strip and the ports at the ends thereof to make such police and sanitary rules and regulations as shall be necessary to preserve order and preserve the public health thereon, and to establish such judicial tribunals as may be agreed upon thereon as may be necessary to enforce such rules and regulations. The President may acquire such additional territory and rights from Colombia as in his judgment will facilitate the general purpose hereof.
>
> SEC. 3. That when the President shall have arranged to secure a satisfactory title to the property of the New Panama Canal Company, as provided in section one hereof, and shall have obtained by treaty control of the necessary territory from the Republic of Colombian as provided in section two hereof, he is authorized to pay for the property of the New Panama Canal Company forty millions of dollars and to

the Republic of Colombia such sum as shall have been agreed upon.... The President shall then through the Isthmian Canal Commission hereinafter authorized cause to be excavated, constructed, and completed, utilizing to that end as far as practicable the work heretofore done by the New Panama Canal Company, of France, and its predecessor company, a ship canal from the Caribbean Sea to the Pacific Ocean. Such canal shall be of sufficient capacity and depth as shall afford convenient passage for vessels of the largest tonnage and greatest draft now in use, and such as may be reasonably anticipated, and shall be supplied with all necessary locks and other appliances to meet the necessities of vessels passing through the same from ocean to ocean; and he shall also cause to be constructed such safe and commodious harbors at the termini of said canal, and make such provisions for defense as may be necessary for the safety and protection of said canal and harbors....

SEC. 4. That should the President be unable to obtain for the United States a satisfactory title to the property of the New Panama Canal Company and the control of the necessary territory of the Republic of Colombia and the rights mentioned in sections one and two of this Act, within a reasonable time and upon reasonable terms, then the President, having first obtained for the United States perpetual control by treaty of the necessary territory from Costa Rica and Nicaragua, upon terms which he may consider reasonable, for the construction, perpetual maintenance, operation, and protection of a canal connecting the Caribbean Sea with the Pacific Ocean by what is commonly known as the Nicaragua route, shall through the said Isthmian Canal Commission cause to be excavated and constructed a ship canal and waterway from a point on the shore of the Caribbean Sea near Greytown, by way of Lake Nicaragua, to a point near Brito on the Pacific Ocean.... (Spooner Act 1902)

There is no record of Morison being involved with battles in the House and Senate between June 4 and June 28. To help promote the Panama cause in Congress he gave three talks on an Isthmian Canal comparing the Panama and Nicaragua routes. The first was to a Commercial Club of Chicago, January 25, 1902 (Morison 1902c). The second was to Reform Club of Massachusetts April 24, 1902 (Morison 1902d). On May 20, 1902 he gave a long talk to the Contemporary Club of Bridgeport, Connecticut (Morison 1902e). All were entitled *The Isthmian Canal.*

He also wrote an article for the ASCE entitled *The Bohio Dam*, in which he proposed a different design from the commission's for the main dam that would dam up the Chagres River to form Lake Bohio. As was common at that time a copy of the paper was published in the *Proceedings of the ASCE* in January so that readers could respond in writing prior to the meeting and orally at the meeting. His Bohio Dam article appeared in the January 1902 issue. He gave his talk on March 5, 1902 and it was published in the *Transactions ASCE* May 1902. In the paper he noted he did not agree with the route, Figure 6, selected by the Commission nor in their design of a dam at Bohio. He testified to the Senate on February 12 1902, "There is one thing that I would like to say before we go. I see that a previous witness has introduced a paper that I prepared on the subject of the Bohio dam. That paper was prepared with a view of bringing the matter before a collection of engineers for discussion, to see what criticisms could be made on what I considered a satisfactory solution of the dam problem at a very much less expense than the Commission's plan. It will be discussed at a meeting in New York on the 5th of next month. When I stated there that I considered the Commission's plan- I have not the paper here or I would give you the exact words- when I stated there that I considered that the Isthmian Canal Commission's plan involved very great difficulties, I certainly did not mean that it could not be done." (Morison 1902b, 19)

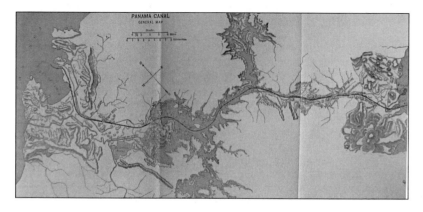

Figure 6. The Panama Route, Isthmian Canal Commission

William H. Burr, the Chairman of the Commission's Panama committee, also wrote a paper for the *Popular Science Monthly* entitled *The Panama Route for a Ship Canal* in August 1902 giving the official design of the Canal at Panama. In it he gave many of the same reasons as Morison in favor of the route.

Morison followed up with an article written on December 23, 1902 entitled, *The Panama Canal*, and published in Volume 35, Issue No. 1 in 1903, in the *Bulletin of the American Geographical Society*. He summarized his position on the routes as follows:

> In no one respect can the Panama route be considered the best. For convenience of access to all ports except those of South America, Tehuantepec is superior, but a summit 700 feet above tide water, too broad to tunnel or to cut through, makes it an impossible canal route. The San Blas route from Mandinga harbor, on the Caribbean side, to

the mouth of the Chepo, in Panama Bay, has the best harbors, and is the shortest line between the two oceans; but the summit is four times as high as that at Panama, requiring a tunnel more than four miles long, which is fatal. The Nicaragua route has the lowest summit and the attractive feature of a great inland lake for its water supply. It is a practical route for a canal; but the distance between the oceans is four times as great as at Panama; it is near the region of maximum volcanic disturbance; there are no existing harbors at either end, and for a distance exceeding the entire width of the isthmus at Panama the Nicaragua Canal must be built through a nearly uninhabited swamp, with a rainfall of over 200 inches a year, where the newly-upturned soil would be full of malaria; and this only brings it to the San Juan River, whose crooked channel must be improved and made navigable. There are three possible routes from Caledonia Bay to San Miguel Bay; the distance between tidewater on each of them is about the same as at Panama; there are good harbors at each end; but all of these routes involve tunnels, although on one of them the summit is but little more than 600 feet high, and on another the length of tunnel can be kept within two miles. Various routes have been proposed between the Atrato River and the Pacific, but all of these involve a considerable length of canal in the delta of a great silt-bearing river and a length of inland navigation much greater than on either the Panama, the San Blas, or the Caledonia routes. While not possessing any single feature better than that of some other route, the Panama route has many less bad features than any of the others. It is for this reason that it has been so often selected as the best trans-isthmian line. (Morison 1903a, 24)

His colleague on the Commission, Emory Johnson also had an article in the same issue entitled *Comparison of Distances by the Isthmian Canal and other Routes*. He wrote, "the Panama route is the more advantageous for the West South American trade, both with Europe and the United States. For the commerce of Europe and the United States with every other Pacific country, with the exception of New Zealand, to which the distances are practically equal, the Nicaragua is shorter than the Panama route." (Johnson 1903, 168)

Morison wrote another an article entitled, *The Summit Level of the Panama Canal* in the June 1903 issue of *Engineering Magazine* (Morison 1903a). In both articles he proposed a route different than the Commission's route between Colon and the Bohio Dam. He also suggested that an additional lock be placed at Tiger Hill making it a three lock approach to the Lake. He thought that the 42 ft. lifts, with two locks at Bohio, was unnecessary and risky. His plan created a new and lower lake between the Tiger Hill Lock and Bohio Lake.

He maintained his interest in the canal until he died suddenly on July 1, 1903 in New York City. He was buried in Peterborough, New Hampshire.

PHILIPPE BUNAU-VARILLA, THE OTHER INDISPENSIBLE MAN OF PANAMA

Bunau-Varilla, Figure 7, was born out of wedlock in Paris, France on July 26, 1859. Not much is known of his family or early life but we next meet him as a scholarship student at the Ecolé Polytechnique in Paris, one of the greatest engineering schools in the world. He graduated in 1880. While there he heard a

Figure 7. Philippe Bunau-Varilla

speech by Ferdinand de Lesseps on the Panama Canal. He like many of the top students was admitted to the Ecolé des Ponts et Chaussées located next door to the Polytechniqué for another three years. It was effectively a graduate school for state engineers. He graduated in 1883 and worked in Algeria and Tunis for a period before being sent to Panama to work on de Lesseps' canal that had began in 1880. He sailed to Panama on October 6, 1884 on board ship with Jules Dingler, the Chief Engineer. Shortly after arriving yellow fever struck again and Dingler's wife, among others, died. In 1885 a small revolution occurred on the Isthmus under Prestan and Azipuru who took Colon and Panama City in bloody battles. It was squashed when the United States put troops on shore at Colon to protect the railroad and docks. Shortly after Colombian troops arrived, and the revolution was terminated. He became Engineer on two divisions and the Port of Colon when one of the engineers left suddenly. In the summer of 1885 Dingler left and was replaced by Maurice Hutin who also left after a short stay. This left Bunau-Varilla in charge of the largest construction job on the face of the earth at the age of 26. According to Bunau-Varilla he put the project back on track and was excavating over 1,4000,000 cubic yards of soil and rock a month in spite of bad weather, floods typhoons, etc. He was sent assistants but they usually died of yellow fever or left within a few months. In late 1886 the company changed its program in favor of large contractors rather than men working directly for the company. At this time Bunau-Varilla was replaced by Leon Boyer, and after surviving a bout of yellow fever he was sent back to France to recuperate.

Boyer himself would die within the month from yellow fever. Bunau-Varilla was asked to return to the Isthmus by Charles de Lesseps as head of his own company to tackle the Culebra Cut. In association with his brother Maurice and M. Antigue and M. Sonderegger, he took over the work of the failed Anglo-Dutch companies. He solved the problem of disposing of waste material from the cut by building wooden trestles perpendicular to the walls of valleys and dumping the soil and rock over the side. This method of disposal kept the dirt cars and trains running on schedule and great progress was made. But even with this progress he knew it would take a long time to advance the cut and provide for a sea level canal. He proposed to de Lesseps that it would be better to build a temporary lock canal with the elevation at the bottom of the cut 140 ft. rather than the -35 ft. that would be required for a sea level canal. The canal could then be opened and generating revenue while work continued in lowering the rock at the summit. When the top level was excavated to the proper level, the top lock gates could be removed and excavation continued at the lower level until all locks were removed. Initially de Lesseps, who was married to a sea level canal, resisted but by later 1887 he agreed to the temporary lock canal idea. Work on the new plan began in early 1888 with Gustav Eiffel contracted to build the lock gates required and Bunau-Varilla estimated the canal could be opened by December 31, 1891. He formed another company to handle the excavations for the locks to be built by Eiffel. Unfortunately de Lesseps and his company were running out of money and needed a major infusion of cash to continue operations. He proposed another issue of lottery bonds but that failed to raise the required money. On December 14, 1889 Bunau-Varilla's dream and the entire company collapsed. A receiver was appointed and one of the first things he did was to form a Commission D'études to recommend a course of action. They reported back on May 3, 1890 recommending a permanent lock canal with a cost of $180,000,000 and seven years to build. Bunau-Varilla then wrote a 173-page book entitled *"Panama-the Past, the Present, the Future"* (Panama: le passé-le présent-l'avenir) that was released on March 20, 1892. In it he gave his plan for a temporary lock canal along with his methods of attacking the Culebra Cut. He presented a financial plan to build the canal and a way to form a new company and take the project out of the hands of the liquidator. The book was filled with photographs of the work and equipment at Panama. He followed this up a 69-page sequel entitled, *Panama- Le Trafic* in September 1892 comparing the Panama and Nicaragua routes. He then went to Russia to get their support for the canal and for a time it appeared they would guarantee the interest on any bonds issued. That deal failed when the French Government failed.

The New Panama Canal Company was formed on October 14, 1894 with a capital of $12,000,000. Its main goal was to continue work on the Isthmus to keep the Colombian Concession alive. It appointed another International Committee of Experts (Comité technique) to study the options available and report back. On November 16, 1898 they came back with a plan for a lock canal similar to that proposed in 1892 by Bunau-Varilla, with the exception that they lowered the summit level by 33 ft. to take into consideration the work that was done in the cut between 1892 and 1898. Bunau-Varilla, however, had been kept out of the loop by the new company as they wanted to keep clear of the stain that was placed on all those who had worked on the canal between 1879 and 1889. After having the report reviewed by another committee for a year, the new plan was publically announced on November 30, 1899. Bunau-Varilla had by this time worked on and in behalf of de Lesseps dream from 1885 to 1899 and of all people in France knew the New Panama Canal Company only existed to sell its rights, etc. to someone else. It is also to be remembered that the Isthmian Canal Commission (The Second Walker Commission) had been authorized by Congress on March 3, 1899.

BUNAU-VARILLA 1899-1902

His first action, as noted, was to meet and talk with Morison, Burr and Ernst (the Panama subcommittee) while they were in Paris. He presented them copies of his two books, and according to himself, saying as noted earlier, "Here is what I have published during the most violent moral storm which has agitated France for many years. If in these books, written for the defense of Panama, you find today a solitary fact or a single incorrect figure you may throw them away. If you find nothing erroneous, adopt the conclusions, which are written therein, because they are the expression of the eternal truth." (Bunau-Varilla 1920, 166) He believed that these meetings began the process of looking at Panama in a different light.

He then used his contacts in the United States to set up a series of presentations on the merits of the Panama route. His first was to the Commercial Club of Cincinnati on January 16, 1901. He followed this with a talk to the Cincinnati Society of Civil Engineers. He then met Senator Mark Hanna who was to become a staunch supporter of Panama and who was the man closest to President McKinley. After another talk in Cleveland he went to Boston for a speech in front of the Commercial Club. He was then back in Chicago with meetings with the National Business League followed by a talk in front of the New York City Chamber of Commerce. While these may have been small venues the local newspapers picked up on the talks and spread his message to many others which, he believed, started the movement in public opinion away from Nicaragua. In order to spread his words even farther he prepared a pamphlet based upon his talks entitled *Nicaragua or Panama?* and published it on March 20, 1901. In its 32 pages plus illustrations he made direct comparisons between the two routes and frequently used information from the previous US Canal Commission's reports. He introduced his pamphlet as follows:

> I firmly believe that when the Truth is advancing nothing can stop it; I firmly believe that its irresistible pressure will overthrow any dam of prejudice erected in order to hold it back.

At the same time I think that individual efforts may largely help the Truth in its progress, by clearing from its path the obstructions of ignorance.

This was my aim when I answered affirmatively to three American friends who invited me to come to this country of free discussion in order to say publicly what they had heard in private conversation with me in Paris.

I am not here as the representative of any private interest; I came to defend a grand and noble conception which gave me several happy years of struggle and danger, and for which I suffered many years of anxiety, during which I do not remember one hour of despair.

It has been to me a great privilege to have the opportunity of exposing to the clear light of day all the irrefutable facts which show that Providence has subjected to a severe test the sagacity of man, by giving apparently to the Isthmus of Nicaragua all good qualities, and to the Isthmus of Panama all defects for an interoceanic waterway, when in reality it has given to the latter and refused to the former all the attributes necessary for the establishment of this natural highway of nations.

I have been happy to speak in this great country, where the first official word of justice for Panama has come from the eminent Isthmian Canal Commission.

My purpose has been attained. I have worked for the scientific Truth on one of those fields, where, as Mr. Carnegie recently and justly said, there is no room for selfish and private aims. (Bunau-Varilla 1901, 3)

He concluded,

The facts that I have related demonstrate the following statements:

I. It is very questionable whether the continuous earthquakes will allow the construction, on the Nicaragua route, with all its indispensable qualities, of a substantial masonry dam, which is the key of the whole Canal.

II. Admitting a dam could be built, the Nicaragua route, whatever may be the engineering skill displayed and the expenses made, will *never acquire* the two most essential qualities necessary to an interoceanic canal, - *continuity of operation* and *security of transit.*

The large ocean steamers will have there to struggle, deprived of any means of resistance, against the combined efforts of the continuous violent gales, of the heavy river currents, of the impediments resulting, of the constant modifications of the depths in the channel, and of the presence of the numerous dredges that will have to keep the way open, and this when they will be steering with great difficulty around an extraordinary number of curves, the short radii of which ought to be, in themselves, considered as incompatible with the navigation of big ocean ships in a narrow channel.

The ships will also have to meet two bad, exposed, and dangerous passages, when going from the Atlantic into the Canal and from the Canal into the Lake of Nicaragua.

One is allowed to say that *continuity of transit* and *safety of navigation will be constantly at the mercy of conflicting elements and beyond the control and prevision of man.*

III. If the experience of four centuries is not a mere word, if the undisputable proofs, written in letters of the continuous violent and increasing volcanic activity in Nicaragua, are not a mere dream, the route over that Isthmus is not only eventually exposed to, but certain, sooner or later, to be the prey of, that uncontrollable power of nature before which flight is the only resource.

If one thinks that not only the enormous cost of such a waterway would be at stake but also the very basis of the prosperity, the wealth of the millions of people who will settle on the west coast of North America, as soon as the construction of the Canal will join them with the other side of the continent and with Europe, one hesitates to calculate the consequences that would result from one of those seismic convulsions, which most probably will be still more terrible in the future than they have been in the past on that part of the Isthmus.

To prefer definitively the Nicaragua route to the Panama route, the unstable route to the stable one, would mean to prefer the stability of a pyramid on its point to the stability of a pyramid on its base when to that stability is attached the prosperity and welfare of a whole continent.

IV. The Panama route, having no winds, no currents (except on rare occasions), no sharp curves, no sediments, no bad harbors, no volcanoes, enjoys to the highest degree the three essential qualities totally wanting for the Nicaragua solution, - *continuity of operation, security of transit, stability of structure.* Outside of that it is three times shorter, will cost much less than the Nicaragua route and is easily transformable into a Bosphorus, the only form that will definitely answer to the world-wide interests to be served by the route, and allow of a passage from ocean to ocean in five hours. (Bunau-Varilla 1901, 3)

Right after publishing his pamphlet he met Senator Mark Hanna for the first time. Hanna was a member of the Senate Committee on Interoceanic Canals and a leader in the Republican Party. Hanna talked with many people about Bunau-Varilla and was very interested in his efforts in behalf of a canal at Panama. Hanna evidently told him that if the Commission's final report came out in favor of Panama he would support it and Bunau-Varilla. After this meeting he also met with President McKinley and gave him a copy of his pamphlet. He even tried to talk with Senator John Morgan but found him to be a "fanatic." He of course sent copies of his pamphlet to the Commission and had an exchange with George Morison on it. They evidently disagreed on the impact of seismic activity on the construction and maintenance of the canal but did not disagree on the potential impact of volcanic activity. He left the United States in April 1901 convinced, or at least hoping, that he had opened the eyes of not only the Commission but Senators, Congressmen, Newspapermen, etc. to the merits of the Panama route over the Nicaraguan route. The *New York Herald* ran a story on April 13 headlined, RESURECTION OF THE PANAMA PROJECT, and on April 18, THE DEFENDERS OF NICARAGUA ARE ALARMED, THE PANAMA CANAL PROJECT ADVANCES RAPIDLY.

Bunau-Varilla went directly to the French people to urge them to pledge their resources towards the completion of the canal for the glory of France. At his own expense he published long articles, using his own unique language, to urge all citizens to urge the French government to undertake the completion of the canal with the support of the citizens. He published several of these articles on April 25, May 10 and December but the government and many newspapers were either quiet or antagonistic towards a renewal of French efforts in Panama considering the financial troubles caused by the collapse of de Lessep's Company. Upon the failure of this effort he knew the only way to complete the canal was to have the New Panama Canal Company sell its rights to the United States for the $40,000,000. Walker and the Commission repeatedly requested the New Panama Canal Company to give them a price for their rights, plans, etc. but, as noted earlier, the only price mentioned was over $109,000,000. Bunau-Varilla knew that the only price that would help to change the mind of the Commission was $40,000,000 that they had estimated so he began a program to force the company to respond to Walker's request with a hard number, and that number should be $40,000,000. He went back to the United States on November 12, 1901 (Theodore Roosevelt has succeeded McKinley at this time) to talk to among other people Senator Mark Hanna and his friend Myron Herrick. On November 21 and 22 the Walker Commission's report and Morison's Minority report were published. The only thing keeping the Commission from accepting the Panama route, at least by Bunau-Varilla's reading of the report, was the price the New Panama Canal Company wanted for its rights. When he talked with Hanna about the report he pointed out the advantages listed by the Commission of the Panama route. He then wrote a long cable to *Le Matin*, his brother's newspaper, stating in part, "In spite of adverse public opinion, and the unpleasant consequences to Nicaragua, the situation may still be saved if the Canal Company abandons all ambiguous diplomacy and dangerous controversy immediately after the publication of the report." (Bunau-Varilla 1920, 208) Partly as a result of this cablegram President Hutin was replaced by Marius Bô and as soon as Bunau-Varilla got back to Paris he met with him in mid-December urging him to inform the Commission that the New Panama Canal Company would accept $40,000,000. This was a large pill for Bô to swallow but the decision was made easier when a cablegram from the United States was published in *Le Matin*, "Senator Hanna announces, on behalf of the Republican members of the Canal Committee, that they are ready to re-examine the question of adopting the route via Panama, should the owners of the French works be disposed to sell their enterprise for forty million dollars." (Bunau-Varilla 1902, 209) On January 4, 1902 the New Panama Canal Company telegraphed its willingness to accept the $40,000,000 offered. In spite of this, on January 9, the House of Representatives voted overwhelming in favor of the Nicaragua route.

BATTLE IN THE SENATE

Morgan stated he wanted to start his hearings on the Hepburn Bill on January 14, but Hanna was successful in delaying them until Roosevelt had the opportunity to talk with the Commission members and their recommendation in light of the French offer. As a result of those discussions, on January 18, 1902 Walker issued, at the urging of President Roosevelt, a supplementary report recommending Panama. Morgan's hearings were discussed in the Morison segment and will not be repeated here. The Interoceanic Canal Committee voted 7-4 to report the Hepburn bill out to the full senate with Hanna, joined by Millard, Prichard and Kittridge, writing a minority report:

1. It is 134.57 miles shorter than the Nicaragua from sea to sea (being 49.09 miles by Panama as against 183.66 miles by Nicaragua).

2. It has less curvature, both in degrees and miles, being but 22.85 miles of curvature as against 49.29 on the Nicaragua, and but 771 degrees for Panama as against 2,339 degrees for Nicaragua.

3. The actual time of transit is less, being but twelve hours of steaming by Panama, as against a minimum of thirty-three hours of steaming by Nicaragua; that is, of one day of daylight as against three days of daylight (for the canal must be navigated by day exclusively at first, and, to a great extent, always, especially by large ships, which chiefly will use it. The Commission's plan does not provide facilities for navigation by night.)

4. The locks are fewer in number, being but five on the Panama to eight on the Nicaragua.

5. The harbors are better, those at the termini of the Panama being good and already used by the commerce of the world, while at the termini of the Nicaragua there are no harbors whatever.

6. The Panama route traverses a beaten track in civilization, having been in use by the commerce of the world for four centuries, while the Nicaragua route passes through an unsettled and undeveloped wilderness.

7. There already exists on the Panama route a railroad perfect in every respect and equipped in a modern manner, closely following the line of the canal, and thus greatly facilitating the construction of the canal, as well as furnishing a source of revenue, and included in the offer of the Panama company.

8. The annual cost of maintenance and operation of the Panama Canal

would be $1,300,000 less than that of the Nicaragua (which sum capitalized is the equivalent of $65,000,000).

9. All engineering and practical questions involved in the construction of the Panama are satisfactorily settled and assured, all the physical conditions are known, and the estimates of the cost reliable, while the Nicaragua involves unknown and uncertain factors in construction and unknown difficulties to be encountered, which greatly increase the risks of construction and render uncertain the maximum cost of completion.

In addition to these facts stated by the Commission are the following, not referred to by them, but which have become of controlling importance, viz:

10. It is recognized that a sea-level canal is the ideal. The Panama Canal may be either constructed as a sea-level canal or may be subsequently converted into one. On the other hand, no sea-level canal will ever be possible on the Nicaragua route.

11. No volcanoes exist on the line of the Panama Canal nor in its neighborhood. On the other hand, the Nicaragua route traverses an almost continually volcanic tract, which has been during the last three-quarters of a century probably the most violently eruptive in the Western Hemisphere. The active volcanoes, Zapatera and Ometepe, rise actually from the waters of Lake Nicaragua.

12. At Panama earthquakes are few and unimportant, while the Nicaragua route passes over a line of well-known crustal weakness. Only five disturbances of any sort were recorded at Panama during 1901, all very slight, while similar official records at San Jose de Costa Rica near the route of the Nicaragua Canal, show, for the same period, 50 shocks, a number of which were severe.

13. As a practical matter the masters of vessels prefer the Panama route for safety, convenience, and shortness of transit, for its less curvature and risks, and for the lower insurance rate by that route.
(Hanna 1902)

These arguments are right out of Morison's and Bunau-Varilla's playbook. Together they won many adherents to the Panama Cause. The effort then shifted to the full Senate.

As described previously in the Morison segment debate on the Spooner amendment to the Hepburn bill took place between June 4 - 19, 1902. Bunau-Varilla sent a copy of his *Nicaragua or Panama?* pamphlet to every Senator along with an article from the *Evening Post* describing the explosion of Mt. Pelée on the island of Martinque on May 6 and how a well known geologist supported his contention about the difference in volcanic activity between Nicaragua and Panama. He also knew that Hanna's speech to the Senate could be the deciding factor in the determination of the Panama vs. Nicaragua debate. He then prepared and convinced Hanna to use a set of

13 visual aides to make the case for Panama. Hanna suggested he put them into a pamphlet under his own name and he would distribute it to every Senator. The pamphlet, under the title *Comparative Characteristics of Panama and Nicaragua*, included the visual aides and brief notes describing the message the reader should obtain from the aide. He had it prepared and printed by June 2 in time for the debate in the senate. Two examples, Figure 8, Figure 9, of his aides are given below.

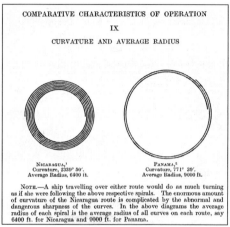

Figure 8. Bunau-Varilla Curvature, 1920, 234

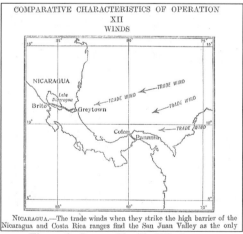

Figure 9. Bunau-Varilla Winds, 1920, 237

The debate in the Senate pitted the entrenched supporters of Nicaragua against a small group of supporters of Panama led by Hanna. It got very personal once it appeared to the Nicaragua men that the Panama men were making progress in converting some members of the Senate who had earlier been supporters of Nicaragua. During the debate word came to Washington that Mt. Momotombo on Lake Managua, located just north of Lake Nicaragua, had erupted on May14. This was exactly what Bunau-Varilla had been warning about but Morgan and his allies did not believe the story and had officials in Nicaragua deny, officially, that any eruption had taken place. It, of course had erupted, but Bunau-Varilla could not prove it given the primitive means of communication with Nicaragua at the time. To make his case he recalled seeing a postage stamp from Nicaragua with a volcano erupting on it. It turned out that it was Mt. Momotombo. He was able to find enough of the these stamps in Washington and mounted them on sheets and sent them to every Senator with the statement, *The Nicaraguan Stamp, (1900 issue) used by the author as proof of the presence of active volcanoes in Nicaragua.* He sent this out on June 12, only a few days before the final vote. He, and the Panama Cause, was now starting to get support in newspapers such as the *Washington Star* and the *New York Sun*. On June 19, 1902 the Senate voted in favor of the Spooner amendment.

The House of Representatives had even more rabid supporters of Nicaragua led by Hepburn. As noted earlier a conference committee was formed with members of the House and Senate to resolve differences in the original Hepburn bill passed by the House and the Spooner amended plan from the Senate. After heated debate the conference committee agreed to submit the Spooner amended bill to the entire house for approval. Bunau-Varilla then sent his pamphlet and stamp to each member of the House of Representatives. The House approved the Spooner amended Hepburn Act by a vote of 260-8 on June 26, 1902. The President signed it on June 28, 1902.

SUMMARY AND RELATIONSHIP BETWEEN MORISON AND BUNAU-VARILLA

With the bill signed into law by the President the two indispensible men had succeeded in changing the hearts and minds of the Isthmian Canal Commission, the President, the Senate, and the House of Representatives away from the "American Route" to the failed French route. They of course received help from others such as Mark Hanna and President Roosevelt and even from William Nelson Cromwell, the lawyer that had been retained by the New Panama Canal Company to lobby Congress to accept the Panama Route, but it was these two men, generally working alone, but at times working in a coordinated manner that succeeded.

It is clear from the record that they had great respect for each other's technical abilities but also for each other's dogged determination to fight for what they perceived as the right solution for an Isthmian Canal. Morison, for instance, in his Bohio Dam paper named a large island in Lake Bohio after Bunau-Varilla writing, "It will be a beautiful body of water, and in it will be an island of about 400 acres, which I have proposed to call the Island of Bunau-Varilla in honor of the brilliant Frenchman who never despaired of the completion of the Panama Canal and to whose untiring energy we owe so much." (Morison 1903) Bunau-Varilla in his 1920 book wrote of Morison, "In the Isthmian Canal Commission of 1899, it was his iron will, based upon Scientific Truth, which was the unconquerable obstacle on which was shattered the dogma of Nicaragua to which the whole of the Commission, himself included, was at the outset devoted." (Bunau-Varilla 1920, 457)

EPILOG

Bunau-Varilla's contribution to the Canal was not over, as he was a key player in the revolution of Panama from Colombia in November 1903. He was named Minister Plenipotentiary for the new Republic of Panama. He negotiated the Hay-Bunau-Varilla Treaty which gave the United States the title to a strip of land 10 miles wide in perpetuity with the right to act as if it were sovereign in return for a lump sum payment of $10,000,000 and a yearly payment of $250,000 starting nine years after the signing of the treaty.

The United States built the Canal between 1904 and 1914. Morison died in 1903. Bunau-Varilla, though living until 1940, never saw the canal he promoted so effectively in 1900, 1901 and 1902. These two men were truly "Indispensible" to the creation of the Panama Canal. Without them the following "Indispensible Men", William Gorgas, John Stevens and George Goethals would never have had the chance to complete the canal.

REFERENCES

Ammen, Daniel, 1880, The American Inter-Oceanic Ship Canal Question, L. R. Hamersly & Co. Philadelphia

Bunau-Varilla, Philippe 1892, Panama-the Past, the Present, the Future, (Panama: le passé-le présent-l'avenir), March 20, Masson, Paris

Bunau-Varilla, Philippe 1892, Panama, Le Trafic, December, (A Supplement to Panama: le passé-le présent-l'avenir), Masson, Paris

Bunau-Varilla, Philippe 1901, Nicaragua or Panama, Pamphlet

Bunau-Varilla, Philippe 1902, Comparative Characteristics of Panama and Nicaragua, Pamphlet

Bunau-Varilla, Philippe 1920, Panama: The Creation, Destruction and Resurrection, New York

Davis, Charles H. 1867, Report of the Interoceanic Canals and Railroads between The Atlantic and Pacific Oceans, Washington, Government Printing Office, Senate Document Volume 2, No. 62, 39th Congress, 1st Session

Evening Star 1902, The Isthmian Canal..., January 17, 1

Hanna, Mark 1902, Minority Report, Contained in Isthmian Canal: Speech in the Senate of the United States, Wednesday June 11 by Charles Warren Fairbanks

Humphreys, Andrew 1876, Report of the Interoceanic Canal Commission, Senate Ex. Document 15, 46th Congress, 1st Session, Vol. 1

Isthmian Canal Commission Act, Section 3, March 3, 1899, Part of the Rivers and Harbors Bill House of Representatives

Johnson, Emory 1903, Comparison of Distances by the Isthmian Canal and other Routes, Bulletin of the American Geographical Society of New York, Volume 35, No. 1.

Ludlow, William 1896, Hearings on House Bill 35 on the Nicaragua Canal, to Committee on Interstate and Foreign Commerce, April 28, 136, 54th Congress 1st Session, Document No. 315)

Morison, George 1899, Circular Letter of May 1, 1899 and Answers Thereto, New York State Committee on Canals, 1899-1900

Morison, George 1895, Presidential Address, Transactions ASCE, Volume 33

Morison, George 1901a, New York Herald, November 22

Morison, George 1901b, Letter to President Roosevelt, December 10

Morison, George 1902a, Statement of Geo. S. Morison before the Subcommittee on Interoceanic Canals, United States Senate, Government Printing Office, Washington, February 12.

Morison, George 1902b, The Bohio Dam, Transactions of the American Society of Civil Engineers, Vol. XLVIII, No. 2, May 1902, pp. 235-258 (Also Proceedings ASCE January 1902)

Morison, George 1902c, The Isthmian Canal, Address to the Chicago Commercial Club, January 25

Morison, George 1902d, The Isthmian Canal, Address to the Massachusetts Reform Club, April 24

Morison, George 1902e, The Isthmian Canal, Address to the Contemporary Club, Bridgeport, Ct. May 20

Morison George 1903a, Lake Bohio and the Summit Level of the Panama Canal, The Engineering Magazine January, 497-508

Morison, George 1903b, The Panama Canal, Bulletin of the American Geographical Society of New York, Volume 35, No. 1.

Spooner Act 1903, Acts of Congress Relating to the Panama Canal, Treaties and Acts of Congress relating to the Panama Canal 1917, 26.

Walker, John G. 1899, Report of Nicaragua Canal Commission, 1897-1899, The Lord Baltimore Press, Baltimore, Maryland

Walker, John G. 1900, Preliminary Report of the Isthmian Canal Commission, November 30, 43, 44, Senate Document No. 5, 56th Congress, 2nd Session

Walker, John G. 1901, Report of Isthmian Canal Commission, 1898-1902, Senate Document No. 222, November 16, 57th Congress, 1st Session

Walker, John G. 1902, Supplemental Report of Isthmian Canal Commission, Senate Document No. 123, Jan. 20, 57th Congress, 1st Session

The American Engineers that built the Panama Canal

J. David Rogers, P.E., P.G., M.ASCE[1]

[1] K.F. Hasselmann Chair in Geological Engineering, Missouri University of Science and Technology, Rolla, MO 65409, ph (573) 341-6198; fax (573) 341-6935; email: rogersda@mst.edu

ABSTRACT

When the United States took over title of the French canal franchise in Panama in 1903 they approached the project with vigor and confidence, treating it as an enormous railroad engineering project. A large part of the eventual success of the United States in building a canal at Panama came from avoiding the mistakes of the French, whose leadership had proven too inflexible. From the outset the Americans employed third party oversight and a knack for innovate solutions on a broad number of challenges which, like the French, they did not foresee.

By 1907 the various excavation problems led American engineer John Frank Stevens to redesign the project, using a series of three locks at either end to lift ships 85 feet and transit across man-made Gatun Lake. In 1908 control of the project passed to four Army Corps of Engineers officers and a Navy civil engineer, who completed the project in August 1914, excavating 225 million cubic yards of material at a cost of $22 million below budget, despite battling landslides for the previous 10 months. The project was the jewel of an emerging American empire, and its contributions to world health and sea-born commerce were without precedent.

AN ENGINEERING LEGACY

As a boy I was fascinated with collecting stamps, and they taught me much about geography and history, but especially, our American legacy. The only group of stamps that ever honored the contributions of American civil engineers were those issued by the American Canal Zone authority, between 1915 and 1979. If you knew someone who "lived in the zone" during that bygone era, you may have caught a glimpse of one of these. They were distinctly American in their design, with the standard motifs of the 19120s, 30s, and 40s. Most of the men they portrayed were public figures for a single decade, that of 1904-14, while the Panama Canal was being constructed and yellow fever and malaria were being all but eradicated.

I cherished these special stamps as a unique part of American history, but knew little of the men portrayed in them. As a philatelist, I knew that the completion of the Panama Canal had been celebrated in commemorative issue of five stamps in 1915, and I knew that George Goethals had been one of the principal figures in the canal's construction, as well as President Theodore Roosevelt, because the U.S. Post Office issued a commemorative stamp featuring them (Figure 1) on the 25th

anniversary of the Canal's opening, in 1939. But, I didn't know a thing about the rest of the faces pictured in Figure 2.

Figure 1. Left – Stamp issued by the U.S. Post Office in 1939 to commemorate the 25th anniversary of the completion of the Panama Canal featured the images of Theodore Roosevelt and George W. Goethals, the project's Chief Engineer. Right – Canal Zone stamp featuring the Goethals Memorial at Ancon Hill in Panama.

Figure 2. Canal Zone stamps commemorating the contributions of those men that built the Panama Canal included, from top left: John F. Wallace, John Frank Stevens, George Washington Goethals, Harry H. Rousseau, Harry F. Hodges, William L. Sibert, David D. Gaillard, and Sydney B. Williamson.

That all changed when I read David McCullough's classic book "The path between the Seas," which appeared in 1977, while I was attending graduate school in civil engineering at the University of California, Berkeley. At that stage in my life I didn't have time to carefully digest the contents of this excellent resource, but I did scan it sufficiently to know that it described many of the civil engineers who were instrumental in making the isthmian canal across Panama. I had read just about everything David McCullough ever wrote, beginning with some wonderful essays he published in American Heritage in the 1960s. So, I was determined to read his book on the Panama Canal someday, when I could devour every morsel without distraction.

That opportunity arose in mid-1991 when I was given an opportunity to serve in the Canal Zone as the Navy's Assistant Officer in Charge of the Patrol Squadron Detachment-Panama in June 1991 on drug interdiction duty with Joint Task Force 4 in Key West, for the U.S. Southern Command. I willingly accepted the assignment because I knew it would provide that long awaited opportunity to read about and see the Panama Canal up close, where I could feel and sense the exciting stories I read.

I was not disappointed. As a geological and geotechnical engineer, I had heard much about the landslides along the canal from my mentor Professor Ralph Peck, one of my Berkeley professors, J. Michael Duncan, and Robert L. Schuster of the U.S. Geological Survey. These men served on the Geotechnical Advisory Board for the Panama Canal Authority, dating back to the late 1960s (Peck), late 1970s (Duncan), and mid-1980s (Schuster).

The successful completion of the Panama Canal in the early 20th Century was a very real testament to American engineering prowess and ingenuity. The men pictured above were simply those in responsible charge, who were able to accept most of the credit and public adulation. But hundreds of other, now largely unknown, American engineers made crucial contributions to the projects ultimate success, and many more, working for the Army Corps of Engineers, continued to keep the canal open in the face of oppressive landslide problems in the interim between the canal's completion in 1914 and its turnover to the Panama Canal Authority (Autoridad del Canal de Panamá) in 1999.

During my duty in Panama I spent every spare moment I had in the library of the Panama Canal Commission, the American entity that ran the canal and was then in the process of turning over responsibility for all operations to the Panamanian government (between 1979-99). The library was filled with thousands of photographs, which have since been transferred to the U.S. National Archives. These pictures provide graphic depiction of the many surprises that challenged the American engineers, their struggles to understand the mechanisms responsible for the problems, and the solutions they employed to solve the problems, not always successfully, the first time around.

The construction organization set up by the Americans in 1904 was plagued and hampered by high turnover during the first four years. The project was literally saved by those few determined individuals who chose to stick it out. These were the engineers who were learning from their mistakes. Each year's experience on the Isthmus tended to make them much sharper and more aware of those factors that could get them, and the project, into trouble. The longer they remained the more savvy they became. Their legacy lives on in the Panama Canal, an engineering system that was built from the ground up with safety of operation as its foremost goal. All of the world has benefited from that effort.

What follows is the story of the civil engineers. The selection of Panama as the canal site, and the lock canal with a summit elevation of 85 feet above sea level was a knock-down-drag-out fight between the most reputable civil engineers in America between 1899-1906. I have provided biographical profiles of each of these men, so the reader can appreciate their technical expertise and the manner of their practical experience. That part of the story is filled with personalities, politics, and intrigue, and may bore the reader a bit, because as engineers, we find those subjects rather boring. The key in understanding those battles was the ability of a handful of engineers to convince President Theodore Roosevelt of their "minority view," and of Roosevelt's dogged determination to always do "*what was right*," and not just "*what was popular.*"

The second part seeks to describe the principal engineers and Dr. W.C. Gorgas, whose role was crucial to the successful completion of the project, but was controlled by engineers who did not sympathize or believe in what he was trying to do. In this arena, the civil engineers charged with oversight of Gorgas's duties nearly sabotaged the whole project. The remainder of the narrative describes how the engineers tackled the various problems with logistics and supply, with meddlesome bureaucracy, with construction problems and labor shortages, and with the perplexing geology of the Isthmus, which almost trumped the American effort.

One of the seldom contemplated facts about the canal construction are the absence of any photos taken during rain storms, which were frequent. In those days rainfall wasn't compatable with outdoor phtography, so viewing photographs of the various construction activities one seldeom realizes that these only depict the most favorable conditions, not necessarily the average conditions, which were typified by water, mud, and sweat.

THE FIRST ISTHMIAN CANAL COMMISSION (1899)

On June 10, 1899, President William McKinley appointed an Isthmian Canal Commission (ICC) to examine which route the Americans might pursue across the Central American Isthmus (Figure 3). The original commissioners were John G. Walker, Peter C. Hains, and Lewis M. Haupt.

John G. Walker. The commission's president was retired Rear Admiral John G. Walker, who though he wasn't an engineer, had participated in numerous nautical and geographical surveys. He was a graduate of the U.S. Naval Academy in 1856 and had served as secretary of the Navy's Lighthouse Board from 1873-78. From 1881-89 he had commanded the Navy's Bureau of Navigation, which oversaw the preparation and upkeep of nautical charts, celestial and magnetic navigation aids, and oceanographic data collection. From 1896 he chaired the Lighthouse Board and directed national efforts to find sites in southern California suitable for basing the expanding Pacific Fleet. In 1897 he retired from the Navy, and had previously been selected by McKinley to head up the Nicaraguan Canal Commission (often referred to as the "First Walker Commission," because he subsequently chaired two different Isthmian Canal Commissions).

Figure 3. The Isthmian Canal Commission appointed by President McKinley in June 1899. Front row, from left: George S. Morison, Samuel Pasco, James G. Walker, Alfred Noble, and Emory Johnson. Rear row, from left: Lewis M. Haupt, William H. Burr, Oswald Ernst, and Peter C. Hains. Walker, Haupt, and Hains had previously comprised McKinley's Nicaragua Canal Commission in 1897-99 (National Museum of American History).

Peter C. Hains. Peter C. Hains was an Colonel in the Army Corps of Engineers. He graduated from West Point in 1861 at the beginning of the Civil War, joining the Corps of Topographical Engineers, which were absorbed into the Army Corps of Engineers in 1863. After the war, he served in the Lighthouse Bureau and designed the tidal basin in Washington, DC, which drained the marshes around Georgetown. This service was interrupted by his appointment as a brigadier general of volunteers during the Spanish American War. He was respected for his contributions in hydraulics and foundation design (Hains, 1894; 1896).

Louis C. Haupt. Louis Haupt had served as Professor of Civil Engineer at the University of Pennsylvania, and was the son of legendary civil and railroad engineer

General Herman Haupt, who had commanded the nationalized federal railways system during the Civil War, then as General Manager of the Pennsylvania Railroad. The younger Haupt was an engineer graduate of West Point in 1867, who taught at Penn for two decades (1872-1892) before he began consulting on canals in New Jersey and on the Ohio-Lake Erie Ship Canal. He was noted in particular for his work on improving harbor and port facilities, and the design and operation of efficient dredges (Haupt, 1898, 1905). Among his noted contributions were patents for the adjustable deflecting shield for maintaining depths of flow in constricted channels and the "Reaction Breakwater," which allowed channels to be excavated naturally by concentrating ebb tide flows across offshore bars. He had previously served on Admiral Walker's Nicaragua Canal Commission from 1897-99.

Oswald H. Ernst. Lt Colonel Oswald Herbert Ernst was born in 1842 in Cincinnati. He graduated from West Point in 1864 during the Civil War and took a commission in the Corps of Engineers, serving in the Union campaigns of 1864 in Georgia. After the war he assisted in the construction of coastal fortifications protecting San Francisco Bay, then commanded an engineer company at Willet's Point, NY. He then served as an engineering instructor at West Point from 1871-78, when he authored the text *Manual of Practical Military Engineering* in 1873. His next duty involved the improvement of navigation of the Osage River, then along the middle Mississippi, between the mouths of the Illinois and Ohio Rivers. In May 1882 he was promoted to Major and assigned to the Mississippi River Commission. This was followed by a tour supervising river and harbor improvements in Texas, before rejoining the Mississippi River Commission in 1888. From 1881-89 he also served on the Engineer Board on River and Harbor Improvements, Bridge Construction. In 1889 he was given charge of maintaining public buildings and grounds in Washington, DC, also serving on the Lighthouse Board from 1892-94. In March 1893 he was named an Aide-de-Camp to President Grover Cleveland and Superintendent of Cadets at West Point. He was promoted to Lt. Colonel in March 1895, and remained at West Point through the spring of 1898 when War was declared on Spain. He was commissioned Brigadier General of Volunteers in June, commanding the First Brigade of the 1st Division in Puerto Rico during the latter half of that year. He was serving as Inspector General of Cuba when he was named to the ICC, with his rank reverting to Lt Colonel of Engineers in the regular Army. He was subsequently promoted to Colonel in February 1903, and to Brigadier General upon his retirement from the Army, in June 1906.

The other engineers comprising the commission. Because of the expanded duties demanded of the ICC to investigate and evaluate a series of competing routes in Central America (Figure 4), several additional members were added to the commission. These included a number of eminent civil engineers, including hydraulics engineer Alfred Noble of Chicago; renowned bridge engineer George S. Morison of New Jersey; and William H. Burr, a Professor of Civil Engineering at Columbia University. The other members were former Senator Samuel Pasco of Florida, Wharton School of Business transportation expert Professor Emory Richard

Johnson, Lt. Colonel Oswald H. Ernst, and Lt. Commander Sidney A. Staunton of the Navy served as Secretary of the Commission.

Figure 4. Map illustrating the five principal canal routes (shown as solid black lines) investigated by the first Isthmian Canal Commission in 1899-1901. The dashed lines show the locations of previous surveys by others (Walker et al., 1902).

George S. Morison. George Morison was an influential figure in the civil engineering profession of the late 19th Century (Figure 5). As a consulting engineer based in New York City, he had worked on an array of high visibility projects, mostly for large railroads and shipping companies. He graduated from Harvard at the age of 20 in 1863 and its law school in 1866, before changing course and determining to make himself an "expert bridge engineer." To achieve this goal he apprenticed himself to Octave Chanute, and the two collaborated on the swing span Kansas City Bridge, the first permanent bridge over the Missouri River. From there he designed a number of record steel truss spans, including the Union Pacific Bridge over the Missouri River at Council Bluffs (the longest truss span in the USA), and culminating with the Memphis Bridge in 1892, the first permanent span across the lower Mississippi River. Morison had a reputation for excellence and his good opinion was valued by most of the civil engineering profession. He had served as President of the American Society of Civil Engineers in 1895, then headquartered in New York City. Morison's professional clout, more than any other single human factor, would play a crucial role in the surprising decision to build an American canal in Panama.

Alfred Noble. Alfred Noble (Figure 5) received his civil engineering degree from the University of Michigan in 1870 and worked on the canal and lock at Sault St.

Marie and on important bridges across the United States before establishing his own consultancy in Chicago in 1894. He was called on the design the hydraulics works associated with the Chicago Ship and Sanitary canal in the late 1890s, up to that time, the largest civil works project in American history. He had previously served on the Nicaragua Canal Board appointed by President Grover Cleveland in 1895, where his varied expertise proved valuable in estimating the means by which hydraulics would control the design of lock gates and the day-to-day operations of any canal. In 1903 he served as President of the American Society of Civil Engineers, at the time America was vigorously pursuing acquisition of the French workings in Panama. In 1905 he was appointed by President Roosevelt to the International Board of Consulting Engineers charged with determining whether a locked or sea level canal should be constructed across Panama. He then served as a consultant to the ICC on the foundation treatment of Gatun Dam and Locks, in 1908-09. His professional clout can be gaged from the length of his memoir in the ASCE Transactions of 1915, which, at 63 pages, was the longest ever published by the society.

William H. Burr. William Hubert Burr (Figure 5) was born in Watertown, CT in 1851 and received his undergraduate training in civil engineer at Rensselaer Polytechnic Institute, graduating in 1872. He joined the Phillipsburg Manufacturing Co., then engaged in constructing wrought iron bridges, mostly for railroads. In 1874 he joined the City Waterworks Department in Newark, NJ, then returned to his alma mater in 1875 as an instructor, and was named Professor of Rational and Technical Mechanics the following year, penning a notable article on the distribution of stresses in eyebar elements of bridges (Burr, 1876). In 1884 he left Rensselaer to become Assistant to the Chief Engineer of the Phoenix Bridge Co. near Pittsburgh, and was subsequently named the firm's general manager, supervising the design and construction of many important bridges (Burr, 1885, 1887), including a camel-back truss across the Ohio River in Cincinnati, with a record center span of 555 feet, completed in 1888.

In 1891 he accepted the position of vice president of the Sooysmith & Company of New York, the foundation engineering company that had developed the "Chicago Caisson." The following year he took a position with as Professor of Civil Engineering at Harvard, and then moved to Columbia University in 1893, where he became the founding chair of their civil engineering program, and where he remained until his retirement in 1916. During his time at Columbia he was a consultant on numerous projects in the New York area, and was very active in the American Society of Civil Engineers.

In 1899 he was appointed to the Isthmian Canal Commission and was assigned roles in evaluating both the Panama and Nicaragua routes. It was the first of three important commissions he would serve on that ultimately determined the nature of the American design for the Panama Canal. Throughout the process of vetting the various schemes and designs, he remained a staunch supporter of a sea level canal.

Figure 5. Left - George S. Morison had served as President of ASCE in 1895 and was one of the most respected consulting civil engineers in New York at the turn of the 20th Century (ASCE). **Middle - Alfred Noble** had the longest association advising the United States government on the various schemes for canals in Nicaragua and Panama, between 1895-1909 (Bentley Library, University of Michigan). **Right - Professor William H. Burr** founded the civil engineering program at Columbia University in 1892 and served as a consultant on many high visibility projects (Columbia University).

The commission report of November 1901. The ICC was charged with making credible engineering feasibility studies of every possible route across the Central American Isthmus, to see if any of the "alternative" to the American scheme for Nicaragua might be competitive, in terms of cost and years to completion. Based on all of the problems the French had suffered in Panama over the previous 20 years, nobody expected their path to be selected. The ICC divided itself into five committees, each designed to take the lead in examining specific subjects. These were: 1) the Nicaragua route (Noble, Burr, and Hains); 2) the Panama route (Burr, Morison, and Ernst); 3) other possible routes (Morison, Noble, and Haines); 4) the investigation of industrial, commercial, and military value of an interoceanic canal (Johnson, Haupt, and Pasco); and 5) various aspects of rights, privileges, and franchises (Pasco, Ernst, and Johnson).

The ICC worked feverishly, completing their report on November 20, 1901, just two months after Theodore Roosevelt had succeeded McKinley as the nation's new president. Seven of the commission's eight members recommended that "the most practicable and feasible" route was through Lake Nicaragua, but the redoubtable George Morison voiced his dissent. Then, he penned a personal letter to Roosevelt outlining the substantiation of his opinions regarding the Panama route as being the most favorable from an engineering standpoint. He then reminded the president that if the French could be coaxed into selling their rights to it for a reasonable sum, the completion of the French canal would only cost $109 million, far less than the ICC's estimate of $200 million to build a Nicaraguan canal. The escalating factor was how much would have to be paid to the French concession, as well as the Columbian government, to secure the franchise owned by the Compagnie Nouvelle du Canal de Panama in Paris. Morison pointed out that even if the United States paid the French $40 million and the Columbians $10 million, the aggregate total of $159 million was $41 million less than a canal through Nicaragua. He also pointed out that the vessel transit times would be lessened through a shorter canal (by a factor of three times because of increased length and the number of lock passages), the cost of maintenance would be much less (because fewer locks would have to be maintained), and it could be completed several years sooner because the Americans would be inheriting the incomplete French excavations across Culebra. Roosevelt recognized cogent apolitical reasoning when he saw it, and he was immediately persuaded by the logic of Morison's arguments.

Morison's letter to President Roosevelt. What occurred next was a surprise to almost everyone. President Roosevelt accepted the ICC report on November 20th, digested its contents and penned his own response two days later, on the 22nd. He shocked everyone by announcing that he favored Morison's minority opinion. His announcement was a political ploy intended to prompt the French to reduce their demand for their canal holdings, and their concession to build the canal in Panama. Politics then took over the process, doing what it does best: assuaging public opinion and fomenting foreign policy.

On Friday, January 3, 1903, the powerful and influential Senator Mark Hanna, former President McKinley's closest friend and ally, and a member of the Senate's Committee on Interoceanic Canals, suddenly announced that he was "impressed with the superior advantages of the Panama Route." A go-between named Walter Wellman, who was a reporter for the Chicago Record-Herald in Washington, DC, was tapped to inform the French canal syndicate that the Senate Committee would probably accept an offer of $40 million, but no higher, and that they must act promptly. They did, offering their concession for exactly $40 million. Admiral Walker received their cable and walked it to the office of Secretary of State John Hay later that same day. Suddenly and unexpectedly, everything had changed.

On January 9th the House of Representative voted 308 to 2 in favor of a Nicaraguan canal. The Chairman of the Senate Committee on Interoceanic Canals, John Tyler Morgan of Alabama, saw the House vote as a providential opportunity for him to nail down his life dream of constructing an American canal across Nicaragua. But, when the Senate's Committee met on January 16th, it was Senator Hanna who placed the ICC report on ice, requesting that the senate committee table the ICC report on Nicaragua until the offer from the French syndicate could be considered, since it held the potential of saving $41 million. Morgan's hands were tied, he couldn't go on record as being opposed to such colossal savings, but he remained confident that the ICC would stand by their 7 to 1 vote against Panama. Senator Hanna then announced that the President was asking Admiral Walker and the ICC to re-convene for the purpose of issuing a supplemental report, which would consider the economic aspects in light of the new French proposal.

Walker would later state that the ICC moved quickly, in deference to the President's orders (Morris, 2001). Two days later, on January 20th, the ICC issued a unanimous finding recommending that the United States purchase the French holdings and concession and complete the canal they started across Panama. The supplemental report was mostly a comparison of figures: the Panama route was 135 miles shorter than Nicaragua; there would be fewer locks with less curvature; the transit time would be 12 hours for Panama versus 33 hours for Nicaragua, there was an established railroad already in place in Panama (95% of the then anticipated excavation would be moved via rail), and two artificial harbors would have to be constructed at either end of a Nicaraguan canal, but only one at Panama. The bottom line now was that a Nicaraguan canal would actually cost about $45.6 million more than one at Panama, even after accounting for the initial outlay of $50 million. In essence, all of this was a recitation of George Morison's minority report which hardly drew any press just five weeks earlier.

The original American scheme (1903). The ICC feasibility study for an American canal across Panama between Cristobal and Balboa in January 1903 was based on a number of important assumptions (Walker et al, 1902). These included the following: The ICC assumed that a locked canal would be the least expensive to construct on top of the incomplete French excavations made between 1882-89 and 1894-95. Their assumption envisioned a locked canal lifting vessels to a central pool at an elevation of 90 feet above sea level. The canal was to be 35 feet deep, with a submerged bed at least 150 feet wide. The canal would be 49 miles long, and the highest cut would be 286 feet high, at the Continental Divide near Culebra.

Their plan assumed that the American canal would follow the Chagres River inland for a distance of about 17 miles to an earthen dam approximately 100 feet high (at the same location where the French had planned to construct a similar embankment dam 75 feet high). This was called the Bohio Dam, and George Morison penned an elegant description of it in the ASCE Transactions, which drew considerable interest from the society's members (Morison, 1902). The dam site was a natural one, at the first bedrock narrows of the Chagres River. Two sets of locks, each with a 45 foot lift, were proposed to be constructed in the bedrock knob forming the dam's left (southwestern) abutment. Each lock was to be 85 feet wide and 735 feet long. Ships passing through these locks would thereby be lifted 90 feet to the level of Bohio Lake, through allow the vessels to travel another 14 miles before entering an eight-mile stretch of increasingly massive cuts that would take them past the continental divide just south of Culebra.

From there Morison's team had sited one 45-foot lock at Pedro Miguel and the other at Miraflores on the Pacific side (at the sites were slightly shorter blocks were subsequently built). Ships would traverse the last 10 miles along a dredged channel at sea level that would take them into the Pacific Ocean. The volume of required excavation would be somewhere between 42 and 54 million cubic yards, with five million of that being dredge tailings from the approach channels. The estimated cost was about $144 million, and the massive project would likely take eight to ten years to complete, once the construction force was mobilized in Panama. No mention was made of the costs of providing sanitation and drainage, to combat typhoid, yellow fever, and malaria, among other tropical maladies for which Panama was so infamous.

SECURING THE CANAL ZONE (1903)

Negotiations with the Columbians. The Spooner Act of June 28, 1902 authorized the United States to purchase of assets owed by the French Compagnie Nouvelle du Canal de Panama, provided that a treaty could be negotiated with the Columbian government, of which Panama was then a state. The balance of 1902-03 was spent in negotiating a suitable agreement with the Columbian government for the right to build and maintain a canal across their lands for a period of 99 years. Three Columbian emissaries were dispatched in succession to Washington, DC to hammer out an agreement with Secretary of State John Hay. The last of these was Dr. Tomas Herran, the son of Columbian General Pedro Alcantara Herran, who had negotiated the Bidlack Treaty with the United States in 1848, which had guaranteed United States transit rights across the Panamanian Isthmus. As a boy, the younger Herran had accompanied his father to those negotiations, and later graduated from Georgetown University, so he spoke fluent English and had friends in Washington, DC. On January 21, 1903 Hay issued an ultimatum to Herran, informing him that President Roosevelt had become so exasperated with the negotiations that the Columbians could either agree to the $10 million in gold with an annual rent of $250,000 being offered to them, or the United States would terminate negotiations and begin dealing with Nicaragua (in 1921 the United States paid the Columbians an indemnity of $25 million, along with an apology for acquiescing to the Panamanian rebellion). The following day Herran signed what came to be referred to as the "Hay-Herran Treaty," and the Americans presumptuously breathed a sigh of relief. But, the Columbian government in Bogota balked, confident that they could broker a better deal for themselves, now that the Americans had committed themselves to the French Compagnie Nouvelle's Panama concession.

The Americans felt that they had executed a legal agreement between the two governments, but the Columbian legislature refused to recognize its legality. Months of haggling ensued. In June 1903, President Roosevelt announced that he "*would do everything in his power to secure the canal route through Panama*" because, the United States had expended "*millions of dollars on engineering studies*" (McCullough, 1977). He was convinced that Panama presented the "best route" for a canal. On August 12th the Columbian Senate rejected the Hay-Herran Treaty by a unanimous vote, sending a clear message to Washington that they wanted more compensation. The deal Roosevelt and Hay were trying to broker was already the largest real estate venture in history. The $50 million price tag just a to gain the rights to excavation was just slightly less than the combined sums of the four largest land acquisitions the United States had ever made: the Louisiana Purchase for $15 million, the Gadsden Purchase for $10 million, the Alaska Territory for $7.2 million, and the recent purchase of the Philippines for $20 million. Roosevelt's patience was stretched to the brink, and he considered the Columbians to be in violation of all rules of civility and diplomacy. The two sides were rapidly losing patience and respect for one another. In late August, Roosevelt hinted that the Unites States might be faced with armed conflict, the administration feeling justified to construct a canal with international access through the tenants of the Bidlack Treaty.

Establishment of the Republic of Panama. Through the summer of 1903 a number of prominent Panamanians, wary that their national government in Bogota would never toss a peso of the American gold in their direction, should a suitable concession be negotiated with the United States, began gathering together to discuss their options. A number of Americans were invited to the secret meetings including the United States Consul General at Panama City, Hezekiah A. Gudger, and two Army Corps of Engineers officers working for the Isthmian Canal Commission, Major William M. Black and Lieutenant Mark Brooke (Figure 6). Black had a distinguished career as a military engineer who could solve difficult problems. He perfected the design of cellular cofferdams to exhume raise the sunken battleship *USS Maine* from Havana Harbor in 1911-13. He eventually became Chief of Army Engineers during the First World War with the rank of major general.

Figure 6. Two Corps of Engineers officers who played a vital role in the early American surveys of the canal route and in the bloodless Panamanian revolt were, from left: Major William Murray Black (West Point Class of 1877) and Second Lieutenant Mark Brooke (West Point Class of 1902). General Black served as the Army's Chief of Engineers during the First World War, while Brooke retired as a Colonel of Engineers in 1932 (USACE and USMA).

The idea of declaring themselves independent of Bogota was freely discussed among the conspirators and with the American officers present, whom they assumed to be official envoys to Washington. Dr. Manuel Amador was dispatched as the group's secret emissary to William Nelson Cromwell, the American attorney representing the French Compagnie Nouvelle du Canal de Panama in New York. From there on, the conspiracy to separate from Columbia with the implied promise of American political and military backing, was cautiously played out. Some $250,000 to finance the revolution appears to have come from "French sources," likely some of the principal stockholders of the Compagnie Nouvelle, eager to see the American assumption of their stagnant holdings in Panama. They had now waited almost 20 months since the $40 million deal had been negotiated with the Americans, whose hands were tied by the apparent intransigence of the Columbian government.

Herran notified government officials in Bogota that *"Panamanian separatists were in Washington,"* trying to negotiate a deal with the Americans, and urged them once again to sign the Hay-Herran Treaty, but to no avail. Clandestine negotiations continued, which included private meetings between Philippe Bunau-Varilla, the Director General of the Compagnie Nouvelle, and President Roosevelt. By mid-October suitable arms had been smuggled into Panama to allow an armed revolt, and Roosevelt was kept informed through a pair of American Army officers dispatched to Panama to assess the situation. But, the Panamanian insurgents would not act without some sort of approval and support from Washington. The Americans responded that it would be legally and politically difficult to openly support a new regime in Panama if their proposed coup became stained with the blood of innocent officials, who they might be tempted to ambush in such an uprising.

Warned of impending trouble, President Roosevelt dispatched several warships and a contingent of Marines to Panama to protect American interests and property in the event of an armed rebellion. The American gunboat *USS Nashville* (Figure 7) was the first to arrive, dropping anchor in Colon round 5:30 pm on November 2nd. 6-1/2 hours later the Columbian gunboat *Cartagena* arrived at Colon, carrying almost 500 troops.

The next morning the Columbians disembarked their troops and marched to the station of the Panama Railroad, demanding to be taken to Panama City, where the territorial capitol and governor were. Their commander, General Juan Tobar and 15 of his senior aides were intercepted by Colonel James Shaler, General Manager and Superintendent of the Panama Railroad, who contrived a scheme to spirit them off in a special one-car train across the isthmus. The plan was for them to be received by Columbian officials in Panama City, including General Esteban Huertas, whose loyalty and that of his troops stationed on the Pacific side of the Isthmus had been secretly purchased by the junta seeking to declare independence. That evening around 6 pm, the revolutionaries seized Columbian Governor Obaldia, Tobar, and his aides, and declared their independence. The next day Huertas' troops were each paid $50 in Columbian silver, and five days the general received $65,000 for his part in the coup, with his officers receiving lesser, but generous amounts.

Captain John Hubbard

Figure 7. Navy Commander John Hubbard was an 1870 graduate of the Naval Academy (shown at right as a Navy Captain in 1908) commanded the American gunboat USS *Nashville* (PG-7), shown at left, which carried eight 4-inch rifles. The American ship and its crew played a pivotal role in assisting the bloodless coup that culminated in the establishment of the Republic of Panama. Hubbard was promoted to Rear Admiral in 1909 (US Navy).

The following morning, on the opposite side of the Isthmus, the Columbian Gunboat *Cartagena* unexpectedly hoisted anchor and sailed away, apparently fearful of the American guns from the gunboat *Nashville* zeroed in in her. This left the almost 500 Columbian soldiers of the Tiradores battalion without any means of support or retreat. Their young commander, Colonel Eliseo Torres, was told that the Panama Railroad could not convey them across the isthmus because the Commanding officer of the *Nashville*, Commander John Hubbard, had ordered Shaler to preclude either Panamanian or Columbian forces form using the railroad, to maintain its neutrality (but there were no "Panamanian forces" on the other side of the isthmus, only the Columbian soldiers whose loyalty had recently been purchased). 40 armed sailors, with extra ammunition and supplies, were strategically placed around the railroad station and the warehouse, to "protect American property" (the Bidlack Treaty of 1846 stipulated that the United States would guarantee "free and uninterrupted transit across the isthmus" on any American owned railroad). After several hours, Junta envoy Porfirio Melendez met privately with Torres, offering him a handsome bribe for cooperation, and warning him that additional American military forces had been dispatched to Colon to sustain the revolution (although this was a ruse, it turned out to be true). Torres refused to be intimidated, as he had seen no evidence yet of an armed uprising by the local populace. He demanded to be taken by train to confer with his superior General Tobar, who was being held in Panama City. A tense stand-off ensued, but neither fired at the other. Finally, Torres agreed to withdraw his forces to Monkey Hill if the Americans would withdraw their naval militia from the warehouse to the *Nashville*, and that he be escorted by rail to Panama City to confer with Tobar.

As the new republic's officials awaited Torres arrival in Panama City, they informed their high-ranking Columbian prisoners that the United States fully supported their coup, had provided the funds and the necessary weapons, and would provide military support, if necessary. This wasn't factually true, but the American naval presence in Colon seemed to support the assertion. Torres then arrived, but Tobar refused to order him to either remain in Panama or to depart, leaving it up to the colonel to do as he saw best. He returned to his troops on Monkey Hill overlooking Colon later that evening.

On the morning of November 5th, Torres brought his troops down off Monkey Hill, complaining that the mosquitoes had savaged his troops. Commander Hubbard responded in kind, again landing his force of 40 armed sailors to protect the Panama Railroad terminal. The standoff continued through the day, but Torres began to worry about how he was going to sustain his forces with the *Cartagena's* untimely departure the previous day. During the day's negotiations, Melendez and Shaler continued working on Torres. Shaler was addressed as Colonel, in deference to his military service in the American Civil War, on the side of the Union. He began warning Torres that "5,000 American troops were due to land at any time" (this was also a ruse, but turned out to be partially true, when 400 Marines arrived that very evening). Around 5 pm Torres agreed to embark the soldiers on the Royal Mail Steam Packet *Orinoco*, which had arrived two days prior. Torres demanded that he be given $8,000 to leave. On the telegraphed word that funds would be forthcoming from Panama City, Shaler ordered $8000 in $20 American gold pieces from the railroad's safe in Colon, which were passed to Torres in two canvas bags around 6 pm. 20 minutes later the auxiliary cruiser *USS Dixie* was sighted approaching the harbor, after having been dispatched from Kingston, Jamaica with a load of Marines. Her sighting hastened the sense of urgency in embarking Torres' soldiers. As the Columbian troops queued up on the wharf to board the *Orinoco* for the voyage across the Columbian Gulf to Cartagena, the ship's purser informed Torres that he must be paid in cash before allowing any of the troops to board. He demanded a bit more than L1000 pounds British Sterling (about $3192 US), which could not be found in Colon because the railroad's safe has just been emptied. The purser finally agreed to embark the soldiers if Commander Hubbard and Colonel Shaler would sign a bank voucher to the Packet Line, which they did.

Just after 7 pm the *Dixie* anchored in the harbor as the *Orinoco* was still loading the Tiradores Battalion. The Columbian soldiers sailed away at 7:35 pm, and 45 minutes later Marine Captain John Lejume with 400 American Marines disembarked from the *Dixie* to maintain order in Colon until the new government could establish law and order. The new Panamanian flag was then handed to Major Black as a gesture of gratitude for the American assistance, and as the flag unfurled, the crowd cheered, and a new era began - for both nations. The new Republic of Panama was established as a direct consequence of the greatest civil engineering enterprise ever undertaken up until that time. Lejume was promoted to major on March 14, 1904 and given command of the Marine Battalion billeted at Camp Elliott near Empire, which were the first American troops stationed on the Isthmus.

Transfer of ownership to the United States. On the morning of May 4, 1904, Lieutenant Brooke (Figure 6) walked into the old Hotel de la Compagnie in Panama City. After brief greetings and exchanges of pleasantries with officials of the French Compagnie Nouvelle du Canal de Panama (formed in 1894), he read a document that was presented to him, and following the instructions of War Secretary William Howard Taft, he signed the agreement, accepting for the United States of America all of the property and equipment of the New French Canal Company. This receipt is written and signed in French, English, and Spanish. Brooke's first action was to take down the French tricolor which had flown from the hotel mast since January 1, 1880 and to replace it with the American star and stripes (Duval, 1947). The American effort to construct an international ship canal in Central America had at last, officially commenced.

THE SECOND ISTHMNIAN CANAL COMMISSION (1904)

The Spooner Act required that the Isthmian Canal Commission (ICC) appointed by the president should thenceforth consist of seven members, at least four of whom must be engineers; and of those, at least one Army and one Navy engineer. One troublesome aspect of the new act was the provision that *all* seven members were to have equal authority. This was intended to preclude any system of political spoils or favoritism from overseeing what promised to be the largest expenditure ever approved by the federal government. At face value, it seemed fair handed and unbiased. It proved, however, to be unworkable.

On March 3, 1904, President Roosevelt appointed a new slate of ICC commissioners (Figure 8), which included a few familiar faces, most prominently, retired Rear Admiral John G. Walker, now 68 years old. He had recently supervised the feasibility studies of various canal routes (Figure 4), which included detailed surveys of the French excavations.

General George W. Davis. The Army representative would be retired Major General George W. Davis, who had just completed a term as military Governor of Cuba. He would be the only member of the new commission who actually resided in Panama, serving as the first Military Governor of the new Canal Zone, a strip of land ten miles wide along the proposed centerline of the canal (Figure 9). Davis was a former infantry officer (he never served in the Corps of Engineers), but had distinguished himself in 1885 by devising a system of elevating machinery that hoisted the marble dimension stones to complete the Washington Monument. He proved himself an able administrator during the Spanish American War as a Brigadier General of Volunteers, and was promoted to Colonel of Infantry in the regular Army in October 1899. In 1900 he was detached from regular duty to serve as vice president and general manager of the Nicaraguan Canal Construction Company, receiving promotions to brigadier general in February 1901 and, was elevated to major general upon his retirement from the army at age 63 in July 1902. His promotions to flag rank in 1901 and 1902 were undoubtedly connected to his important role in the American bid for a Nicaraguan canal, whose proponents lobbied

Congress to fund as a national undertaking, similar to the federal land grants used to stimulate construction of the first transcontinental rail lines (between 1850-71). The idea failed to garner sufficient political support because of the lack of reliable surveys and hydrologic data necessary to develop engineering plans. Davis had been chosen to secure such data so that suitable plans could be prepared and the costs reliably estimated. He would be paid a salary of $10,000 per annum.

Figure 8. Members of the second Isthmian Canal Commission in Washington, DC in the spring of 1904. Left to right, seated: General George W. Davis, Admiral John G. Walker, Frank J. Hecker, and William Barclay Parsons. Standing: William H. Burr, Benjamin M. Harrod, and Carl E. Grunsky.

The other Commissioners. The rest of the commissioners named by the President included William Barclay Parsons, C.E.; William H. Burr, C.E.; Benjamin M. Harrod, C.E.; Carl E. Grunsky, C.E.; and Detroit businessman Colonel Frank J. Hecker. They would each be paid an annual salary of $7,500 (U S Senators received $5000 per year at that time). Hecker had served as a sergeant in the Union Army during the Civil War before founding the Peninsular Car Works of Detroit in 1879, and then the Peninsular Car Company in 1884. He came to Roosevelt's attention while serving as Colonel and Chief of the Army's Division of Transportation during the Spanish American War, which oversaw the care and transport of Spanish prisoners of war. He volunteered to serve as the Commission's eyes and ears down in Panama, departing in May. At age 58, he also felt the strain of his exertions, and he resigned his seat on the commission on November 16th, 1904, returning to his home in Detroit. The Commission issued a sterling letter of appreciation for his efforts.

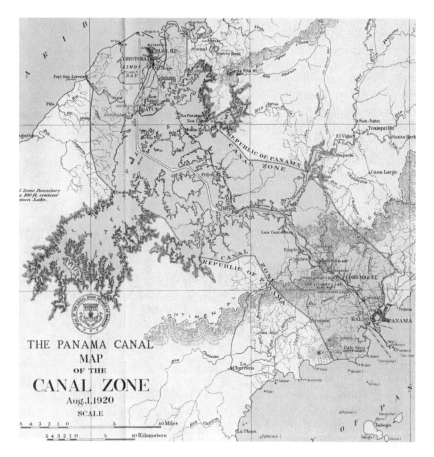

Figure 9. Map of the Panama Canal and the U.S. controlled Canal Zone, a ten mile wide strip of land stretching across the isthmus, along with those lands inundated by Gatun Reservoir (NAS, 1924).

Parsons, Burr, Harrod, and Grunsky were all practicing civil engineers, who had published widely and possessed considerable expertise with the design and construction of waterworks and hydraulics. The group would undoubtedly have included George S. Morison, whom everyone, including President Roosevelt, considered to be the "Panama expert," but he had died unexpectedly on July 1, 1903. The commissioners fulfilled their duties from stateside offices in Washington, DC and New York City.

William Barclay Parsons. William Barclay Parsons, Jr. (Figure 10) was a native of New York City and the great-grandson of a British naval officer whose ship had foundered off the coast of Long Island during the War of 1812. Born in 1859, he went by his middle name Barclay. At the age of 11 he was sent to Torquay, England for four years of schooling, which included private tutors on trips to France, Germany, and Italy. In 1875 Parsons enrolled in Columbia College, and upon his graduation in 1879, he then entered the Columbia School Mines (precursor to the present School of Engineering), graduating at the top of his class in 1882. He secured a position with Erie Railroad, becoming roadmaster of the Susquehanna Division, then moved to the Greenwood Lake Railway. During this time he published two design manuals: one on *Railroad Turnouts*, and the other tiled *Track: A Complete Manual of Maintenance of Way*. On January 1, 1885 he started his own consultancy in New York City, and was soon joined by his brother Harry deBerkeley Parsons. In 1891 Barclay Parsons was deputy chief engineer of the newly formed New York Rapid Transit Commission, and became the chief engineer in 1894. He went to Europe to glean information and data on underground railways in operation there, which convinced him of the utility of employing electric traction motors over steam power. The engineering plans were forwarded to the state capitol in Albany where they languished, waiting for monies to be authorized for construction.

Figure 10 Three of the principal figures comprising the Committee on Engineering of the second Isthmian Canal Commission, appointed on March 4, 1904: Left - William Barclay Parsons of New York City (Library of Congress); Middle – Carl E. Grunsky of San Francisco (author's collection); and Right – Benjamin M. Harrod of New Orleans (ASCE). The fourth member of the committee was Professor William Burr of Columbia University, who had also served on the first Isthmian Canal Commission.

When the Spanish American War erupted in April 1898, Parsons quickly volunteered his services to raise an Army Engineer regiment, and took charge of a training camp in Peekskill, NY. The war was short-lived, but before he could lead his men to the conflict, Parsons was approached about leading the efforts of an American syndicate to construct a rail line between Hankow and Canton, China, across Hunan

Province. It was an area yet unexplored by western engineers. Parsons departed with five assistant engineers in the summer of 1898. He succeeded in personally leading the survey of the treacherous line, a distance of over 1000 miles. During his voyage home he penned a book about China, one of the first on the subject in English.

Parsons returned to Manhattan in March 1900, just in time to witness groundbreaking for underground rapid transit system he had labored to design for so many years. The system was constructed by the Interborough Rapid Transit Company, much of the line being in novel cut-and-cover design pioneered by Parsons. In 1903 he had sparred publicly with fellow civil engineer Alfred Noble over the latter's preliminary design of the Steinway (later known as Queensboro) Tunnel, which Noble later completed with Parson's approval, in September 1907. The first subway line opened for business on October 27, 1904, while Parsons was serving on the ICC, from its office in New York City. In late 1904 Parson resigned his post with the Rapid Transit Commission and once again established his own consultancy, called Barclay Parsons & Klapp. He consistently favored a sea level canal across Panama, despite the 20 foot tide differential.

Carl E. Grunsky. Carl Ewald Grunsky (Figure 10) was born near Stockton, California in 1855, during the California Gold Rush. His boyhood ambition was to become a physician and he graduated from Stockton High School at age 15. He immediately took a position as an elementary school teacher and taught for one year before departing for Germany with his younger brother in 1872. Upon his arrival he enrolled in the Realschule in Stuttgart studying medicine for two years. By 1874 he realized that his meager savings would be insufficient to complete medical school, so he decided to study engineering at the Stuttgart Polytechnikum in Wurttemberg, one of the outstanding engineering schools in Europe. In 1877 he graduated at the top of his class and returned to his native California the following year.

In 1878 he accepted a position with the State's Engineer's office, where he steadily advanced himself from Topographer to Assistant, then Chief Assistant Engineer (from 1878-88), working with State Engineer William Hammond Hall, C.E. It was during this time that he developed the skills later expressed in his textbook *Topographic Stadia Surveying*. In 1884 the California Legislature passed the Wright Act, which set forth the legal tenets for the establishment of local irrigation districts, which required topographic and hydrologic surveys by a competent engineer, which were all reviewed by the state engineer. Once established, irrigation districts provided a legal means of securing water rights.

With this new market for consulting engineers, Grunsky founded his own consultancy in 1888, specializing in irrigation, waterworks, and sewerage problems. He was appointed to the Examining Commission on Rivers and Harbors for California in 1889 and 1890; the Sewerage Commission of San Francisco in 1892-93, and as Consulting Engineer to the San Francisco Public Works Commission from 1893-95. He was active in a number of professional societies, serving as President of the Technical Society of the Pacific Coast in 1893. In 1898 he joined the American

Society of Civil Engineers (ASCE). In 1900 he was named the first City Engineer of San Francisco, and was the man most responsible for initiating that city's designs on securing the waters of the Tuolumne River, which was completed many years later.

Grunsky was serving as San Francisco City Engineer when President Roosevelt tapped him for the Isthmian Canal Commission in early March 1904. He was the only commissioner ever appointed from the western United States. The citizens of San Francisco threw a complimentary banquet in his honor at the Palace Hotel on March 15th, whose guests were provided with a 62 page testimonial (Complimentary Banquet, 1904).

After a year of service on the ICC, he became as advisor to Secretary of the Interior Ethan A. Hitchcock, and as a consulting engineer to the newly formed Reclamation Service, under Frederick H. Newell. In 1908 he reopened his consulting office in San Francisco, which he continued to operate until his death, in 1934. In 1910 he received ASCE's Norman Medal for his article *The Sewer System of San Francisco, and a Solution of the Storm-Water Flow Problem* (Grunsky, 1909a). In 1910 he published a discussion on the article *Water supply for the Panama Canal* in the ASCE *Transactions* (Grunsky, 1910).

Grunsky was also known for his valuable contributions to the field of economics, expressed in his books *Valuation, Depreciation, and the Rate Base* (1917; 1927); and *Public Utility Rate Fixing* (1918), which became national standards. He served as President of the California Academy of Sciences from 1912-34, President of the Commonwealth Club of California in 1920, President of the Pacific Division of the American Association for the Advancement of Science in 1924, President of the American Society of Civil Engineers in 1924, and as President of the American Engineering Council in 1930-31.

Benjamin M. Harrod. Benjamin Morgan Harrod (Figure 10) was born in 1837and raised in New Orleans. He attended prep school in Flushing, NY before matriculating through Harvard University, graduating in 1856, intending to pursue a technical or constructive profession. After a year of preparatory work he was employed as a draftsman in the United States Engineers Office then engaged in the design and construction of fortified works and light houses along the Gulf Coast, between the Mississippi River and the Rio Grande. During this time he was appointed Assistant Engineer.

In 1859 he started his own engineering and architecture business in New Orleans. When the Civil war erupted in 1861 he was commissioned as a Lieutenant in an artillery regiment of the Confederate Army, but detailed to perform engineering duties. He served as a brigade and division engineer in the construction of fortifications around New Orleans and Vicksburg and was captured in the Siege of Vicksburg in July 1863. After a prisoner exchange, he was commissioned as a Captain of Engineers in Virginia and engaged in the defense of Richmond and

Petersburg until the surrender of Virginia forces at Appomattox in April 1865, when he mustered out with the rank of major.

Harrod resumed his consultancy in New Orleans in 1865, and was thereafter referred to as "Major Harrod," in respect of his wartime service. In 1877 he was appointed Chief State Engineer of Louisiana and joined the American Society of Civil Engineers. In his position, he became much engaged in the construction of a system of earthen levees to protect regions of that state from annual flooding by the Mississippi River and its distributaries. In 1879 he was appointed as one of two civilian engineers on the newly formed Mississippi River Commission, established by an act of Congress to survey the Mississippi River and its tributaries, and to devise schemes to improve these channels downstream of the mouth of the Ohio River to the Head of Passes in the delta. From 1888-92, Harrod served as the City Engineer of New Orleans, then as Advisory Engineer for the drainage, sewerage, and waterworks systems of that city, from 1897-1902. In 1897 he served as President of the American Society of Civil Engineers.

In 1903 he was one of three delegates representing the United States at the International Congress on Navigation convened in Dusseldorf, Germany. In 1904 he was appointed to the second Isthmian Canal Commission, where he continued to serve until 1907. He then returned to New Orleans working as a consulting engineer until his death in 1912.

Organization of the Commission. The new commissioners all met one another for the first time on March 22, 1904 in Washington, DC. Admiral Walker served as the Chairman, and Carl Grunsky was selected as the secretary *pro tempore*. Whenever Admiral Walker was not present, Grunsky served as the commission's temporary chairman. The record shows that Grunsky offered more resolutions than any of his fellow commissioners, and they consistently sided with him on any technical issues that required a vote (Isthmian Canal Commission, 1908). This tendency did not bode well for poor Dr. Gorgas, as described later.

The commission's first debate was in regards to the compensation they should offer to the chief engineer of the canal project. General Davis felt that a salary of $25,000 per annum was excessive, insofar as only the President of the United States received more remuneration from the federal government ($35,000). Davis introduced a motion to reduce the offer to $20,000. Carl Grunsky responded that Wallace was the only civil engineer who had substantive managerial experience with one of the nation's principal railroads, overseeing the design and construction of every manner of improvement, which involved all manner of structures (as Vice President and General Manager of the Illinois Central Railroad). Grunsky then reminded the commissioners that Wallace had stated that he would not leave his current position for anything less than $25,000. The commissioners then voted six to one (with Grunsky) in favor of offering Wallace the higher amount (the commissioners were being paid just $7,500 per annum). Wallace was interviewed by the Commissioners in Washington on May 5th and offered the position by President Roosevelt the next day, which he accepted.

Early on, the 1904 commissioners also agreed that standard civil service employment procedures should be employed in the Canal Zone, so the favoritism and the system of political spoils would be discouraged as much as possible. The ICC committees were responsible for reviewing requisitions and expenditures.

On May 10th, General Davis and Major Black departed New York for Panama, along with Navy Paymaster E.C. Toby, and several others who were in the vanguard of the small group of Americans that would supervise the canal project for the next 10 and a half years. On May 17th, Commissioner Frank J. Hecker departed for the Isthmus to supervise the matters of receiving, listing, and storing the enormous quantity of property, and to decide what of it was salvageable and what was beyond likely repair. He remained on the Isthmus till the following November.

The Executive Committee supervised the general business of the Commission and its policies; the supervision and management of the Panama Railroad; and recommended staff appointments. The Committee on Engineering was responsible for seeing that proper reports were made and submitted to the Commission; reviewed reports by the Chief Engineer (Wallace); examined all requisitions for engineering supplies; passed on applications by the Chief Engineer for increases to his staff; and recommending these for approval or disapproval to the Executive Committee.

On October 25th, the Commission was subdivided into five Standing Committees: An Executive Committee comprised of Parsons, Grunsky, and Admiral Walker; a Committee on Finance comprised of Hecker and Harrod; Committee on Legislation comprised of Harrod and Hecker; a Committee on Engineering comprised of Burr and Parsons (who remained in New York); and the Committee on Sanitation comprised on Grunsky and Burr. In addition, a Committee on Engineering Plans was formed comprised of all the civil engineers (Burr, Grunsky, Harrod, and Parsons), who would *"study and prepare the general plans of the Canal works to be submitted to the Commission"* (Isthmian Canal Commission, 1908). Major Harrod was elected chairman of this group, likely because he was senior to the other members.

Compensation of Army and Navy officers. One of the Commission's first actions on March 22, 1904, was to employ the two Army Engineer officers in Panama, Major William M. Black and Second Lieutenant Mark Brooke. They were put on the payroll of the Isthmian Canal Commission and charged with "collecting engineering data on the Isthmus" and making monthly reports to the commission on the progress of construction. As would be the custom throughout the canal project, officers were to be paid salaries 50% above their respective military grades. Major Black's new salary would be $4,500 per year, while Brooke was to be paid $2,250 per year. The Commission adopted this policy at their meeting in Colon on April 17th, while they were visiting Panama. Army and Navy officers detached for duty on the Isthmus would be eligible for quarters, medical attendance, medicine, as well as travel expenses, to and from the United States. Curiously, on August 10, 1904, when Colonel Gorgas requested that 10,000 pounds of freight, the normal allowance for military officers changing stations in the United States, be granted for the transfer of his household goods from New York to Panama, his request was denied by the Commission on the grounds that "*it not being the policy of the Commission to grant such allowances to its officers and employees.*"

The Commission took their first inspection tour of the canal between April 5th and 20th, taking a steamer from New York to Colon and return. Doctors Gorgas, Ross, and LaGarde accompanied the group to make a reconnaissance of the Isthmus and develop a scheme for sanitation (described later). During this same visit, they asked Major Black to return to Washington with them so they would have someone on their staff that was familiar with the terrain and the engineering problems in Panama. The next officers placed on the Commission's payroll were Surgeon Colonel Gorgas at $7,500 per annum, who would serve as Chief Sanitary Officer and the Director of Hospitals in Ancon. His principal assistants Dr. John W. Ross and Dr. Louis A. LeGarde were to be paid $7,000 and $6,000 per annum, respectively. Ross came down with malaria and returned to the mainland in late January 1905.

Peer review. In August-September and November 1904, the ICC commissioners once again traveled to Panama for inspection tours (in November they were accompanied by War Secretary Taft). Daily meetings were convened for almost a month in Ancon, on the Pacific side, where evening trade winds graced the knoll rising about 300 feet above the Gulf of Panama. This was also the location where Gorgas was building the project's principal hospital.

Wallace returned to Washington, DC the following month for more conferences. Wallace was not a man used to the rigors of political combat. He had always tried to work within the confines and restrictions of those placed in authority over him. This was the flaw in his character for the Panama assignment, because its political nature required a chief engineer who could battle with the politicos in Washington for what he desperately needed, regardless of protocol (Stevens, 1927).

Report of the Committee on Engineering Plans. As described above, in late October 1904, a Committee on Engineering Plans was formed to "*study and prepare the general plans of the Canal works to be submitted to the Commission.*" This group (Harrod, Burr, Grunsky, and Parsons) began compiling data in the hopes of reaching a consensus on the type of canal that should be constructed by the Americans. On January 17, 1905, Burr and Parsons sailed for Panama to make a detailed assessment of the situation there, while Grunsky and Harrod remained in Washington, DC so that a quorum of the Commission would be able to continue the transaction and approval of countless requisitions and hiring decisions.

Burr and Parsons focused much of their efforts collecting and assessing topographic surveys, hydrologic data, metrological data, and foundation conditions at the various dam sites that were then being explored using borings. They also met with General Davis and Chief Engineer Wallace, and were apprised of the progress on the canal construction. Their goal was to determine if the best type of canal was still the concept developed by the previous ICC in 1902-03, calling for a locked canal with a high water elevation of 90 feet above sea level across the Continental Divide at Culebra.

When all seven members of the ICC reconvened in New York for their 80th meeting on February 23rd, Burr and Parsons made the case for construction of 8,000 lineal feet of breakwater (involving 3.4 million yards of riprap) to protect the anchorages in Limon Bay at Colon and Cristobal, and to dredge the harbor to a depth of 30 feet, considerably enlarging its freight handling capacity and wharf frontage.

Recommendations for a sea level canal. Burr and Parsons spent the remainder of that session describing their recommendations for three canal schemes, which had been forwarded to the Commission in writing on February 14th. The inclusion of a sea level scheme evolved from their discovery that the average monthly cost of excavation at Culebra had fallen below 50 cents per cubic yard, based on data gleaned from the three new Bucyrus steam shovels that were working the faces (and 17 more were on order). These machines could excavate 1,000 cubic yards per day, far more than assumed by the previous ICC in its report of 1902 (the French ladder excavators could remove up to 1,400 yards per day, but only in shale and soil, not rock).

The 1903 ICC report had assumed an average cost of rock excavation to be 80 cents per cubic yard. This *reduction in unit cost* of excavation had the potential to shave $15 million off the budgeted cost of the project! Burr and Parsons cited this windfall, and then assuaged: *"This large saving in cost makes a fundamental reconsideration of canal plans by this commission essential. It is obvious that this actual reduced cost of excavation justifies a reduction in elevation of the summit level of the canal by a correspondingly greater volume of excavation."* (Isthmian Canal Commission, 1908).

Their report laid out three conceptual designs: 1) a locked canal with a summit water elevation of 60 feet above sea level for $178 million; 2) a locked canal with a summit water elevation of 30 feet above sea level for $194 million; and, 3) a sea level canal for $230 million. This assumed that approximately 96 American steam shovels could excavate 30 million cubic yards of rock per year, averaging 1,000 cubic yards per day per machine. Their estimates included 20% for contingencies.

They concluded by recommending that their scheme for a sea-level canal should be adopted, with a tidal lock on the Pacific side. They recommended that the sea level canal should be constructed with a minimum bottom width of 150 feet, not less than 35 feet deep (similar to those recommended by the first ICC in 1903). They suggested that, as an additional precaution, calculations should be undertaken to undertake feasibility studies of incorporating a minimum water depth of 40 feet throughout the canal's length, to account for the increasing draft of steel battleships, or "dreadnaughts," then being considered by the world's great navies.

Their recommendations also included a caveat on the dimensions of ship locks, if required, which advocated a minimum width of 100 feet in lieu of the 85 feet then planned (a width of 110 feet was eventually chosen), and that the lengths of each lock be increased to 1000 feet (a length of 1100 feet was later selected).

Recommendation for a dam at Gamboa. Burr and Parsons then described the increasing difficulties of constructing the earthen dam at Bohio that had been proposed by Morison (1902). This finding was based on disappointing news gleaned from the additional exploratory borings made in 1904, which showed a buried channel extending as much as 139 to 172 feet below sea level, making it impractical, if not impossible, to carry the embankment down to the desired bedrock. They recommended that the fickle flows of the Chagres River be controlled by an earthen dam at Gamboa, the same site that the French had initially explored in 1881 (a concrete gravity dam 220 feet high was eventually constructed 8.5 miles upstream of this location, near the village of Alhajuela in 1931-35).

One remarkable assumption by Burr and Parsons was that the Gatun dam site was "*unsuitable*" because bedrock lay 140 to 173 feet below sea level, too deep for the placement of a masonry core wall of an embankment dam. Masonry core walls extending down into the underlying bedrock was an accepted method of securing an effective seepage cutoff at that time, which had been employed for the Sudbury Dam in Massachusetts and Pilarcitos Dam in California (Stearns, 1902). But, there were numerous embankment dams that employed cutoff trenches backfilled with fine grained material that were often extended deeper using steel or timber sheetpiles. In the case of the North Dike of the Wachusett Reservoir for the Boston Water Works, a cutoff trench 50 feet deep was excavated beneath the embankment, which was 200 feet above the underlying bedrock (Stearns, 1902).

The Bohio dam site was similarly deemed unsuitable because of deeply incised channels in the bedrock as much as 158 feet below sea level. Morison (1902) had compared the foundation conditions as the Bohio dam site with those handled successfully under the North Dike of the Wachusett Reservoir, including geologic sections through both sites. Why Burr and Parsons chose to ignore the articles by Morison and Stearns remains a mystery. They assumed that the old French dam site at Gamboa was the only workable solution for harnessing the Chagres River, where a masonry core earth dam could be constructed to a crest elevation of 200 feet above sea level to garner sufficient flood storage. They recommended that work on this dam should commence immediately.

In light of the events that subsequently transpired in the coming months, the assertions of Burr and Parsons regarding the unsuitability of the Gatun dam site seem rather premature. What they couldn't visualize was such an enormous earthen dam (6,400 feet long and up to 2,300 feet wide), without a masonry cutoff wall, similar to the North Dike of the Wachusett Reservoir, which Stearns (1902) had described in his lengthy discussion of Morison's 1902 article on the Bohio Dam. It would be the world's largest embankment dam by volume, retaining the largest man-made lake in the world (110 square miles of surface area), equipped with the world's largest gated spillway, and connected to the largest locks in the world. Gatun Dam, spillway, and locks were colossal undertakings for their time, the likes of which very few engineers could envision or appreciate, even some of the best engineers in the world!

Epilogue on the 1905 sea level scheme. Carl Grunsky (1909b) stated that none of the technical data collected by Burr, Parsons, and Davis in Panama was ever transmitted to Washington, DC to be included in the appendices of the ICC Proceedings (although the report by Burr, Parsons, and Davis was included). Nor did said data accompany any progress report of Chief Engineer Wallace. Wallace's annual report (conveyed to the Commission over a month after he left, on July 31st) only included the comment "*that no temporary or tentative plan be adopted that will interfere with the final adoption of the "sea-level plan."* The shortcoming in all of this was the absence of any credible scientific data or reliable cost estimates for how much the sea level scheme would cost and how long it would take to complete, in comparison with a locked canal, were never aired credibly for later boards or commissions to pass judgment on. Grunsky (1909b) later assuaged that he did not believe that the Culebra excavations could be advanced to such great depths with an average slope of "3 on 2" (vertical to horizontal), or 56° from horizontal (shown in Figure 11).

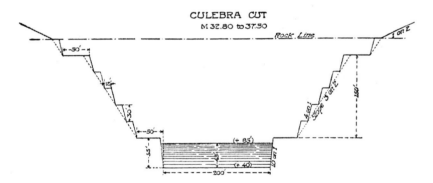

Figure 11. Maximum section at the Continental Divide at Culebra illustrating the intended excavation profile for a lock canal in early 1905, which proved overly optimistic (NAS, 1924).

In March 1905, Benjamin Harrod proposed that the recommendations of the Committee on Engineering should be adopted by the Commission in favor of the sea-level canal, but Grunsky did not agree because he felt they had not properly estimated the cost or time-to-completion differentials. So the matter was referred to the

Committee on Engineering Plans, of which Grunsky was a member, for "further examination." Before the ICC Committee on Engineering Plans completed their formal review, the Commissioners were asked to tender their resignations by President Roosevelt on March 29, 1905. On April 5th, Roosevelt appointed an International Board of Consulting Engineers (described below). Although Burr and Parsons remained adamant in their support of a sea level canal, eight months later, Harrod was persuaded to favor the locked canal scheme, as recommended by the minority report of this new board.

JOHN F. WALLACE – FIRST CHIEF ENGINEER

Professional background. John Findley Wallace (Figure 12) was born in Fall River, MA in 1852, but he grew up in Illinois because his father founded Monmouth College of Illinois in 1856, an institution sponsored by the United Presbyterian Church. Young Wallace attended Monmouth as an undergraduate, but declined to take the required theological courses, such as Hebrew and Greek. At the age of 19 he departed to take a position with the Army Corps of Engineers working on navigation improvements at Rock Island Rapids, along the Mississippi River. This work kindled his interest in civil engineering and he returned to Monmouth, completed his degree and enrolled himself at the University of Wooster to study engineering, while supporting himself as a tutor.

After working two years for the Corps and three years as a county surveyor, he found a position as civil engineer in charge of route surveying, construction, and operation of a new railroad line between Peoria and Keithsburg, IL, which was subsequently absorbed by the Iowa Central Railroad. Wallace's abilities were noted by the Iowa Central, and in 1883 he was named Chief Engineer of Construction. During the succeeding three years he designed and built transfer facilities for the railroad at Keithsburg, across the Mississippi River, a feat of considerable note in those days for such a young man.

In 1887 he was hired by the Atchison, Topeka & Santa Fe Railway (AT&SF) as a bridge engineer for an extension of their transcontinental line between Chicago and Kansas City, across the Missouri River at Sibley, MO. This bridge was designed by Octave Chanute, who became President of ASCE in 1891. The Sibley Bridge involved the sinking of caissons under compressed air in a river noted for swift, treacherous currents with a high sediment load, as well as changing flow levels. The bridge was one of the outstanding projects of that era, chronicled in an article by Wallace for the ASCE *Transactions* (Wallace, 1889).

From 1889-91 Wallace worked with future ASCE President Elmer L. Corthell in Chicago, supervising the construction of a joint entrance into that city by the AT&SF and the Illinois Central Railroad. During this same interim he served as Corthell's associate on the foundation construction of the Merchant's Bridge over the Mississippi River in St. Louis, and a proposed bridge over the lower Mississippi at New Orleans (Wallace, 1894). He also penned an oft-cited piece on the various methods that should be employed when advancing borings to bedrock in river channels, as a discussion accompanying S. M. Rowe's article on the Red Rock Bridge over the Colorado River, then the longest cantilever bridge in the Americas (Wallace, 1891).

Figure 12. John Findley Wallace as he appeared while serving as the first Chief Engineer of the American effort to construct the Panama Canal in 1904-05. He had served as President of ASCE in 1900 (Jackson & Son, 1911).

By 1891 these high visibility projects and articles landed him an appointment a Chief Engineer of the Illinois Central Railroad (ICR), where he set about reconstruction of that line's terminal facilities in Chicago, in anticipation of the expected upswing in rail traffic for the 1893 Columbian Exposition and World's Fair. Wallace was influential in the line's decision to elevate their tracks in downtown Chicago, becoming the first railroad to do so (Wallace, 1906). During the 1890s Wallace oversaw numerous improvements of the ICR system, which included improved rail yards and terminal facilities in St. Louis and New Orleans, as well as outside Chicago. In 1898 Wallace was promoted to the ICR's executive department, where he served as assistant vice president, assistant general manager, and general manager. During his time with the Illinois Central (1891-1904) he oversaw construction expenditures of more than $100 million, a sum of unimaginable scale to civil engineers of that era (Molitor et al., 1922).

In 1899 he helped found the American Railway Engineering Association, serving as its first president. The following year (1900) Wallace was elected

President of the American Society of Civil Engineers. He also served two terms as president of the Western Society of Engineers.

Wallace's Selection as Chief Engineer of the Panama Canal. John Wallace was serving as Chief Engineer and General Manager of the Illinois Central when President Theodore Roosevelt appointed him Chief Engineer of Construction for the Panama Canal on May 6, 1904, after conferring with his new Isthmian Canal Commission. Wallace's appointment coincided with that of Army Surgeon Colonel William C. Gorgas as Chief Health Officer. Wallace's salary was to be a staggering $25,000 per year, $10,000 more than the Illinois Central was paying him and more than any other federal employee, with the exception of President Roosevelt (until 1907 William Gorgas would be only be paid $4,000 per year, the wage of an Army colonel). General Davis arrived at the Canal Zone on May 17th, while Wallace and Colonel Gorgas arrived on June 28th. The great enterprise lay before all three, but only Gorgas would see it through to completion.

Challenges facing Wallace in 1904-05. On May 9th, an executive order by Roosevelt placed the ICC under the newly appointed Secretary of War William H. Taft, who had recently served as Territorial Governor of the Philippines. Under the new arrangement, the chief engineer (Wallace) and chief health officer (Gorgas) would reside in Panama during the project, but all requests for expenditure had to be reviewed by all seven members of the ICC back in New York and Washington, DC. Even the hiring of any employee making more than $1,800 annually had to be approved by the entire commission. Admiral Walker had hundreds of requisitions sitting on his desk each day, waiting for his approval. Their unwieldy oversight of each expenditure, regardless of how small, was a political compromise required by the Spooner Act.

When Wallace arrived in Colon the challenges facing him were almost without precedent. Everywhere he looked there were dilapidated buildings (2,149 according to McCullough, 1977) and rusting metal castings, pipes, rails, and all manner of mechanical equipment. The path of the French workings was littered with rusting French and American-built earth moving equipment. Many of the dredges and barges long since sunk in the shallow water, providing mute testimony to the dismal failure that preceded the optimistic American entry (Figure 13). In 10+/- years of digging the French had succeeded in excavating about 80 million cubic yards of earth and rock, above ground and below water. They had employed rail muck cars with capacities of only 4.6 to 5.25 cubic yards on narrow gage lines and 6 to 9 cubic yards on their standard gage side-dumping rail cars. Of these excavations, only about 30 million yards were useable in the high level lock canal eventually selected by the Americans in mid-1906. Near the Continental Divide at Culebra the French had succeeded in excavating 163 vertical feet of a cut intended to become more than 368 feet high (Figure 11). But, much of the French workings would not be used by the Americans because their final alignment varied considerably from what the French had proposed, in large measure because of how they decided to handle the unpredictable Chagres River.

Figure 13. One of the mammoth floating dredges abandoned by the French in 1895. Being made of more corrosion resistant iron (instead of steel), most of them were easily repaired and put back in service by the Americans in 1904-05 (Bishop, 1916). The first American dredges were not placed in operation until July 1907.

Maltby's Dredges. In January 1905, the Commission hired Frank B. Maltby as their first Division Engineer for Dredging and Harbors, at a salary of $5,000 per annum (Maltby, 1945). Originally from Pittsburgh, Maltby had acquired his expertise in dredging while working as an Assistant Engineer for the Army Corps of Engineers in the Midwest, under Captain Hiram M. Chittenden. Around 1899 he was named Superintendent of Dredging Operations on the Mississippi River below Cairo, Illinois, for the Mississippi River Commission. In this role he was responsible for overseeing the maintenance of the navigational channels of the lower Mississippi River, which included various improvements to expand the ports of southern Louisiana. He was by recommended Major Harrod, who being from New Orleans, was familiar with Maltby's abilities (Maltby, 1945).

As their first task, the ICC Committee on Engineering Plans asked Maltby to propose designs for sea-going dredges that could be used to improve and significantly expand the port facilities at Cristobal, upon which the entire project depended, as the great majority of stores, equipment, and personnel flowed from the Atlantic terminus (Figure 9). He also directed all dredging operations in the Canal Zone. Upon viewing all of the sunken wrecks littering the area around Colon, Maltby began directing the restoration and repair of the abandoned French dredges. This surprised everyone because many of these machines had lain underwater for as much as 15 years! Maltby discovered that the high grade of Scottish iron used to fabricate the old dredges was far more resistant to corrosion in saltwater than the lighter weight steel that was rapidly overtaking the American market (Maltby, 1945). Within three months, he succeeded in restoring five of the old Scottish-built ladder dredges (Figure 17), which he manned with Greeks. Maltby established himself as one the most respected division engineers on the Isthmus, guiding and directing a flotilla of dredges that would later complete the Culebra Cut.

The ICC gradually enlarged its flotilla of rehabilitated floating dredges, and Frank Maltby was pressed to recruit marine engineers, dredge operators, boiler techs, and deck hands capable of keeping their hydraulic monitors working and the centrifugal pumps from becoming clogged with debris (Figure 14). Maltby was kept busy traveling back and forth between the Isthmus and Washington, DC, working on the specifications for the various kinds of dredges he thought would be most appropriate for the canal project. In June 1907 the first American built dredge finally arrived, built to Maltby's precise specifications. The number of dredges gradually increased, from five in 1906 to 27 in 1912.

Figure 14. One of the floating 20-inch suction dredges used to excavate the eight mile channel between Colon and Gatun on the Atlantic side of the canal. More than 220,000 lineal feet of ductile iron pipe were shipped to Panama to convey dredge tailings as far as 12 miles from their source.

Resurrecting rusty abandoned equipment. The rusting hulks of French and Belgian locomotives and muck cars were everywhere (Figure 15), all specially configured for the Panama Railroad's unusual five-foot gage, using 50 to 60 pound per yard steel rail (the weight of a three foot section), imported from England or Belgium. The European rolling stock was of a smaller size and capacity than their

American counterparts, employing single axels at either end, while the Americans routinely employed dual trucks (twin axels) with leaf or coil springs; capable of hauling 50 to 100% more load per rail car. Despite these limitations, by the close of 1905 (1-1/2 years after Wallace's arrival) over 100 of the diminutive French and Belgian steam engines that could be refurbished were back in operation so that excavation work might "proceed with all haste" (Figures 16 and 17).

Figure 15. Rusting French and Belgian locomotives at Empire, near the eastern end of the Culebra Cut, as seen in 1905 (Library of Congress).

Identifying the needs. In mid-1904 there were only about 500 canal workers on the isthmus, who were making slow progress on the excavations around Culebra. These men had been inherited from the second French company, which was obliged to "continue working" in order to retain their franchise with the Columbian, and now Panamanian governments. Wallace's first job should have been constructing housing for the work force, spread out over 50 miles. But, first he needed to hire an array of office professionals, including surveyors, civil and mechanical engineers, doctors, nurses, orderlies, telephone linemen and operators, secretaries, accountants, dispersing agents, draftsmen, and cartographers. He needed railroad men of every type, including yardmasters, engine foremen, foremen helpers, switchmen, switchmen helpers, work train conductors, trainmen, locomotive enginemen, switch enginemen, foremen, engine drivers, firemen, watchmen, laborers, switch tenders, mechanics, boiler technicians, machinists, and shop and foundry workers.

Figure 16. View of the cut at Culebra Cut at its deepest point around April 1905, near the Continental Divide. This shows the French equipment: the single axle muck cars and Belgian steam engines, all using single track benches.

Figure 17. The largest of the six Scotch-built ladder excavators left by the French is shown here, alongside the smaller French muck cars, pulled by a diminutive Belgian locomotive, in early 1905. These machines were only capable of excavating 50 cubic yards per hour (Library of Congress).

In October, ICC Paymaster E.C. Toby was promoted to Chief of the Department of Material and Supplies, which he completely re-organized. Foremost among his concerns was the provision of additional wharfage at La Boca on the Pacific side, where lumber shipments from California and Washington were piling up. He also sought to entice skilled blacksmiths, boiler-makers, machinists, coppersmiths, moulders, pattern-makers, mechanics and "other labor of this character" to come down to Panama to maintain equipment and make necessary repairs and alterations, because much of the older machinery was beginning to wear out break down with increasing frequency.

Tropical tramps. A large number of the workers Wallace had received in Panama during his year there were young and inexperienced raconteurs, or men who "had a past" that they sought escape from in America, such as black listed (scab) laborers, alcoholics, or those seeking to make a lot of money in a short period of time, which were referred to as "tropical tramps." The great majority of these men were unmarried and they had to brave the poor reputation for death and disease attributed to Panama for so many decades. As working conditions slowly improved on the isthmus, more skilled labor and higher quality professionals were enticed to immigrate to the Canal Zone, and this can be seen in the increasing percentage of married men with their wives and families. But, that was after Wallace's departure.

Karner recruits West Indian laborers. By November 1904, Wallace had 3,500 men working in the Canal Zone, but he was short on laborers, of all types. He desperately needed more help. On November 15th 1904, the ICC resolved to hire an agent "*who shall investigate the sources, character and supply of labor in the West Indies, and how much of said labor is available for the purposes of canal construction*" (Isthmian Canal Commission, 1908). Nobody suitably qualified in the view of the Commission applied for the position, so it went unfilled, for awhile.

In late December 1904, War Secretary William Howard Taft, Chief Engineer Wallace, and British Ambassador to Panama, Hugh Mallet, set off for Kingston, Jamaica to recruit unskilled West Indian laborers. Mallet arraigned an audience with the British Governor, Sir James Sweetenham. The Americans had to negotiate financial agreements with the governing authorities of the English protectorates to guarantee the safe passage of such workers to and from Panama, so that their government wouldn't have to bear such expense. Thousands of West Indian workers had been stranded in Panama in the late 1880s and 1890s when the French operations went bankrupt. For Jamaica, this meant a deposit of five pounds English Sterling for each worker transported to Panama (about $16 US), for a minimum period of 500 days for a starting wage of 10 cents per hour. The workers would be shipped on the Royal Mail Steam Packet Company steamers, and not on American vessels. For all these efforts, very few Jamaicans initially volunteered their services, as the name Panama conjured up many bad memories of the past. In the end almost 90,000 Jamaicans would make the trip to Panama to work on the Canal between 1905-14, after initial fears were overcome and better living conditions were established on the Isthmus.

On each leg of this initial foray to Jamaica, John Wallace was plagued by seasickness, which required several days of recuperation after each landfall. Upon his return to work in Panama, he turned to his trusted Assistant Engineer William J. Karner, an Illinois Central colleague he had brought from Chicago. In fact, Karner had made the voyage to the Isthmus with Wallace in late June 1904, and the two men had lived in Wallace's home until Sadie Wallace arrived that fall. Karner had been preparing daily and monthly progress reports, which included engineering economics assessments, such as tracking labor and average unit costs (Karner, 1921). In October 1904 he came down with malaria, and was taken to Ancon Hospital, where he had recuperated sufficiently by January 1905 to have come back to work. When Wallace returned from Jamaica, he appointed Karner as the new Special Disbursing Agent (Chief Recruiter), knowing that it paid considerably more than he was making as an assistant engineer and it would allow Wallace to stay off the high seas.

In mid-January 1905, Karner set off for Barbados, where the American counsel scheduled an audience with the British Governor, Sir Gilbert Carter, whose wife was an American. Karner the civil engineer had never done anything like this before, but he was a quick study, and Lady Carter taught him the rudiments of diplomacy (Karner, 1921). He quickly established a recruiting office in Bridgetown, with similar ground rules to those posed unsuccessfully in Jamaica the previous month; 500 days minimum stay at 10 cents per hour, with safe passage on English steamers. Barbadian laborers were only making the equivalent of 25 cents per day, but they didn't seem particularly interested in risking what they had for something they knew nothing about. All of the West Indian workers had to be at least 20 years old, subject themselves to medical exams by doctors hired by the Americans, and inoculations prior to their being accepted for shipment to Panama.

But, when Karner departed Bridgetown on January 26th, he only had 16 Barbadian men with him, sailing for Colon. On January 26th, Commissioner William H. Burr (on duty in Panama) prepared a special report outlining the acute shortage of unskilled laborers, which they had hoped to hire at rate of 1,000 men per month. From October 1st to January 26 they had only succeeded in attracting 800 laborers. Burr stressed the need for more recruiting by Karner, with "*ample funds to meet whatever expenses he incurs in securing labor.*" Fortunately, while visiting Bridgetown, Karner had struck a deal with the local agent of the Royal Mail Steam Packet Company, Samuel E. Brewster, which would increase his business and his share of each worker shipped to Panama. Brewster understood the local culture, and he began working the local men who were poor and unemployed, enticing them to take a chance on Panama. When Barbadian men began returning from the Isthmus with money and respectable store-bought clothing, they attracted the local ladies, leading to a rush by everyone else to catch the next steamer for Panama.

By the time the canal was completed, 45,000 of the 200,000 inhabitants of Barbados would make the journey to Panama, about 40% of the island's adult male population, and 7,000 would remain in Panama permanently. This exodus increased the cash flow of the British Colony by $300,000 per annum. Similar recruiting efforts were made elsewhere in the Caribbean, attracting British and French Caribbean workers from Jamaica, St. Lucia, Trinidad, Antigua, Martinique, and St. Vincent. Their starting wage was just 10 cents per hour (80 cents per day).

Other recruiters worked in Europe, and despite the resentment of the Spanish government, some 7,000 peasants from the Biscayan Provinces of Spain were recruited to work in Panama, for a starting wage ($1.60 per day) double that given to the Caribbean workers. The starting pay for American workers was $5.20 per day.

Problems with the supply system. A representative example of the mindless bureaucracy created by the ICC was the approval process that required unanimity of the commission's seven members. One of the most badly needed materials was common pipe; iron pipe for water supply and terracotta pipe for sewer lines. These were essential components for the living quarters being raised across the wide expanses of the project, spread over 50 miles. In July 1904, Wallace issued requisitions for prodigious quantities of water and sewer pipe, including only those sizes and types that he was confident most American manufacturers kept in stock. To better insure their prompt delivery, Wallace took the additional precaution of including specifications for both eastern and western standards for such pipes. Commissioner Grunsky was from San Francisco, so favored the western standards, which he had helped develop, while Commissioner Parsons was from New York and Commissioner Harrod from New Orleans. Instead of acting promptly to approve Wallace's request for pipe, the virtues of eastern versus western pipe specifications became a technical discussion and debate, further delaying any hope of the unanimous approval required before the requisitions could even be cabled to the United States! It was finally decided to shelve the issue until Wallace could be interrogated about the matter during the visit of the Engineering Committee in August 1904.

When the issue was finally taken up with Wallace, the commissioners learned that Wallace had issued both specifications in hope of gaining a more prompt approval, and that he had no particular preference. What he needed was large quantities of pipe "as soon as possible," even in small lots if that's all that were available. Grunsky convinced his fellow engineers that the western pipe standard was "superior," and so should be the only one used on the project, and this resulted in many months delay in getting potable water supplies and sewerage conveyance installed on the isthmus. The Rio Grande Dam and Reservoir had been completed in September 1904, but the first load of water pipe did not arrive by steamer until March 1905. It was a very inefficient system of supply, the likes of which can only be created by a committee of people with equal rank with nobody in charge.

The all-important steam shovels. In early August 1904, Admiral Walker and Commissioners Burr, Harrod, and Grunsky returned to Panama, where daily conferences were convened at Ancon Hill. In his view, Wallace's success or lack thereof seemed to be judged by his monthly progress reports to the ICC in

Washington, who wanted to know "how many cubic yards of earth had been excavated." Wallace's organization had made significant progress in ascertaining what sorts of equipment was best suited to the massive excavations proposed.

By early August his staff had settled on the largest track-mounted steam shovels then available, behemoths weighing between 70 and 94 tons (depending on the model) built by the Bucyrus Steam Shovel and Dredge Company of South Milwaukee, Wisconsin (Figure 18). 24 of these monsters had been used in the construction of Chicago's Sanitary and Ship Canal, which involved the excavation of 43 million cubic yards of earth and rock between1894-1900 (Layton, 2012), and they were being used to excavate the New York State Barge Canal, the largest domestic project in the United States during the first decade of the century. In the hands of a skilled operator, these shovels could excavate about 150 cubic yards per hour, provided they were working with an unlimited string of rail-mounted muck cars.

On August 18th, the Commission was meeting in Ancon and they approved the purchase of one 70-ton Bucyrus shovel for $8,800 and two 95-ton shovels for $24,500. These were the largest equipment purchases to that time, which were shipped from New York. In October, Wallace ordered eleven more Bucyrus shovels, and the following spring, a dozen more (Isthmian Canal Commission, 1908). On October 22nd, the Commission approved the purchase of five 70-ton shovels and six 95-ton shovels from the Bucyrus Company for $117,250, the low bidder.

Steam shovel engineers (operators) were among the highest paid skilled laborers on the project. In 1904 they received $190 per month ($2,280 per year) and $1 for each 1000 cubic yards of excavated material loaded by their shovels in excess of 25,000 cubic yards per month, as a performance incentive. The largest shovels averaged between 150 and 225 cubic yards per hour, which meant they were capable of excavating 30,240 to 45,360 cubic yards per month (assuming eight hour shifts six days per week). In this way most of shovel operators earned bonuses of between $5 and $20 per month.

In the spring of 1905 two new steam shovels (either 70-ton or 95 ton size) arrived on the Isthmus each month, and this stream of continued throughout 1906 and 1907. By the end of 1908 there would be 77 Bucyrus and Marion steam shovels working Culebra Cuts, filling muck cars as fast as they could be positioned.

Organization and progress reports. In regards to taking stock of what he had on the Isthmus to work with, Wallace was receiving assistance from General Davis, who had arrived at Colon on May 17th, 1904. He immediately assumed direction of all administrative and executive work as the Managing Representative of the ICC. Commissioner Frank J. Hecker arrived on May 24th and "began supervising the canal work proper," inheriting the organization set up by the New French Company. During the August 31st meeting, Governor Davis suggested that an Executive

Figure 18. The first of 77 Bucyrus 70-ton and 95-ton steam shovels arrived on the isthmus in November 1904. The 95-ton example, shown here, hoisted a five cubic yard bucket, which allowed it to excavate between 150 and 225 cubic yards per hour, highest of all the above-ground earth movers on the canal.

Committee should be formed in the Isthmus comprised of himself, Commissioner Heckler, and one other commissioner to be named by the chairman, whose occupancy could be rotated between members of the ICC.

This committee would review requests down in Panama, and make recommendations to the full Commission in Washington, DC. This new procedure was approved by a vote of 5 to 1. The problem was that Chief Engineer Wallace, although the second highest paid public servant in the land would remain subordinate to the new Executive Committee, as well as the full Isthmian Canal Commission. The cumbersome statutes of the Spooner Act were just beginning to be felt.

During the August 31st meeting of the Commission in Ancon they enlarged the responsibilities of the chief engineer to include the sewerage and waterworks systems of Panama City and Colon, and all engineering works associated with those cities, as well as the canal project. They also asked Wallace to provide the Commission with monthly progress reports as well as an annual report, due on July 31st of each year.

In mid-March 1905 the canal building organization was overhauled and restructured into eleven major departments, based on their respective tasks:
1. Office of the Chief Engineer
2. Quartermaster Department
3. Office of the Division Engineers (who oversaw the technical activities within each of their sub-specialties, such as construction, surveying and mapping, port and harbor facilities, etc.)
4. Excavation Department
5. Mining Department
6. Track Department
7. Transportation Department
8. Spoil Dumps Department
9. Camp and Building Department
10. Water Service Department
11. Sanitation Department

Tropical weather. One of the vexing problems with "management by numbers" was the tropical weather in Panama, which varied considerably across the isthmus. The rainfall was considerably heavier on the Atlantic side, at Cristobal, where most of the stores, construction supplies, and workers arrived and had to be off-loaded and trans-shipped to their respective destinations. On average Cristobal receives about 130 inches of rainfall during a seven months season (typically, between April and October). Culebra receives about 100 inches, while Panama City, on the Pacific side, receives only 70 inches per year. Despite the lower cumulative rainfall totals, there are only three or four dry months on the Pacific side each year, although no months are entirely free from precipitation. The lowest rainfall recorded on the isthmus occurs within the sheltered interior behind the Azuero Peninsula, where they only receive 50 inches of rainfall.

The problem was torrential downpours, the likes of which North Americans and Europeans were completely unfamiliar. These tropical storms played havoc with the Chagres River, which could rise 20 feet over night, flooding everything on the coastal plains below Gamboa, and bringing construction activities to an abrupt halt. One essential activity that was continually at the mercy of the rain were the hundreds of test borings at key locations along the proposed canal, where major structures, such as locks and dams, were to be located. These included locations and along the Culebra Cut, between Obispo and Miraflores, where the deepest excavations were proposed. In these areas the drillers kept "losing the holes" to flooding and cave-ins (these were predominately "wash borings" using drag bits, long before mud rotary technology was developed in the 1930s).

The lack of progress during the wet season appeared ominous to those watching in Washington, and it seemed as though each storm brought a new round of set-backs that required repairs and reconstruction of many essential elements, such as bridges, waterworks, and dikes. Nobody trusted the water in Panama, fearful of typhus outbreaks, which were both frequent and deadly, so the provision of drinking water took precedence over all other activities. The work areas required near-continuous maintenance of the drainage systems, and these activities could only be performed *after* the storm waters had receded, preventing crews from working on other essential activities. It took the first year and a half to figure out how many men and machines were needed just to maintain the support infrastructure in various parts of the project through the rainy season.

Railroad ties, lumber, and coal. Prior to the American arrival, all of the ties for the Panama Railroad had been fashioned from lignum vitae trees, a very heavy tropical hardwood. This was necessary in order to preclude their destruction by wood eating ants (spruce or pine ties would only last a year). *Lignum vitae* only grow in the West Indies and certain areas of tropical Americas. The trees are of diminutive size, so an entire tree was required to produce each tie (two sizes of ties were used on the canal: 7" x 9" x 9' and 6" x 8" x 8'. The wood was so hard and dense that holes had to be drilled in them to allow rail spikes to be driven into them, and one tie was required for every 18 inches of track. Despite these limitations, orders were placed for more than 338,000 rail ties in lots of 25,000, but they could not be harvested quickly enough, so iron tie rods were attached to conventional timber ties from the Pacific Northwest and covered with bitumen (tar) to dissuade the ants, and these sections were most often employed as temporary sections of "panel track."

In late January 1905, the requests were increased to lots of 50,000 to 100,000 ties of 7" x 9" x 9' dimension and one hundred sets of longer switch ties. The Commission could only locate two lots of 10,000 rail ties. Wallace then ordered 50,000 ties made of red cypress, feeling that either this or redwood would offer increased resistance to rot (they were employed on mountainous lines at high elevations in the United States subject to increased moisture and snow). In early February the Commissioners approved the purchase of 50,000 nine foot ties and 4,800 of the longer switch ties from Brown & Co. in New York. If they purchased lots of 10,000 ties or more, Brown & Co. would furnish them to the ICC for 68 to 77 cents apiece, depending on the size. These orders were soon followed by a request for 2,500 tons of 75 pound (per yard) steel rails (200,000 lineal feet of rail), followed by an order for 3,500 tons of 70 pound rail in March 1905.

In many of the material specifications Wallace's intimate association with the American Railway Engineering Association (AREA) proved valuable. He quickly recommended adoption of AREA material specifications, which American railroads began to employ in the early 1900s to establish minimum materials standards for items like rails, tie plates, switch frogs, and timber rail ties. Recognizing the need to increased resistance to dry rot, new "Panama Standards" were added to existing AREA standards for acceptable grades and types of lumber, which included the minimum number of dry (dark) and wet (light) season growth rings for each rail tie, to better insure the rejection of porous, green, or lightweight timber which was more prone to water absorption and dry rot. As it turned out, many of these "Panama Standards" remained in the AREA Specifications manuals lasted long after the canal was completed (Panama Canal Rule, 1915).

In October 1904, Wallace asked for 2.7 million board feet of lumber and eleven more Bucyrus steam shovels. He was badly in need of housing, warehouses, machine and maintenance sheds. The lumber order was so large it was tackled by Commissioner Heckler and Grunsky, who sought competitive bids from sources in the United States. The Bellingham Bay Improvement Co. in Washington submitted an attractive bid for $19.14 per thousand board feet, provided the government could offload 75,000 board feet per day from their freighters. Grunsky and Hecker declined to accept Douglas fir for drop siding on buildings because they feared it wouldn't stand up to the tropical weather. They decided to purchase 150,000 board feet of drop siding and 400,000 board feet of general lumber from the Continental Lumber Co. of Houston, who submitted the lowest bid for Southern Yellow Pine. Early in 1905 they began ordering lumber from J. J. Moore & Co. in San Francisco, with an initial shipment of 3.2 million board feet for $63,181, to be shipped in either American or foreign vessels.

Competitive bids for coal had resulted in prices of $5.75 per gross ton delivered to Cristobal, and $6 per gross ton to any location along the Panama Railroad, as far as Panama City. During 1905, the ICC ordered increasing volumes of coal, beginning with 3,000 tons during the first quarter, followed by 5,000 in the second, 9,300 in the third, and 12,500 tons in the 4th quarter. At the end of his term in late June 1905, Wallace had only succeeded in constructing 150 new buildings and renovating 336 old French structures. It was a poor showing for a project that envisioned something between 25,000 to 50,000 workers.

An unwieldy work force. Wallace had supervised the design of numerous rail projects for the ICR, but the construction had been parceled out to private contractors in hundreds of separate contracts. Railroads hadn't managed the actual construction of their lines since the early days of transcontinental railroad construction (between 1861-86), when each mile of rail brought huge tracts of public land into the ownership of the rail lines. During Wallace's career railroads hired engineers on a case-by case basis to lay out new lines with acceptable grades and curvature, and only retained them to oversee construction of specific structures that might require specialized expertise, such as the sinking of caissons using compressed air or the erection of false work around active rights-of-way. In those days bridge iron or steel superstructures were erected by their fabricators, dominated by seven bridge manufacturers clustered along the banks of the Ohio River near Pittsburgh.

In this regard Wallace's challenges in Panama were without precedent. His first major requisition was to purchase 1,000 rail dump cars with capacities of 20 to 40 cubic yards each. This request was approved by the Commission on September 3rd, but subject to receiving three competitive bids. 500 dump cars were ordered two days later. The Commission also requested alternative bids for transport of said rail cars to the Isthmus, and set up on tracks at Cristobal after shipment. On October 20th, Wallace requested the Commission purchase 1,000 dump cars and 500 flat cars, his largest requisition to date. On the 24th, the Commission approved the order of 500 flat cars from the American Car & Foundry Co., but requested changes in specifications for bids on the dump cars, rejecting all those that had been offered. The next day, Wallace presented a proposal from the Standard Steel Car Company to furnish 200 flat cars with the modified specifications for just $875 per car, with an option to purchase 500 more within 90 days. This revised bid was approved, and the cars began to arrive at Cristobal on April 1, 1905.

In January 1905, the Commission approved the purchase of 12 Ingoldsby patent dump cars produced by the Pullman Co. of Chicago. Wallace requested the purchase of the latest dump cars that employed an air dumping device, which operated off the train's compressed air braking system. Desperate for more flat cars, Wallace then asked the Executive Committee to purchase 100 steel flat cars for the Panama Railroad, which could be used interchangeably with the construction flat cars. All of the locomotives and railroad rolling stock had to be specially configured for the Panama Railroad's unique five foot track gage (since the Civil War the United States had employed a "standard gage" of 4 feet 8-1/2 inches).

Wallace continued to request the services of key men he had worked with previously, such as F. B. Harriman, a Division Superintendent with the Illinois Central, William J. Karner, a civil engineer with the Illinois Central in Chicago, E. A. Courtney of the Illinois Central, W. E. Angier of Thebes, Illinois as a bridge engineer, and his nephew, supervising architect M. O. Johnson.

Fairness versus efficiency. On August 18th, the Isthmian Canal Commission unanimously passed a resolution that *"all purchases of machinery, materials, and supplies would only be made after "due advertisement, inviting public competition"* unless there was some sort of emergency. Wallace still couldn't get the water pipe he needed for basic sanitation and water supply to the workers housing, or anywhere else. Essential supplies and materials were requested through the proper channels, then somehow lost or forgotten in the many layers of approval that each expenditure required (in the wake of Admiral Walker's forced resignation in late March 1905, more than 160 unfilled requisitions were collected from his cluttered desk).

Wallace soon found that he was battling the ICC's cumbersome approval process for more control and streamlining of the requisition process, more of a say in accepting or denying quality workers, and organizational and logistical problems that seemed to get worse with each passing month. Although he had held numerous positions of responsibility within engineering societies and organizations, he was not an inspiring leader.

The 3,500 workers on the project by November 1904 never saw much of him and as their ranks continued to swell, the signs of things going more and more out-of-control began to emerge. Work crews were suddenly shifted from one activity to another, often in locations far from where they had been working. These sudden shifts were usually made without any explanation and began to hamper morale. Wallace was just trying to keep all of the plates spinning, largely by himself.

In the fall of 1904 Wallace learned that exploratory borings at the Bohio Dam site along the Chagres River indicated that competent bedrock lay more than 168 feet below sea level, a revelation that surprised all of the engineers. With each passing month he was discovering more "changed conditions," which reminded everyone of the disastrous problems that had broken the French years before. Wallace was of the mind to build a sea level canal because, in the end, it would offer the greatest utilization for ship traffic, allowing a maximum number of vessels to pass through the isthmus in any 24 hour period. This predilection towards efficiency was typical of those men associated with American railroads at the time, because theirs was a very competitive business, with people being hired and discharged more or less continuously.

Appealing to Secretary Taft and the President. Wallace became increasingly frustrated with all of the red tape. He was a gentleman engineer, not a fighter. He bid his time, and waited for an appropriate opportunity to air his concerns privately to Secretary Taft. He returned to Washington for a visit in September 1904. When he returned in November he was accompanied by his wife Sadie, just in time for Taft's first visit to Panama, a 10-day inspection tour.

Taft was accompanied by his wife Nellie and they stayed with the Wallace's at the Chief Engineer's residence in Panama City (Figure 19). The two couples got

Figure 19 – War Secretary William Howard Taft and his wife Nellie (Helen) playing cards while traveling to Panama for his first trip to the Isthmus in November 1905. During their 10 day visit they stayed with John and Sadie Wallace at their home in Panama City (Library of Congress).

along well, and during the train rides between the various stops along the line of construction, Wallace began making his case for restructuring the unwieldy Isthmian Canal Commission. Despite its good intentions, it was slowing strangling the project and engendering a foreboding sense of frustration with the administration and the bureaucracy in Washington, which was sure to run the project over-budget. Wallace pushed Taft for centralized control by his own office, down in Panama; otherwise *"the work just wouldn't get completed until one or two presidential administrations into the future."*

Wallace also politely expressed his preference for a sea level canal to Taft, which differed in concept with the locked canal of 1903, approved by Roosevelt and the first ICC. Wallace pointed out that the kingpin structure of that scheme was the massive earthen dam at Bohio, which now appeared unworkable. When the Tafts departed he seemed convinced of Wallace's recommendations, and Wallace began a regular correspondence with Taft which succeeded in bringing about major changes to the canal project.

Taft began conferred with President Roosevelt throughout late November and December 1904, appraising him of the various problems with the existing plans, and the unwieldy nature of the ICC's current structure, governed by Admiral Walker, who was now almost 70 years old. Wallace waited hopefully while Roosevelt took matters into account and began formulating an improved organization. In January 1905 Roosevelt announced that he was going to restructure the ICC. On March 29th he asked for the current commissioner's resignations and between April 1st and 3rd Taft announced the changes to the Commission structure and members.

New Commission structure. In late March 2005, Wallace and Sadie had traveled to New York and he was in Washington on April 3rd when Taft announced that the ICC would henceforth be guided by an executive committee of just three men: railroad executive Theodore P. Shonts, the new head of the Commission; ICC general counsel Charles E. Magoon (who had moved to Panama in late July 1904) would replace General Davis as the new Military Governor; and Wallace, whose title became "Chief Engineer." Wallace had, in fact, recommended Shonts to Taft; they were both graduates of Monmouth College who had attained considerable success in the engineering and operation of railroads. In a surprise move, it was announced that Shonts would be paid $30,000 per annum, $5,000 more than Wallace. In late March, Wallace and Sadie returned to the States for ICC meetings in Washington with Taft. Taft urged Wallace to return to Panama with haste to prepare for the arrival of Shonts and Magoon, but he demurred, taking an extended vacation to see family in Illinois. His absence from the isthmus stretched almost to two months.

The two new members of the ICC's reorganized Executive Committee converged on Cristobal, arriving by May 25th. The new structure gave Wallace the ability to promulgate requisitions for supplies and qualified personnel in a more timely manner (described under "Third Isthmian Canal Commission, below). Taft and Roosevelt were hopeful that transfer of decision making power from Washington to Panama would invigorate the construction process, and get the project up and going, with more tangible results. What they underestimated was the general panic that swept the isthmus each time a deadly outbreak of yellow fever broke out.

Discouragement and panic. In the late summer and early fall of 1904 Wallace witnessed his personal assistant, William Karner, taken by malaria, followed by his valet, then by his personal cook, all inherited from the French occupation and believed to be more immune to the tropical diseases. Wallace initially doubted the scale of effort Colonel Gorgas and his sanitation department was exercising to combat malaria and yellow fever, which was requiring 35% of the project's annual budget in the early going (described below, under William C. Gorgas).

The incidence of yellow fever was running about 30 cases per month, but this was expected to worsen during the coming summer months. In April, Governor Davis came down with malaria and was unable to continue. While Wallace was away in April and May an outbreak of yellow fever had struck the employees of the Administration Building in Ancon. One of those quickly taken was the Canal Zone (CZ) Supervising Chief Architect, 29-year old M. O. Johnson, who was Wallace's nephew, who had worked for him at the Illinois Central Railroad in Chicago. In early July 1904 Wallace had enticed him to come to Panama to work on the canal, at a salary of $3,600 per year. He allowed Johnson to be buried at Corozal in one of the Maxwell sheet metal caskets he had shipped to Panama for himself and Sadie, in case either of them should fall victim to disease.

The next to fall was Wallace's auditor, then the wife of Wallace's personal secretary, followed by one of the executive male secretaries. During May 1905, 60 people from the administration building were transferred to a vacant structure next to the hospital at Ancon. Panic swept through the leaderless halls of the project, and in less than two weeks over 200 CZ employees resigned and boarded steamers bound for New York. Surgeon Colonel Gorgas issued orders to keep all of the screens closed on the administration building, and had his sanitation workers repeatedly spray the building with insecticides (Gorgas ordered more than 120 tons of Park Davis' insect powder). There were now 4,000 men working on the canal, of which, about 700 were Americans, and 1800 Jamaicans.

The Wallace's returned at the end of May, during the height of the panic. More than 1,000 people had been admitted to the Zone hospitals. By the end of May, only seven of the 18 clerks who had been working for Wallace in the administration building when he departed in late April were still on the job. The remainder had been stricken with fever (Lewiston Evening Journal, 1905). In April, May, and June 1905 about three-quarters of the white work force fled Panama.

It was too much for Sadie Wallace to bear. What good was $25,000 per year when everyone was dying or resigning? During their recent travels Wallace had received some feelers about a respectable corporate position in New York. So many resignations followed that the ships were filling with returning passengers as fast as they could be loaded. The clincher was most probably the stunning news that one of the canal workers had been diagnosed with bubonic plague. Wallace had been back on the isthmus just three weeks when he suddenly cabled Taft, informing him that he had "complicated business" which necessitated his returning to New York, immediately.

Resignation and embarrassment. Wallace's met Taft at the Manhattan Hotel in New York on June 25th. Taft brought attorney William Nelson Cromwell "as a witness" to the events that were to be discussed (Cromwell had represented the French Compagnie Nouvelle du Canal de Panama in the sale of their holdings to the United States). What Wallace didn't know was that Taft had been tipped off by Governor Charles E. Magoon about how Wallace intended to negotiate a higher salary; by threatening to resign (he had informed Magoon that he had employed this technique to his advantage with the Illinois Central Railroad). It is doubtful that salary was the motivator so much as it served as a respectable means of escape. The entire episode was out-of-character for Wallace, considering his behavior before and after that fateful day.

Their meeting was not amiable. Wallace opened with the news that he had been offered a "very attractive position" in the states, which would pay $50,000 to $60,000 per annum (more than double his current salary). Taft reminded Wallace that he was receiving 67% more compensation from the ICC than he had from the Illinois Central ($15,000 per annum), and that he had gone to great lengths to restructure the ICC in such a manner as to accede to the concerns Wallace had voiced over the previous year. Taft demeaned Wallace for abandoning his nation in the hour of crucial need, asserting *"For mere lucre you change your position overnight without thought of the embarrassing position in which you place your government by this action. By every principle of honor and duty you were bound to treat this subject differently."* Taft's words were carefully scripted and majestic, like some sort of court order (he was a former judge and would later serve as Chief Justice of the United States). Wallace responded by offering to remain in Panama in some capacity that would allow minimal upset to the construction schedule, but Taft demanded his immediate resignation. This was passed to President Roosevelt the next day, who made the formal announcement on June 28th.

What Wallace didn't realize was that the interaction had been carefully choreographed by Taft to protect the Roosevelt Administration from political fallout. A stenographer had been strategically positioned in an adjoining room recording everything that was said between the two men, like a legal proceeding. Wallace's choice of words suggest that he did not appreciate how his personal character would be compromised in the press to save the political face of the project. A typed transcript of their interchange was passed to the newspapers two days later, after Roosevelt had publicly accepted Wallace's resignation. The transcript soiled Wallace's professional reputation, at least for a time. No significant offers were immediately forthcoming, nor was it ever demonstrated that such a substantive offer ever really existed.

Several intimates blamed Wallace's dark mood on the fact that Shonts had been brought in at his recommendation, but placed over him, with a higher salary ($30,000), and that he alone viewed himself as necessarily *"the man in charge."* This may have figured into his decision, the abruptness of his about-face coming off a two month break must have been triggered by something more ominous. Wallace and Sadie had just lost a cherished relative who they had enticed to follow them to Panama, and they were now living a nightmare. They wanted out, immediately. The excuse offered by Wallace was the probably most honorable excuse he could conjure up at that moment.

His character demeaned in the newspapers, the Wallace's moved to Chicago where he found employment as a "confidential advisor" (his compensation not being made public) to the Chicago & North Western Railway for the design of their new terminals there. The following year (1907) he accepted the presidency of Westinghouse, Church, Kerr and Co. in New York, which may have been the position that enticed him off the isthmus the previous year. There he supervised the firm's design and construction of terminal and dock facilities, as well as hotels, office, and factory structures. He remained with them for about six years, supervising expenditures averaging about $10 million per year. In 1913 he returned to Chicago as an advisor on an array of projects, and for a time served as Chairman of the Chicago Railway Terminal Commission. In 1915, he became an independent consulting engineer and served on the boards of various companies and public commissions, mostly in Chicago and New York. On July 3, 1921, Wallace suffered a fatal heart attack at age 68 in Washington, DC, while attending Senate hearings in connection with the Board of Economics and Engineering, of which he was a member. In 1946 the Canal Zone issued a 25 cent stamp featuring Wallace (Figure 2).

THE ENGINEERS AND DOCTOR GORGAS

A military career. Many have argued that America's greatest contribution to come out of the Panama Canal was in the arena of sanitation, combating yellow fever and malaria. Surgeon Colonel William C. Gorgas (Figure 20) was the only Army officer who remained in the Canal Zone from the onset of the project in 1904 almost to its conclusion, in late 1913. Born in October 1854, he was eldest of six children of career Army officer Josiah Gorgas (West Point Class of 1841) and Amelia Gayle, the daughter of a former Alabama governor. Though a native of Dauphin County, Pennsylvania, the elder Gorgas sided with the Confederacy in 1861 and was appointed Chief of Ordnance by Jefferson Davis at the outbreak of hostilities, which Amelia and her son Willie witnessed because they were living at the Charleston Armory when Fort Sumter was attacked in April 1861. He was a lad of 10 when the family was living in Richmond with Confederate generals making frequent visits to see his father, a general in the Confederate Army.

Willie dreamed of a military career. After the war the family moved to Brierfield, Alabama, where his father tried to operate an iron manufacturing plant, which ended in bankruptcy. In September 1868, his father accepted a position run at the new University of the South in Sewanee, Tennessee, an Episcopal institution. The following July Willie enrolled in the preparatory department of the new college at the age of 14-1/2. Willie grew with the new school, and he received one of the first bachelor's degrees in August 1875. Willie was determined now to attend West Point, like his father. His father was now Vice Chancellor of the new college and he tried to dissuade his eldest son from the pomp and circumstance of a military career, which would have him shifting from one military post to another in a strict system of seniority that frustrated the brightest young officers. Josiah encouraged Willie to become an attorney, like his mother's brother in New Orleans. Willie moved to New Orleans and spent a year studying law, before his father permitted him to at last apply for admission to West Point. In this regard he received many favorable letters of recommendation, but he was unable to secure an appointment during the summer of 1876 and on October 4th he would turn 22, the upper age cutoff for plebe cadets at that time.

What he learned was that the Army needed doctors for their medical corps. His father approved of his attending medical school, hoping it would dissuade him from pursuing a military career. In September 1876, Willie began his medical studies at Bellevue Medical College of New York City, where he would be known by his initials "W.C." The next three years were difficult, requiring him to take out loans. He surprised himself by gaining a keen enthusiasm for his studies, which was noted by one of his professors, Dr. William H. Welch, who later became President of the American medical Association and an executive with the Rockefeller Foundation. He received his M.D. in 1879, and decided to do his internship at Bellevue Hospital. He joined the Army as a physician and was commissioned as a First Lieutenant in June 1880 (Figure 20), at a monthly salary of $133.

In August 1882, Gorgas was sent to Fort Brown, near Brownsville, Texas, where a number of people had been inflicted by an outbreak of the dreaded yellow fever. One of those struck with the disease was the sister-in-law of the post commander, Marie Cook Doughty of Columbus, Ohio. Gorgas soon fell ill to the disease as well, and the two convalesced together and fell in love. They married two years later, in September 1885. Being survivors of yellow fever, both were thereafter immune to the dreaded disease that would ultimately shape their destiny.

Studying yellow fever in Cuba. The Spanish American War erupted in late April 1898, and Gorgas was promoted to Major in July 1898. The war was quickly won and the following January, Gorgas was appointed Chief Health Officer of Havana, Cuba (Figure 21). Havana had been suffering from outbreaks of yellow fever and malaria during the recent American occupation. It was during this assignment that he met Cuban physician Carlos J. Finlay, an 1855 graduate of Jefferson Medical College in Philadelphia. Gorgas soon learned of Finlay's novel theory that yellow fever was transmitted by the Stegomyia (now called Aedes aegypti) mosquito, which he had been studying since 1881. Gorgas initially rejected his theory, reasoning that there were over 3,000 species of mosquitoes, and the chances of singling out just one particular type seemed preposterous. Gorgas busied himself supervising the improvement of general sanitation, believed by most people at that time to be the cause of yellow fever and malaria. Despite these efforts the incidences of yellow fever increased markedly, frustrating Gorgas.

Figure 20. Left – W.C. Gorgas as he appeared in 1880, when he received his commission as an Army Surgeon (National Library of Medicine). Middle – Captain Gorgas in 1895, when he was stationed at Fort Barrancas in Pensacola (US Army). Right - Surgeon Colonel William Crawford Gorgas as he appeared during construction of the canal.

In May 1900, the Army dispatched a Medical Board to Cuba to study the yellow fever epidemic (Figure 21). This board was led by Surgeon Major Walter Reed, assisted by Dr. James Carroll, Dr. Aristides Agramonte, and Dr. Jesse W. Lazear. They began working with Gorgas, Dr. Henry Rose Carter of the Marine Hospital Service, and Dr. Finlay. Carter had been studying yellow fever in Mississippi and had concluded that it was carried by particular species of mosquito, which only transmitted the disease after a period of incubation lasting 10 to 14 days. The synergy of the group soon grew and a systematic series of tests were conceived to prove or deny the mosquito theory, which began in August 1900. Only two of the 11 people bitten by Finlay's mosquitoes developed yellow fever, one of these being Dr. Carroll. A few weeks later Dr. Lazear allowed himself to be bitten by one of Finlay's Stegomyia mosquitoes, which had bitten a yellow fever patient within three days of his becoming ill, and that 10 days had then passed before it had bit Dr. Lazear (all of this data being recorded and Gorgas). Lazaer immediately came down with yellow fever and, despite the group's best efforts, died on September 25th.

Upon Lazear's death Walter Reed immediately returned to Cuba and set up a comprehensive series of tests using Stegomyia mosquitoes and yellow fever patients, paying newcomers to Cuba (who had never previously been infected) $250 to participate in the study. In a controlled structure Reed deposited filthy rags, clothing, and bedding of yellow fever victims, carefully excluding any mosquitoes. A portion of the subjects were placed in this horrid environment, in near continuous contact with the vomit, blood, and excrement of yellow fever victims, but none of them came down with the disease. The only subjects who did contract yellow fever were those bitten by Stegomyia mosquitoes that had been allowed to bite yellow fever sufferers more than 10 to 14 days previously, as assuaged by Dr. Carter.

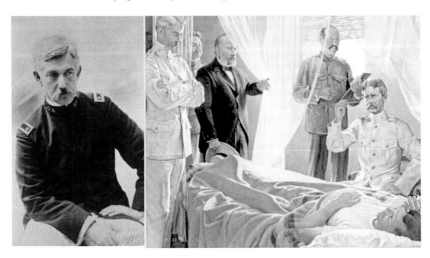

Figure 21. Left image shows Gorgas as a major while serving as Chief Health Officer of Havana in 1899-1902. Right - Painting by Robert Thom showing members of the Yellow Fever Commission in Havana, from left: Major W.C. Gorgas, Dr. A. Agramonte, Dr. Carlos J. Finlay, Dr. James Carroll, and Major Walter Reed (University of Michigan).

The incubation theory was then tested using syringe injections, taking the blood of someone who had contracted yellow fever within the previous 72 hours and injecting a tiny amount of that blood into a healthy subject. Within 10 to 14 days that subject contracted the fever. The germ was also found to be submicroscopic, unlike the malaria parasite. If the infected blood was heated to 55 degrees Centigrade and then injected into a subject, they did not contract the disease. Gorgas and the other doctors were convinced that they now knew the means by which the deadly yellow fever was transmitted. The abrupt eradication of the yellow fever epidemic in Havana brought Reed, Carter, Carroll, Finlay, and Gorgas international acclaim as physicians. Reed died of peritonitis in November 1902, while serving in Washington, DC. He was just 51 years old. His leadership in the battle against yellow fever and malaria would prove invaluable in the American effort at Panama.

Selection as Chief Health Officer for the Panama Canal. Prior to the President's appointment of the second Isthmian Canal Commissioners in early 1904, the American Medical Association (AMA) had lobbied the White House to appoint a chief sanitary officer as one of the commissioners. Disappointed that their recommendation was ignored, Dr. William Henry Welch, Dean of the Medical College at Johns Hopkins University, led the AMA's newly formed Committee on Medical Legislation, which petitioned the White House to meet with President Roosevelt. After several hours of discussions they succeeded in convincing him that if the United States was committing itself to the Panamanian Isthmus, the enterprise should first seek to tackle the sources of disease that had crippled the French and plagued the American military in the Caribbean during the recent war with Spain and the military occupations that followed (Gorgas and Hendrick, 1924).

Roosevelt's decision was no doubt influenced by his recent military experiences in Cuba, where malaria and yellow fever dealt far greater harm to the American troops than Spanish bullets. Roosevelt's close friend and mentor General Leonard Wood (the first regimental commander of the famed Rough Riders regiment) had been elevated to the role of Military Governor of Cuba after the cessation of hostilities because of his training as an Army Surgeon.

Roosevelt instructed Admiral Walker to *"find the very best medical man in the country"* to supervise the sanitation work in Panama, but to consult first with Dr. Welch (McCullough, 1977). Welch's choice was Gorgas, whom Welch had mentored during the former's internship at Bellevue Hospital in New York in 1879-80, before he entered the Army. In 1902 Gorgas was transferred from Havana back to Washington to continue working on protocols for eradicating tropical disease in anticipation of the Americans constructing a transoceanic canal in Panama or Nicaragua. In 1903 then Major Gorgas was promoted two ranks, skipping the grade of Lieutenant Colonel, to full Colonel and Assistant Surgeon General in the Army's Medical Corps, by a special act of Congress. This was in appreciation of his work in Havana controlling infectious diseases.

On April 10th 1904, Roosevelt appointed Colonel Gorgas as Chief Health Officer of the Canal Zone, and in late April-early May Gorgas led a group of medical experts on a tour in conditions in Panama, along with Admiral Walker of the rest of the Isthmian Canal Commission. This group included John W. Ross, Medical Director of the U.S. Navy, Captain Cassius E. Gillette of the Army Corps of Engineers, and Major Louis A. LaGarde of the Army's Medical Corps. They inspected the line of the French canal workings and prepared a plan for the sanitation of the new American Canal Zone, as well as the cities of Colon and Panama City (Gorgas, 1904). The results of these observations and the group's recommendations were then presented by Gorgas to Admiral Walker in Washington.

Gorgas immediately issued a number of requests for window screens, sanitary cloth, disinfectants and such that he knew were in short supply, but was rebuffed by Walker, who instructed him to *"go on with your party to Panama, look the situation over and see what you need. Then make out an order for it. We'll see that you get it promptly*" (Gibson, 1950). This was Gorgas' first clue that his new assignment might be more frustrating than he might have hoped. On June 4th 1904, Gorgas and Marie sailed for Panama and on June 30th he was officially given charge of the new Sanitary Department. He would live in Panama longer than any of the other senior managers, remaining there till late October 1913.

Early battles with engineers over sanitation. When he moved to Panama Gorgas was already known in medical circles for his pioneering research in Florida and Cuba, working with fellow Army physician Walter Reed, who led the team that postulated and confirmed the theory that yellow fever is transmitted by the female anopheles mosquito in 1900. When he arrived on the isthmus he soon learned that in addition to malaria and yellow fever, there were numerous cases of tuberculosis, cholera, smallpox, and diphtheria, as well as a few cases of bubonic plague. During the height of the French efforts to construct a canal (1882-99) yellow fever and malaria had killed an estimated ten to twenty thousand people (Gorgas, 1915).

The challenges Gorgas faced in Panama were without precedent, but he accepted his new responsibilities as Director of the Sanitary Department with zeal and determination. His primary goal was to abate the widespread transmission of yellow fever and malaria by controlling the mosquitoes that conveyed those diseases, at a time when there was still considerable skepticism regarding such measures, which many did not believe. During the summer of 1904, Gorgas began submitting a string of requisitions asking for the same sorts of supplies he had employed in Havana, without any forthcoming approval from the seven members of the ICC, who didn't consider his requests to be urgent because no apparent outbreaks of disease had been reported (Gibson, 1950). That began to change on July 12th, when the first case of yellow fever was reported, with the victim succumbing two days later. This was followed by a second case two weeks later.

On August 3, 1904, Admiral Walker, Commissioners Harrod, Burr, and Grunsky arrived in Ancon to view what progress was being made on the canal. They would remain until September 10th. Gorgas took the opportunity to press his case for more drainage and sanitation issues the commission, but was rebuffed by Carl Grunsky of the Committee on Sanitation (Figure 22). He informed Gorgas that before his requests could be approved the commission needed to perfect a suitable plan of organization for the entire project, which took another 3-1/2 weeks. A review of the minutes of the Commission's meetings reveals that just about everything Gorgas requested in 1904 was either denied or markedly reduced (Isthmian Canal Commission, 1908). It's a wonder he held on and persevered against such harsh treatment. It appears that in the early going the ICC Commissioners exercised very little respect or regard for Gorgas, despite his record of accomplishment (and there were but a handful of Surgeon Colonels in the entire Army at that time). On August 13th, Gorgas submitted a request to increase the salaries of some of his senior medical staff, including an increase of his own salary from $7500 to $10,000 per annum, while the Commission was meeting in Ancon. This proposal, like so many others, was promptly rejected.

On August 28th, the ICC adopted Grunsky's proposed reorganization of the project, which subordinated Gorgas beneath all seven members of the commission. Grunsky advised Governor Davis to refrain from allowing Gorgas control of draining all standing water and spraying oil across Colon or Panama City until such a time that diplomatic issues could be worked out with those cities that were technically under the sovereignty of the Panamanians. This precluded access to many of the areas where mosquitoes were breeding for another four to five months, during which time yellow fever was able to spread into the Canal Zone, reaching crisis proportions by the spring of 1905 (described previously).

In large measure this failure to act was because Admiral Walker, General Davis, and Grunsky were convinced that something as simple as mosquitoes could not possibly be responsible for the spread of such deadly diseases as yellow fever or malaria (Gibson, 1950). Grunsky was not approving numerous requisitions being promulgated by Gorgas. One example was a request for 4,000 pounds of "old newspapers," which was denied because he thought it was reading material for hospital patients, and kindly pointed out that local newspapers could be fetched for far lower costs, than shipping such bulk by steamers, etc. In point of fact, Gorgas wanted the newsprint to seal cracks and crevices of structures prior to fumigation with sulfur and pyrethrum, so more cables passed back and forth, causing more delays.

Part of the problem was that Gorgas was not a fighter, but the string of denials became increasingly vexing and eventually taxed his patience (in Havana his immediate superior had been General Leonard Wood, who being a physician, gave him everything he requested). When funds for construction of a research (pathology) laboratory and the purchase of badly needed ambulances were denied by Grunsky, a frustrated Gorgas turned to the American Medical Association (AMA). In February 1905, the AMA sent one of its former presidents, Dr. Charles A.L. Reed of Cincinnati, down to Panama to make an "independent investigation," which was submitted to Secretary of War William Howard Taft on March 1st (Reed and Taft were both from Cincinnati and knew one another well). A six-page summary of his report was also published in the *AMA Journal.* It was a scathing critique of Carl Grunsky, asserting that engineers *"without proper medical training"* were exercising oversight on Gorgas' activities, which were meant to save lives. Reed termed Grunsky's denials of requisitions as *"unnecessary and unreasonable restraints,"* displaying *"petty, almost despicable antagonism."* The AMA team was outraged that Gorgas's every move was being second-guessed by individuals without formal medical training. What they did not know was that Grunsky originally studied medicine in Germany before switching to engineering, and that this background and his role as the senior member of the Committee on Sanitation may have caused him to overstep sensible bounds of professional etiquette. It was the only dark mark on Grunsky's professional career, which included service as ASCE President in 1926

(Civil Engineering, 1934; Marx et al., 1935). Grunsky and the other members of the Second ICC were requested to tender their resignations by President Roosevelt March 1905, shortly after the AMA report appeared.

Gorgas's methods on trial. In late May 1905, the new Military Governor of the Canal Zone arrived, Charles E. Magoon (Figure 22). Neither he nor the Chairman of the new ICC Executive Committee, Theodore P. Shonts (Figure 22), believed that yellow fever was transmitted by mosquitoes. Like their predecessors, they questioned the enormity of expenditures being requested by the Sanitation Department, despite the fact that 38 patients had been admitted into the CZ hospital in Ancon the very month of their arrival (May 1905), and 22 had died during the first five months of the year. Then, as described earlier, things got worse. In June, the number of yellow fever cases increased to 62, with 19 deaths recorded. Worker morale began to plummet, and Chief Engineer Wallace and his wife suddenly departed, his rumored resignation being confirmed on June 28th.

Figure 22. Left – San Francisco civil engineer Carl Grunsky chaired the ICC's Committee on Sanitation (author's collection). Middle – Charles E. Magoon was the second Military Governor of the Canal Zone, who arrived in late May 1905. (Jackson & Son, 1911). Right – Railroad executive Theodore P. Shonts was appointed Chairman of the new Commission and oversaw the three man executive committee that approved all expenditures after mid-1905 (Jackson & Son, 1911).

Shonts' first move as the new chairman was to reorganize the Sanitary Department, replacing Gorgas with someone whose theories on disease prevention would be more in line with the status quo. Shonts' recommendation was approved by War Secretary Taft and passed onto President Roosevelt for his approval (Taft's approval of the request has always baffled historians because as he was a close friend of Dr. Reed, who was also from Cincinnati). Shonts recommended an osteopath friend who had "*been in the south*" and had "*seen yellow fever*" (Gibson, 1950). This candidate was rejected, but Shonts found another physician who was adamantly opposed to the "mosquito-transmission theory."

Roosevelt rejected these candidates, but favored the appointment of Dr. Hamilton Wright, who had achieved considerable acclaim performing similar sanitary work in the Straits Settlements of Malaysia. Unsure what to do, Roosevelt recalled his meeting with Dr. William H. Welch of Johns Hopkins the previous year, asking him if Dr. Wright might be a suitable replacement for Gorgas. Though respectful of Wright's abilities, he replied that Gorgas was the world's foremost expert on yellow fever, and that no one had battled the disease with more efficiency. Welch's forceful declaration gave the President pause. He then turned to Dr. Alexander S. Lambert of New York City, one of his hunting buddies, who had previously lobbied for Gorgas to be made a full-fledged member of the ICC. He asked Lambert to visit him at his home on Long Island. Roosevelt enjoined the doctor to speak frankly on what he thought of the controversy Gorgas was causing amongst the engineers in Panama. They talked details into the night. Lambert eventually succeeded in convincing Theodore Roosevelt that the traditional theories about bad odors, swamp gas, rotting vegetation and such were not based on any scientific evidence, but simple association with the environments that commonly bred mosquitoes, and that it was a particular type of mosquito (the female Stegomyia) that Gorgas had zeroed in on, which explained why sudden outbreaks of the disease could occur in a particular area, but never on the confines of a hospital, if the latter was provided with window screens. Lambert encouraged Roosevelt to let the engineers take care of engineering and the physician Dr. Gorgas take charge of the medicine, or risk the workers dying like flies, as had happened to the French. Roosevelt thanked Lambert and promised him he would keep Gorgas in Panama.

Theodore Shonts was then summoned to Washington for a private audience with the president. Roosevelt informed him that he was rejecting his request to discharge Colonel Gorgas and that he sought for both parties to cooperate with and be respectful of each other's technical abilities. He instructed Shonts to henceforth approve Gorgas' requisitions for supplies and to *"give him everything he wants,"* and let him, thereby, *"accept the burden for who lives and who dies in the great undertaking."* Shonts never rejected another request from Gorgas as long as he was on the ICC.

The seven step process of providing sanitation. Worker morale plummeted with the yellow fever outbreak of May and June 1905, followed by the sudden departure of Chief Engineer John F. Wallace. His replacement was the mercurial John F. Stevens, a self-taught civil engineer who had personally supervised the construction of more miles of railroad in Canada and the United States than any other man, before, or since (Budd, 1944). Stevens was pragmatic, judging men by the results they achieved, and little else. Upon his arrival Stevens detected defeatism and low morale among the workers, especially in the American ranks.

Unlike his predecessor, Stevens accepted Gorgas's premise that mosquitoes were an enemy of the project that needed to be eradicated. If infectious diseases took hold of any significant proportion of the work force, Stevens realized the low morale could easily doom the mammoth project.

Stevens ceased work on the excavations at Culebra and focused first on improving the living conditions, which included housing and sanitation. He encouraged Gorgas in his zealous battle with the mosquitoes. Gorgas was finally free to implement the same sorts of preventative measures he had used in Havana to rid the isthmus of yellow fever and malaria. He began by dividing up the Canal Zone into 25 sanitary districts, each between 15 and 35 square miles. A lead sanitary inspector was then assigned to each district, some containing only a few hundred residents, while others contained as many as ten thousand.

Gorgas implemented a seven-step program (Gorgas, 1904), which included drainage, trimming of brush and grass, oiling of ponds and swamps, larviciding, drinking of quinine, liberal application of window screens, and killing adult mosquitoes. All of these activities required ongoing maintenance and vigilance (Figure 23).

Figure 23. Left – drainage channels were excavated and maintained around all living areas (Panama Canal Authority); Middle – Worker swabbing a lined drainage ditch with oil to dissuade growth of mosquito larvae; and right; Worker spraying oil onto ponded water in a drainage ditch (Library of Congress).

In early 1905, Gorgas oversaw a work force of about 400 Sanitary Department workers, virtually all of whom were unskilled laborers from Jamaica and Barbados. By the end of 1905 that figure had swelled to a force of almost 1,000 workers. As it had been in Havana, the initial focus was on the provision of drainage, which needed to be improved just about anywhere they looked. Their central aim was to do whatever was needed to preclude ponding because mosquito larvae only bred in stagnant water (Figures 23 and 24). All pools within 200 yards of villages and 100 yards of any houses were drained. Subsurface drainage was preferred wherever possible, to prevent pools. Where the grades were gentle or the structures clustered close together, his workers would line the drainage ditches with concrete, or install buried drain tiles. Covers were placed over water barrels, roof gutters were sloped to preclude ponding, and low areas in filled to preclude ponding (a major problem with the areas receiving fill spoils). Drainage inspectors were assigned to each sector to make sure the drainage integrity was maintained.

Figure 24. Spreading larvicide along a roadside ditch to combat mosquitoes. These wagons and their spreaders increased in size as the project was constructed.

Once the living areas were spruced up and the runoff was conveyed to suitable points of discharge, the sanitary department moved farther afield, to begin draining the infamous Panama swamps, which dominated the coastal river plains on both sides of the isthmus, but over a much broader area between Colon and Gatun.

As shown in Figure 25, all brush cover was cleared and grass over 12 inches tall cut within the same zone as the drainage (within 200 yards of villages and within 100 yards around individual homes). The belief at that time was that mosquitoes would seldom venture over 100 yards across dry, open areas. Subsequent research has shown that the average female mosquito lives within one to two miles of their larval water source, but has been found as much as 75 miles distant, probably aided by wind during storms.

Figure 25. ICC offices and residences of the senior managers were re-positioned by John F. Stevens to this hilltop in Culebra in the fall of 1905, closer to the center of the work. The Sanitary Department cleared the brush around these hamlets, cut the grass between buildings, maintained drainage, and fumigated structures to root out mosquitoes.

In those locations where drainage by gravity flow was not possible, oil was dropped on the water surfaces to kill mosquito larvae. Some areas were inaccessible because of their location, such as the temporary benches of the massive excavations over the nine miles of canal across the Continental Divide at Culebra. The floors of these excavations were pervasively congested with steam shovels, muck trains, drill rigs, push carts, air and water hoses, and thousands of workers. In those areas all the sanitation workers could do was to spray crude oil thinned with kerosene on the surfaces of stagnant ponds, which reduced the dissolved oxygen content of the water, killing mosquito larvae (Gorgas had employed this technique with great effect in Cuba).

When the flow of water precluded oiling, the workers resorted to larviciding. At the turn of the 20th century there were no commercial insecticides, so a production facility was set up at Ancon to produce suitable mixtures. In these situations the sanitation workers placed small ash cans (used to remove ash from fireplaces) containing a mixture of carbolic acid caustic soda, and resin. A small hole was drilled in the bottom of the ash can, just wide enough for a wick, through which the mixture would seep downward by capillarity and be taken into whatever pool or watercourse lay below. Drop by drop this deadly mixture would be dissolved in the water and carried downstream, killing mosquito larvae in the stagnant pools.

The prevalent prophylaxis for malaria in that era was a daily dose of quinine. Quinine is a crystalline alkaloid that reduced fevers, quelled muscle pain, and served as a crude anti-inflammatory. Quinine was extracted from the bark of the cinchona tree, which is native to Bolivia and Peru. It was first exported to Europe in 1631 to combat an outbreak of malaria in Rome. Gorgas ordered quinine in prodigious quantities, and it was dispensed to all workers along the line of construction through 21 dispensaries. Quinine dispensers were also positioned on all hotel and canteen mess tables. On average, half the 100,000 workers took a prophylactic daily dose, which came to be known as the "Panama Cocktail."

Measure #6 was screening or doors, windows, and porches (Figure 26). These screens were made of copper wire (to better resist corrosion), and were to be affixed to all structures where people worked or slept (Gorgas ordered $90,000 worth of copper screens in 1905-06). Not everyone believed that screens were necessary, but after the yellow fever outbreak of 1905, resistance diminished. The volume of screening taxed American factory capacity for several years, but it kept flowing to the Atlantic side of the isthmus, and became fashionable in Florida and other parts of the South soon thereafter, as the public became more convinced of the health benefits.

Figure 26. The YMCA Club at Culebra was typical of the structures in the Canal Zone, surrounded by wide porches and wrapped by steel or copper screens, to keep the mosquitoes out.

The last and most important mitigation measure was the eradication of adult mosquitoes. Adult mosquitoes typically remained in tents or houses after feeding on humans, most often in the evening hours or during the night. Gorgas had men specially trained to ferret out adult mosquitoes from these structures during the daytime. These workers would enter a building, seal off all the usual openings, beneath doors, around windows, vents, and so forth, and then fill the rooms with pyrethrum, which put the flying bugs to sleep, causing them to drop from ceilings, drapes, and other places of concealment. Between mid-1905 and mid-1906 Gorgas received 330 tons of sulfur and 120 tons of pyrethrum, the total output of the later for the entire United States.

After fumigation sanitation workers would collect the adult mosquitoes for examination at the Board of Health Laboratory, supervised by Dr. Samuel T. Darling. This activity, probably more than any other, had an immediate and dramatic impact on lowering the incidence of yellow fever and malaria. These treatments continued until no more mosquitoes were detected.

One of the most vexing aspects of tropical development were the copper water pans or Terra Cotta dishes placed around every supporting post beneath every structure and even every woody plant (Figure 27). These were filled to water to preclude the ascendency of voracious wood eating ants, for which the isthmus was infamous. Smaller versions were then used inside structures, to preclude ants from climbing bed posts or table legs. Gorgas's workers soon discovered that the stagnant water placed in these "ant guards" was a prime location for nurturing mosquito larvae. Each pan had to be treated with a drop or two of oil, then periodically refilled and checked for larvae. One of the more expensive techniques for controlling ants and cockroaches was a thin line of powdered boric acid, which became staple method of pest control in the CZ for almost a century thereafter.

Figure 27. Terra Cotta plates filled with water were employed as "ant guards" around plants at the Ancon Hospital (Le Prince, 1916).

Gorgas's sanitation workers often improvised techniques to rid the isthmus of not just mosquitoes, but ants, cockroaches, and rats as well. They even outlawed the domestication of cattle in the lowlands because stagnant water was found in their hoof prints, which were about the size of a cup of water.

During his first year of operation (mid-1904 to mid-1905), Gorgas' expenditures only totaled $50,000. After Stevens' arrival in late July 1905 the expenditures for the next 12 months jumped to $2 million, an increase of 4,000 percent. The results of Gorgas's efforts were tangible. From the record high of 62 cases of yellow fever and 19 deaths in June 1905, only 42 cases were recorded in July, 13 of which were fatal. In August the figures continued to decline to 27 cases and nine deaths. By December there was only one case of yellow fever, which was not fatal. Only one case of yellow fever would be recorded in all of 1906 (believed to have originated elsewhere) and it was the last one recorded during the remainder of the project. In March 1907, Colonel Gorgas was finally appointed to the Isthmian Canal Commission as a full member and his compensation was elevated from $4000 to $10,000 per annum. He continued to shoulder the responsibility for the sanitation work on the canal project (described later).

The campaign against malaria. Gorgas won the war with yellow fever, but continued to battle malaria and pneumonia. In 1897, British Army Surgeon Major Ronald Ross had succeeded in discovering the malaria parasite inside the gut of *Anopheles* mosquitoes in India, which appeared in the December 1897 issue of the *British Medical Journal.* That same year, an Italian physician named Giovanni Battista Grassi and his colleagues had established the developmental stages of malaria parasites in female *Anopheles* mosquitoes. Although Gorgas was not yet convinced of the new "mosquito theories," the results of drainage improvements in the tropics were far too effective to ignore, and these influenced Gorgas's preventive work in Havana.

Gorgas knew from his experience in Havana that the best way to reduce the incidence of malaria was mosquito abatement. The antimalarial work of his Sanitary Department focused on rural areas along the line of the Panama Railroad and canal, between Colon and Panama City. Within this roughly 50 mile strip there lived about 80,000 people within a half miles of the canal, within 30 villages or work camps. When the project commenced in 1904, Gorgas estimated that one-sixth of the population of Colon were suffering from malaria.

In 1906, the Sanitary Department spent just over $2 million and the Ancon Hospital was staffed by 470 people. More than a dozen subordinate medical facilities were built elsewhere along the path of the canal, and all health care was provided to railroad and CZ employees free of charge.

That same year (1906), there were 821 cases of malaria among the employees of the ICC and the Panama Railroad. 224 succumbed to the disease, and the death rate peaked at 16.21 per 1,000 employees in July 1906. 94% of these were blacks, mostly Jamaicans. The following year (1907), Gorgas's efforts improving drainage began to be felt, as the morbidity rate dropped 50%, to 424 cases per 1,000 employees, with the mortality rate falling to 3.92 per 1,000. In 1908, there were 282 cases of malaria per 1,000 employees. By December 1909, the malaria incidence dropped to just 2.58 per 1,000 (about 1.6% of the workforce). By 1913, the morbidity rate decreased to 76 per 1,000 employees and the death rate to just 0.37 per 1,000 employees.

The percentage of the total work force that was hospitalized for malaria illustrates the effectiveness of Gorgas's program. 9.6% of the project's workforce was hospitalized in 1905; 5.7% in 1906; 1.8% in 1907; 3.0% in 1908; and 1.6% in 1909. It was not until 1921 that no fatal cases of malaria were recorded for any CZ employees or their dependents.

Arguments with Goethals. Gorgas's inability to forestall malaria as successfully in Panama as he had in Cuba was a sore point for him to ponder. Gorgas privately blamed Colonel George W. Goethals (Figure 28), who assumed control of the canal project on April 1, 1907 (Gorgas, 1915). In 1908, Goethals re-organized the canal project, placing everything under his direct command, akin to the structure of a field army. Like most generals, Goethals viewed himself as the person who made the key decisions and issued the orders. Any subordinate who disagreed with him was suspected of being disloyal. In a move to seek greater efficiency much of the maintenance work previously executed by Gorgas' Sanitary Department was shifted to the Quartermaster Department, such as grass mowing and clearing brush cover. Gorgas was so cheerful and good-natured that neither Goethals nor Taft assuaged him to be an able administrator, keeping the men "on their toes" and hard at their assigned tasks, from dawn till dusk.

To Goethals the engineer, it made no difference who mowed the grass or cleared the brush; he did not see why such activities should be directed by a physician (Goethals, 1915). Gorgas protested, because his inspectors would no longer control the landscape maintenance. Goethals held his ground, citing a substantial cost savings. He later justified his decision by citing statistics that the incidence of malaria continued to decline across the isthmus during the first six months of his consolidating the landscape maintenance. In his way of thinking, it mattered not who supervised such work, just so long as it was accomplished (Goethals, 1915). In the midst of these petty struggles, in 1909 Gorgas was elected President of the American Medical Association, the first Army physician to hold that office, which reflected the respect he had garnered among his medical peers.

Gorgas's efforts to reduce the incidence of pneumonia were not nearly as successful. In October 1905 26 black workers died of pneumonia. In 1906 pneumonia claimed 413 lives, 95% of these being the black workers, mostly from Jamaica, which outnumbered the whites by a ratio of approximately 3:1. This was thought to be related to substandard billeting of black workers, but housing conditions for the black workers improved each year. By 1913, when there were more than twice as many employees as in 1906, only 47 people succumbed to pneumonia. Some of these problems were later traced to poor diets because of the inflated cost of food on the isthmus. During Gorgas' tenure as Surgeon General of the U.S. Army during the First World War, pneumonia would claim more soldiers' lives (40,000) than any other cause, including combat.

Figure 28. Colonels Gorgas (left) and Goethals (right) at Culebra in 1909. Gorgas was four years older and senior in date of rank, but Goethals had responsible charge of the entire project and nobody seriously challenged him during the last seven years of the project (Nichols Collection-Linda Hall Library).

According to Marie Gorgas (Gorgas and Hendrick, 1924), around 1912 Goethals came to Gorgas and asked him if he felt he could justify the enormous cost of his mosquito abatement program, which was costing the government $3.50 per capita per year for each person living within ½ mile of the canal. Goethals then assuaged that that sum equaled about $10 per mosquito that Gorgas's men killed, which was more than a weeks' wages in those days. This seemed excessive to Goethals. Gorgas smilingly replied that "*one of those $10 expenditures might well have saved Goethals' life, and that his value to the successful completion of the project was worth many millions of dollars!*" Goethals never spoke to Gorgas again about his work on the canal project.

In June 1913, Gorgas was awarded an honorary doctorate by Columbia University and served as the commencement speaker. In 1914, Colonels William Crawford Gorgas and George Washington Goethals were awarded the first Public Welfare Medals issued by the National Academy of Sciences. The canal opened that August and Goethals was named the first civil Governor of the Canal Zone. The next year, Goethals published a book titled *Government of the Canal Zone* (Goethals, 1915). It contained a detail summary of the canal's construction, but glaringly omits any mention of Colonel Gorgas. Goethals chose instead to give credit for eradication of yellow fever to Sir Ronald Ross in India, and Doctors Reed, Lazear, Carroll, Agramonte, and Finlay working in Cuba! It was an intentional slight of Gorgas's contributions, especially in Havana, and a dark stain upon Goethals' character (his treatment of Gorgas has received harsh commentary in dozens of books and articles, especially by those in the medical community). Gorgas, by then serving as the Army's Surgeon General in Washington, DC, penned his own tome, titled *Sanitation in Panama* (Gorgas 1915), which presented his views of the challenges he faced during his 8-1/2 years in the Canal Zone, working under engineers and bureaucrats that seldom agreed with or appreciated him. He singled out Walter Reed, John F. Stevens, and Theodore Roosevelt as the three men most responsible for his success, and blamed Goethals for not supporting him fully in the war on disease and sanitation.

Surgeon General of the Army. In October 1913, Gorgas requested permission from Goethals to depart the Canal Zone to make a medical tour of South Africa at the bequest of the South African Chamber of Mines, concerned about mortality rates of Negro miners succumbing to pneumonia. Goethals approved the request and in late October, Gorgas, his wife Marie, Army surgeon Major Robert E. Noble, and Chief Pathologist Dr. Samuel Taylor Darling departed Panama on the *SS Ancon*, bound for New York. The party then traveled to London to do some research, and continued onto Capetown, arriving in Johannesburg in early December.

It was in Johannesburg that Gorgas learned of his appointment as Surgeon General of the Army by President Woodrow Wilson, to be effective January 16, 1914, with the rank of brigadier general. On March 14th, Congress approved his promotion to major general and his party returned to Washington in April. In August, the canal was completed and the First World War erupted in Serbia, but the United States remained neutral for the next 2-1/2 years. In 1916, Gorgas was entreated to serve as an advisor to the newly organized International Health Board, and he undertook a tour of South and Central America with a number of assistants to advise these countries on how to combat yellow fever.

Gorgas was never inclined towards the administration of subordinates, but he possessed superior bedside manner and remained fascinated by the scientific study of disease prevention after his association with Walter Reed's team in Havana, which he termed the "nadir of his career." In January 1917, he informed the Secretary of War, Newton D. Baker, that he wished to be placed on the retirement list, but the deterioration of relations with Germany altered these plans, and the

United States declared war on Germany in three months later, in April. Gorgas was soon absorbed in the administration of the Army's medical department during the World War; coordinating its activities with the Secretary of War, the Council of National Defense, the General Medical Board, and various Congressional committees. During this trying time he depended on his faithful subordinate from his days in Panama, Colonel Noble, who oversaw administration of the medical department during most of the war.

Death and legacy. General Gorgas was retired on his 64th birthday on October 3, 1918, and Noble was named Surgeon General of the Army's Expeditionary Forces in France. The Armistice was signed on November 11th. Gorgas once again availed himself to work with the International Health Board and was soon tasked with investigating outbreaks of yellow fever on the west coast of Africa. In May 1920, he and Marie (Figure 29) sailed with his staff for London to attend a meeting of the International Hygiene Congress in Brussels. Upon his return to London he suffered a stroke and three weeks later was knighted by King George V in the Queen Alexandria Military Hospital in London. He died a week later, on July 3, 1920.

Figure 29. Marie and William Gorgas in May 1920, shortly before his fatal stroke (Library of Congress).

His knighthood accorded him a state military funeral in St. Paul's Cathedral, which included the band of the Coldstream Guards playing Chopin's Funeral March through the streets of London, the first American ever honored in such a manner. His body was then transported to Washington, DC where it lay in state for four days at the Church of the Epiphany before interment at Arlington National Cemetery (Wiggins, 2005). He was the most celebrated physician of his era, and the first Army Surgeon to serve as the President of the American Medical Association, an honor he believed should have been bestowed on Walter Reed, had he not died prematurely. In 1928 Gorgas was featured on a green 1-cent stamp, the first regular issue by the Canal Zone. In March 1928, Congress passed a joint resolution renaming the new $2 million hospital in Ancon as Gorgas Hospital, which continued operating until October 1999.

THIRD ISTHMIAN CANAL COMMISSION (1905)

As described previously, War Secretary William Howard Taft made a 10 day visit to Panama in November 1904, ostensibly work out some new tariff policies with the new Panama Republic and to take stock of the progress being made on the canal excavations. He was politely informed by Wallace of the increasing frustration with the lack of progress in the collection of basic topographic, geologic, hydrologic, and meteorological information. Wallace's aides asserted that the necessary scientific equipment and personnel had been requisitioned months earlier, but to no apparent avail. Taft listened politely to the critiques by Wallace, which were later supported by Burr and Parsons after their visits in December. When Taft returned to Washington he was convinced that the ICC in its current state needed to be overhauled in such a manner so as to speed up the processing of requisitions for equipment and trained personnel to use it. He briefed President Roosevelt, who was always eager to hear about "progress," and irritated to hear excuses for the lack thereof.

On January 13, 1905, Theodore Roosevelt asked Congress to allow him to restructure the Isthmian Canal Commission, but Congress failed to act immediately on this request. The issue of whether to construct a sea level or locked canal also became a frequent topic of conversation in Washington, as well as the nation's newspapers. While the Senate and House committees debated the topic of the restructuring, Roosevelt began thinking about how best to settle the technical issue of which canal to build, with an eye on the estimated time-to-completion.

As an accomplished naval historian and former Assistant Secretary of the Navy, Roosevelt appreciated the utility of a sea level canal. But, he was also a pragmatist, realizing that the longer the canal project took the more tenuous its political support. Roosevelt perceived the canal as the crowning jewel of his administration, because it represented America's emergence as the preeminent power of the Western Hemisphere. Roosevelt felt that the trans-oceanic canal was essential to America's emergence as a true world power, with a two-ocean Navy uninhibited by a 17,000 miles transit around Cape Horn. He determined that the technical problems must be ironed out as soon as possible so the project could accelerate in an efficient manner, for which American ingenuity and problem solving was internationally renowned.

On March 29, 1905, President Roosevelt accepted the resignations of all but one (Harrod) of the Isthmian Canal Commissioners appointed in 1904 and appointed a third Commission on April 3rd (Figure 30), whose powers were concentrated in a three man executive committee, which would henceforth reside in Panama and retain sufficient authority to approve all requisitions, instead of requiring unanimous approval of all the commissioners.

Figure 30. Meeting of the third Isthmian Canal Commission on February 2, 1906. The commissioners pictured here, are from left foreground in clockwise order: General Peter C. Hains, Admiral Mordechai Endicott, Chairman Theodore P. Shonts, Benjamin M. Harrod, Military Governor Charles E. Magoon, Secretary Joseph B. Bishop, and General Oswald Ernst.

New Executive Committee. The new Executive Committee would be comprised of Chief Engineer John F. Wallace, Chairman Theodore P, Shonts, and Military Governor Charles E. Magoon. Shonts would be paid an astounding $30,000 per year, Wallace $25,000, and Magoon $17,500 (his salary as General Counsel in Panama had been $7,500 per year). This committee was required to meet on the isthmus at least once every three months. The other commissioners were Rear Admiral Mordechai T. Endicott of the Navy, retired Brigadier General Peter C. Hains and Colonel Oswald Ernst of the Army Corps of Engineers, and Benjamin M. Harrod of New Orleans. Hains and Ernst had previously served on the first canal commission and Harrod on the second. These four commissioners would be paid $7,500 per year and, along with Shonts and Bishop (explained below), would remain in Washington, DC. With the addition of Chief Engineer Wallace, there were now eight instead of seven commissioners, including the chairman. These appointments met the legal requirements of the Spooner Act. In September 1905 President Roosevelt named Joseph Bucklin Bishop as the commission's Executive Secretary. His appointment was arguably one of the most important made during the course of the project, after those of Gorgas, Stevens, and Goethals.

Chairman Theodore P. Shonts. Theodore P. Shonts (Figure 31) was born in Crawford County, Pennsylvania in May 1856, where his father was a doctor. The family moved to Iowa just after the Civil War. Shonts attended Monmouth College in Illinois, founded by John F. Wallace's father, graduating in 1876. He then marketed his services as an account to banks in Iowa, promising to standardize and simplify their methods of bookkeeping. Then he decided to study law in Centerville, Iowa, where he became associated with General (later Governor) Francis M. Drake, who had financial interests in a number of railroads. In 1882, Shonts married Drake's daughter Harriet and was given charge of railroad construction of a portion of the Iowa Central Railroad. This led to similar managerial contracts with other rail lines in the Midwest. Through careful investing, he soon became majority holder of the Iowa Railroad, then sold his share and assumed control of the Toledo, St. Louis, and Western Railroad, which he rehabilitated and turned into a profitable concern. He was managing this railroad system when President Roosevelt appointed him Chairman of the Isthmian Canal Commission on April 3, 1905. He continued sitting on the boards of directors of a number of rail lines during his tenure as ICC Chairman, and resigned his post as Chairman on January 22, 1907 to accept the presidency of the Interborough-Metropolitan Company in New York.

Governor Charles E. Magoon. Charles E. Magoon (Figure 31) was born in Steel County Minnesota in December 1861 and received his legal education at the University of Nebraska, being admitted to the bar in 1882. He never married and practiced law in Lincoln, Nebraska, later serving as judge advocate of the Nebraska National Guard. He departed Nebraska to join the War Department in 1899, as legal officer for the newly established Bureau of Insular Affairs. On July 1, 1904, he was named General Legal Counsel to the Isthmian Canal Commission. He accompanied Secretary Taft on his first tour of Panama in September 1904. Magoon's new role would have two hats: as Military Governor of the Canal Zone (CZ), and as the American Minister to the new Republic of Panama.

Magoon was charged with the administration and enforcement of law in the Canal Zone, along with "the direction of all matters of sanitation: the custody of sanitary supplies and all sanitary construction," not only within the CZ, but also in Colon and Panama City, which were a veritable mess. He arrived on the isthmus on May 25, 1905, and after conferring with Colonel Gorgas, was instrumental in supporting the latter's efforts to countermand the yellow fever epidemic then threatening the project (30 confirmed cases in May 1905). In June, there were 62 new cases of yellow fever, and everyone sensed that something bold needed to be done. During his first weekend on the job Magoon ordered the Administration Building to be fumigated, which it was from then on, repeatedly. He then ordered wire screens to be affixed to all structures, an existing rule that his predecessor (General Davis) had not enforced (and Davis came down with malaria and was sent back to the United States).

In a shrewd move Magoon began hiring Panamanian physicians as his sanitary inspectors, and he offered cash rewards for anyone that would report new cases of yellow fever so that the carriers of the disease could be isolated from the Stegomyia mosquitoes in screened rooms. As if by magic, the preventive measures worked, reducing the number of cases to just six by September, with last case on the isthmus being reported on September 29th. He was reassigned as Provisional Governor of Cuba on October 12, 1906. He was replaced by Richard Reid Rogers, the ICC's General Counsel. In November, the President temporarily abolished the office of Governor of the Canal Zone, to give greater autonomy to the chief engineer. This allowed all of the Governor's duties to be carried out by the general counsel, who would serve in the capacity of governor, but without the title, and he was able to remain in Washington.

Executive Secretary Joseph Bucklin Bishop. Joseph Bucklin Bishop (Figure 31) was born in September 1847 in what is now Rumford in East Providence, Rhode Island. He attended Brown University, graduating in 1870, supporting himself as a journalist for the *Providence Morning Herald*. After graduation he joined the New York Tribune, working under Horace Greeley and John Milton Hay, former assistant secretary to Abraham Lincoln. In 1883, he went to work for the *New York Evening Post* working under Edwin Godkin. He established himself as a determined investigative researcher, exposing many Tammany Hall figures involved in crime and corruption in New York City Hall. His association with Theodore Roosevelt began in 1895, when Roosevelt was serving as President of the NY City Police Commission. When Godkin retired in 1899, Bishop moved to the *New York Commercial Advertiser* as chief of editorial writers. He was always supportive of Roosevelt's progressive policies seeking to root out corrupt machine politicians. Bishop provided strong editorial backing of the Roosevelt Administration's role in supporting Panamanian independence in November 1903, and the subsequent canal treaty with the new republic.

On September 1, 1905, Roosevelt named Bishop to the Isthmian Canal Commission as its new Executive Secretary, with an annual salary of $10,000. He was charged with managing the ICC's day-to-day activities from an office in Washington, DC, and ensuring public support for the canal project through press releases and by keeping an official history of the project, which promised to be the largest civil engineering enterprise ever undertaken. Most insiders saw Bishop as Roosevelt's "eyes and ears" on the inner workings and management of the massive project, as their mutual regards for one another was no secret.

In the fall of 1906, Bishop visited Panama to prepare for the President's upcoming visit in November, which was the first time a sitting American president had ever journeyed outside the United States. Congressional critiques of Bishop as an "overpaid press secretary" continued to be aired around Washington each time congressional appropriations for canal construction were debated. In August 1907, Secretary Taft ordered Bishop to relocate himself in Panama to be closer to the work he was supposed to be documenting and away from the partisan political rumblings.

One of his official duties was to listen to and investigate complaints by the workers, which overwhelmed him the first six months in Panama, because of a backlog. Bishop's presence in the Canal Zone was pivotal to the project's success. In September 1907, he began publishing the *The Canal Record*, a weekly newspaper for the Canal Zone, which had enormous impact on the morale and productivity of the canal project (described later). He remained on the isthmus for the next seven years, except for month-long vacations to the states each summer.

Figure 31. Left - Theodore P. Shonts served as Chairman of the new Isthmian Canal Commission and received the highest salary (Jackson & Son, 1911). Middle - Charles E. Magoon was Military Governor of the Canal Zone and Minister to Panama from 1905-06. Right – Joseph Bucklin Bishop served as the ICC's Executive Secretary from 1905-14.

In 1913, Bishop published a book about the canal's construction titled *The Panama Gateway*, which was moderately successful. Bishop departed Panama in the summer of 1914 to resume his literary career, shortly before the canal opened. At Theodore Roosevelt's request, he began working on a biography of Roosevelt in early 1915, which appeared as a two volume set in 1919 and 1920, about a year after Roosevelt's death in early January 1919. Titled *Theodore Roosevelt and His Time: shown in his own letters*, it was the only biography that met with Mrs. Edith Roosevelt's approval during her lifetime (1861-1948). The first book he published on Roosevelt was *Theodore Roosevelt's Letters to his Children*, which was released in 1919. It became Bishop's only best-selling book, insuring his financial security.

During the 1920s he labored to complete a biography of Major General George Washington Goethals, whom he had come to know and admire while working on the Canal Commission in Panama. Goethals died in January 1928 and Bishop passed the following December, at age 81. His son Farham Bishop completed the manuscript, which was released in 1930 as *Goethals: Genius of the Panama Canal*.

INTERNATIONAL BOARD OF CONSULTING ENGINEERS

Frustrated with the inability of his Isthmian Canal Commission to reach a consensus on the vital issue of what kind of interoceanic canal to construct across the Panamanian Isthmus; on June 24, 1905, President Roosevelt appointed another prestigious board to advise his administration. This august body was called the International Board of Consulting Engineers, and their job was to determine whether the United States should construct a sea level waterway or a locked canal. The outcome of this study would have an overwhelming impact on how the construction should proceed, what the project would cost, and how long it would likely take to complete.

This group was also referred to as the "Board of Consulting Engineers" in the press (Figure 32). It was composed of thirteen engineers: five from Europe, six American civilians, and two Army Engineers. These included: retired Major General George W. Davis as chairman, assisted by Army Engineer Captain John C. Oakes as the board's secretary. The American engineers were Alfred Noble, consulting engineer in New York; William Barclay Parsons, chief engineer of the New York Subway; Professor William H. Burr of Columbia University; retired Brigadier General Henry L. Abbot, who had been appointed to the international *Comite Technique* for the French Panama Canal in 1897; Frederic P. Stearns, chief engineer of the Metropolitan Water and Sewerage Board of Boston; Joseph Ripley, general superintendent of the St. Mary's Falls Canal in Sault Ste. Marie; and Isham Randolph, chief engineer of the Sanitary District of Chicago. The foreign engineers were: William Henry Hunter, chief engineer of the Manchester Ship Canal in England (nominated by the British government); Eugen Tincauzer, chief engineer of Germany's Keel Canal (nominated by the German government); Adolphe Guerard, inspector general of public works in France (nominated by the French government); Edouard M. Quellennec, consulting engineer for the Suez Canal Company; and J. W. Welcker, chief engineer of the system of protective dykes in the Netherlands (nominated by the government of The Netherlands). Hermann F. Schussler, Chief Engineer of the Spring Valley Water Co. of San Francisco and J. B. Berry, Chief Engineer of the Union Pacific Railroad were also appointed to the board, but declined to serve because of previous commitments.

President Roosevelt's charge. President Roosevelt invited the distinguished group to his home on Sagamore Hill overlooking Oyster Bay on Long Island. There he instructed them as follows:

"What I am about to say must be considered in the light of suggestion merely, not as direction. I have named you because in my judgment you are especially fit to serve as advisors in planning the greatest engineering work that the world has yet seen; and I expect you to advise me, not what you think I want to hear but what you think I ought to hear.

There are two or three considerations which I trust you will steadily keep before your minds in coming to a conclusion as to the proper type of canal. I hope that ultimately it will prove possible to build a sea-level canal. Such a canal would undoubtedly be best in the end, if feasible, and I feel that one of the chief advantages of the Panama route is that ultimately a sea-level canal will be a possibility. But, while paying due heed to the ideal perfectibility of the scheme from the engineer's standpoint, remember the need of having a plan which shall provide for the immediate building of the canal on the safest terms and in the shortest possible time.

If to build a sea level canal will but slightly increase the risk, and will take but a little longer than a multilock higher-level canal, then, of course, it is preferable. But, if to adopt a plan of a sea level canal means to incur great hazard, and to insure indefinite delay, then it is not preferable. If the advantages and disadvantages are closely balanced, I expect you to say so. I desire also to know whether, if you recommend a high-level multilock canal, it will be possible after it is completed to turn it into or to substitute for it, in time, a sea-level canal, without interrupting traffic upon it.

Two of the prime considerations to be kept steadily in mind are - 1. The utmost practicable speed of construction. 2. Practical certainty that the plan proposed will be feasible - that it can be carried out with the minimum risk. The quality of work and the amount of work should be minimized so far as is possible.

There may be good reason why the delay incident to the adaptation of a plan for an ideal canal should be incurred; but if there is not, then I hope to see the canal constructed on a system which will bring to the nearest possible date in the future the time when it is practicable to take the first ship across the Isthmus; that is, which will in the shortest time possible secure a Panama waterway between the oceans of such a character as to guarantee permanent and ample communication for the greatest ships of our Navy and for the largest steamers on either side of the Atlantic or Pacific. The delay in transit of the vessels owing to additional locks would be of small consequence when compared with shortening the time for the construction of the canal or diminishing the risks in the construction.

In short, I desire your best judgment on all the various questions to be considered in choosing among the various plans for a comparatively high-level multilock canal, for a lower-level canal with fewer locks, and for a sea-level canal. Finally, I urge you the necessity of as great expedition in coming to a decision as is compatible with thoroughness in considering the conditions (Report of the Board of Consulting Engineers, 1906).

Roosevelt then added that he expected them to visit the Panamanian Isthmus in the September or October of 1905 and submit their report to him around the first of January 1906.

Figure 32. The International Board of Consulting Engineers at their inaugural meeting in Washington, DC on September 1, 1905. From left, standing: Secretary John C. Oakes, General Henry L. Abbot, Eugen Tincauzer, Edouard Quellennec, Isham Randolph, Frederic P. Stearns, and Professor William H. Burr. From left, sitting: Joseph Ripley, William Henry Hunter, Adolphe Guerard, J. W. Welcker, Alfred Noble, General George W. Davis, and William Barclay Parsons (Bentley Historical Library, University of Michigan).

Organization of the international board. The board included three members of the first Isthmian Canal Commission back in 1899: General Davis, Professor Burr, and Mr. Parsons, who a few months previous, had submitted a recommendation favoring a sea-level canal. Other members of the board were known to favor locked canals. The members of the board, therefore, naturally fell into two groups, one which was more or less pre-disposed to a sea-level canal, the others to the lock type of canal.

The engineers were appointed to committees which were assigned the task of discussing the canal problem from these divergent standpoints. These included the Sea Level Canal Committee, Lock-Canal Committee, and the Committee on Unit Prices. The board as a whole, however, passed on judging certain features in order that the conclusions thus reached might serve as a guide in determining other features of the project.

In this regard, it was resolved that ship locks should have a usable length of 1,000 feet, a width of 100 feet, and a depth of 40 feet. The board also recommended the type and dimensions of the open canal section, upon which subsequent comparisons would be made to estimate costs. This had never previously been done, the sea level canal always being about a third of the size of a locked canal, because if its greater depth.

In September 1905, the board was given access to an extraordinary body of information and data, including the various schemes prepared by the French between 1880-98, as well as new proposals by an array of reputable engineers, several of which are summarized below.

Gillette's objections to a sea level canal. On September 15th, the board reviewed an article by Major Cassius E. Gillette of the Army Corps of Engineers, titled *"The Panama Canal: Some serious objections to the sea-level plan,"* which appeared in the July 27, 1905 issue of *Engineering News*. Gillette had been part of the advance team sent to Panama with Dr. Gorgas back in April 1904. He reviewed all of the previous work and pointed out that sedimentation from the Chagres River had been more or less ignored in all previous schemes. He concluded that a locked canal with a summit elevation of 100 feet above sea level and laid out on a fairly straight line would be the best choice, to ensure navigational safety and to allow for future enlargement, to accommodate ocean going vessels of increasing size and tonnage. He discarded the dam site at Gamboa and recommended a higher dam be constructed just eight miles from the Atlantic coast at Gatun. This embankment would impound a lake with a surface area of about 100 square miles, capable of absorbing the inflows of all 38 tributaries crossed by the proposed canal alignment. He also recommended that a sheetpile cutoff wall 60 feet deep be driven beneath the core of the Gatun Dam, followed by three-inch diameter uplift relief wells spaced 5 to 6 feet apart, just downstream of the cutoff wall. He envisioned single flights of three locks at either end of the canal, capable of lifting vessels 35, 35, and 30 feet.

The Minority Report of the Board in January 1906, adopted Gillette's recommendations for the massive embankment dam at Gatun, but lowered its lake elevation to 85 feet above sea level, providing an unprecedented 30 feet of freeboard for flooding (a crest elevation of 115 feet). They also adopted his proposed alignment of the channel from Gatun to the canal entrance on the Atlantic side. The major disparities between Gillette and the Lock-Canal Committee were in the estimated costs of construction. Gillette estimated the cost of the great embankment dam at Gatun to be about $2 million, while the Committee estimated those costs to be about $8 million. This considerable difference was attributed to errors on the topographic maps Gillette was using and the Committee's addition of a massive concrete gravity spillway structure.

Gillette resigned his Army commission on December 31, 1905 and became Chief of the Bureau of Filtration for the City of Philadelphia, where a few years later (1913), he received a patent for his invention of a vertical lift canal lock gate that would serve dual purpose as a bridge. He never received any recognition for his contribution to the Panama Canal, which was worthy of mention.

Bunau-Varilla's lock to sea level canal compromise. On September 19th, 1905, the board read and discussed a new proposal prepared by the French engineer Philippe Bunau-Varilla. He advocated a plan for initially constructing a locked canal, followed by above ground excavation and subaqueous dredging to enlarge and deepen the cuts across the Continental Divide until such a time that the locks could be progressive removed (Figure 33 upper). This plan began with a summit water elevation of 200 feet above sea level with a narrow channel, only 150 feet wide, using five locks of 40 foot lift at each end (Figure 33 lower). The scheme was similar to what the French had conceived in 1894, a compromise intended to garner support from advocates of lock and sea level canals. It was envisioned as a means to complete a locked canal at the earliest possible date, leaving the door open for the eventual completion of a sea level canal with a base width of 300 feet, with all of the finished side slopes assumed to remain stable at inclinations of 45 degrees.

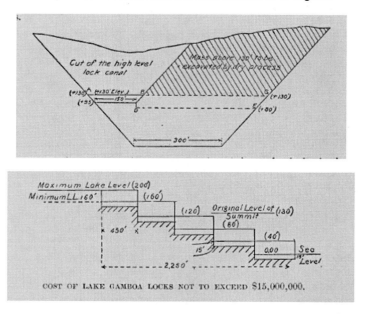

Figure 33. Bunau-Varilla's scheme for transforming a high elevation lock canal to a sea level canal, a compromise intended to garner popular support from both camps of thought (Board of Consulting Engineers, 1906).

Ward's proposal for a massive dam at Gatun. On October 2nd, 1905 the board undertook a thorough review of a recent article titled "*The Gatun Dam*" by Charles D. Ward, which appeared in the 1904 ASCE *Transactions* (Ward, 1904). At the timer Ward was Chief of Maintenance and Extension for the Brooklyn Waterworks. In 1880, he and Ashbel Welch (President of ASCE in 1882) had visited Panama and while touring the Isthmus Ward suggested that a dam approximately one and a quarter miles long be constructed across the lower Chagres River at Gatun, to capture that stream's discharge for a locked canal. The original French plan was to dam the Chagres at Gamboa. When they shifted to a locked canal scheme in 1894, they chose a dam and locks site at Bohio, slightly upstream of that posed by Morison in 1902. The Bohio dam site was eight miles upstream of Gatun.

When the United States took over the canal project in 1903, Ward requested that he be allowed to make the case for a large earthen dam at Gatun, which he presented at the annual meeting of the American Society of Civil Engineers on May 18, 1904. This was followed by the aforementioned article, which first appeared in the ASCE Proceedings, where he gave the proposed dam's dimensions: a length of 6,750 lineal feet with a dam 100 feet high impounding a lake at elevation of 90 feet above sea level (Figure 34). He then provided cost estimates for various alternatives, including a lake half as high (45 feet), if foundation conditions were adverse to a dam 100 feet high.

Eleven years later, after the Panama Canal was completed, one of the international board members, Alfred Noble, would recall: *"The Ward paper created so deep an impression among engineers that one of the earlier examinations made by the First Construction Commission appointed by President Roosevelt was to determine by means of boring the depth to rock and the nature of the overlying earth along the line of the proposed dam. The earth was found to extend to such a depth that the construction of a masonry dam was found to be impracticable, and the project for a dam of this kind was too hastily set aside. There were many, however, who believed with Mr. Ward that a perfectly safe and satisfactory dam could be made of earth, and that advantages of the location from the point of view of navigation were so obvious that the subject was not allowed to drop, and it received favorable consideration from the minority of the Board of Consulting Engineers in 1905"* (Wegmann, 1916).

Noble then added: *"While it is possible to say that the Gatun location would not have been adopted had Mr. Ward's paper not been written, the fact that the first suggestion to this end was dropped and remained unheeded, except by him, for more than 20 years, and that its revival was due to him exclusively, make it most probable that, but for his persistent efforts, the excellent plan of Canal now adopted would not have developed, and either an inferior lock plan or a far more costly sea level plan would have been adopted* (Wegmann, 1916).

Bates' bi-level canal. Another competing plan for a locked canal was presented by American consulting engineer Lindon Wallace Bates, who testified before the board on November 24th, 1905. He suggested an ingenious scheme for terminal lakes at either end of the canal and a higher central lake across the continental divide. The lake on the Atlantic side would be formed behind an earth dam at Mindi, while the lake on the Pacific end would be supported by an embankment dam constructed across the Rio Grande, between Ancon and Sosa Hills. The high lake was to be at an elevation of 62 feet above sea level, formed by damming the Chagres River at Bohio (the same site selected by the French in 1893 and the Americans in 1902). Bates would subsequently pen a scathing critique of the canal's locks in June 1906, predicting the ultimate failure of the canal when the first ship damaged a lock gate and all of the water in Gatun Lake would discharge through the breach (Bates, 1906).

Figure 34. Plan view of the massive embankment dam at Gatun first proposed by Charles D. Ward in 1880, and which he later described at the ASCE annual meeting in May 1904, summarizing his recommendations in the ASCE Transactions (Ward, 1904).

Visit to Wachusett Reservoir. In September 1905, Board Member Frederic T. Stearns was the incoming President of the American Society of Civil Engineers for 1906. He invited the other members of the International Board to take a field trip by rail from Washington, DC up to Worcester, Massachusetts to see the new Wachusett Reservoir along the Nashua River, which had been under construction for the previous five years and was then nearing completion. With a design capacity of 193,340 acre-feet of storage, it was the sate-of-the-art waterworks project in America prior to the Panama Canal. The main dam was concrete masonry rising 228 feet above the deepest level of its cutoff trench, and a height of 144 feet above the stream bed.

Of particular interest to the other members of the Lock-Canal Committee, of which Stearns was a member, was to view the North Dike of the Wachusett Reservoir, by itself the largest earth dam in America at the time. The dike was 65 feet high and two miles long, and contained over 5,500,000 cubic yards of earth fill. A massive cutoff trench 5,245 feet long and 50 feet deep with 45 degree slopes had been excavated beneath the dike, and Cypress sheetpiles had been driven to depths of about 46 to 48 feet beneath this level, flooring in glacial tills 150 feet above the underlying bedrock.

The foundation conditions bore a resemblance to the proposed earthen dam at Gatun in Panama, which they would be visiting in a few days. Unlike Burr and Parsons, the Lock-Canal Committee saw little problem in placing a massive earthen dam on relatively low permeability foundations, without feeling obliged to install a masonry cutoff wall (described earlier). After spending a day and a half touring the project, the members of the international board took a train to New York and boarded the steamer that would take them to Panama.

Reconnaissance of Panama. On September 29th, the international board of engineers boarded the *S.S. Havana* and sailed for the isthmus, so they would be there during the dry season, working under abnormally sunny skies. This proved to be disadvantageous because they failed to witness the torrents of water conveyed by the Chagres during the wet season (Stevens, 1927). They worked from their steamer in Limon Bay (adjacent to Cristobal), interviewing key figures and taking field trips to various parts of the isthmus. On October 10th, they interviewed John Stevens and several of his division engineers at length, aboard the *Havana*. Stevens informed them that he had only arrived 74 days previous, on July 27th, and that he could only provide answers that were "*largely guesswork*" (Stevens, 1935). Nevertheless, he was staunch in supporting a locked canal with a summit elevation of 75 to 80 feet above sea level served by two locks at either end, at Miraflores, Pedro Miguel, and Gatun, with an earthen dam at Gatun.

All of the board's engineers had practiced engineering in more temperate climes, not in the tropics with monsoons. Although the board learned much about the route by personal observation, the biggest shortcoming was that their visit coincided with an unusual period of drought, until the last two days. As a consequence, they were unable to view the typical wet season flood flows, erosion, sedimentation, or landslippage that presented the greatest engineering challenges on the canal. Fortunately, there were other forces of persuasion, and in the end, sane engineering judgment prevailed. They departed Panama on the evening of October 11th and reconvened in Washington, DC on October 20th.

Hearings before the board. While in Panama the board interviewed Chief Engineer John F. Stevens; Frank B. Maltby, Division Engineer at Cristobal (in charge of dredging and harbors); H. F. Dose, Division Engineer at Culebra, who described the monthly excavation quantities and their unit costs; Charles Bertoncini, Chief of Division of Map Making and Lithography (held over from the French regime); and Walter E. Dauchy, Assistant Chief Engineer for Construction.

By November 14th, the Lock-Canal Committee had whittled its candidate schemes down to four possibilities:

> Project No. 1 – Summit level at elevation 85 feet, to be maintained by a flight of three locks at Gatun on the Atlantic side, with one lock at Pedro Miguel and two locks in flight at Sosa Hill adjoining La Boca pier on the Pacific side. Estimated cost, $141,236,000.

> Project No. 2 – Same as No 1 except that on the Pacific side there would be two locks in flight at Pedro Miguel and one at Miraflores rather than at Sosa. Estimated cost $148,272,000.

> Project No. 3 – Elevation at summit level 60 feet, maintained on the Atlantic side by a flight of two locks at Gatun, and on the Pacific side with a single lock at Pedro Miguel and another at Miraflores. For the purposes of control of the Chagres River and to furnish a water supply a dam near Gamboa was included. Total estimated cost, $171,190,000.

> Project No. 4 – Summit level at elevation 60 feet to be maintained by a dam with single locks at Gatun and Bohio on the Atlantic side, and with single locks at Pedro Miguel and Miraflores on the Pacific side, with a dam at Alhajuela. Total estimated cost, $175,929,720.

These estimates included 20% for contingencies.

The entire board then discussed the four lock canal proposals, but failed to gain a majority vote any of the four candidate schemes. Three members voted for Project No. 2, three for Project No. 4, and two members who thought the Atlantic locks should be positioned at San Pablo and Obispo instead of Gatun and Bohio. The rest favored Project No. 1. A motion favoring Project No. 1 was made by Mr. Welcker, which seemed to be the consensus opinion, but no vote was taken.

On November 15th, the sea level canal schemes were described and discussed by the entire board. The consensus opinion was that such a scheme would entail 12 or 13 years of construction if no complications (such as landslides, etc.) were encountered. At the November 18th meeting, the Sea-level Committee agreed that the minimum channel width should be 200 feet at sea level, narrowing to 150 feet at a depth of -40 feet (Figure 35). They also recommended that a double tidal lock be constructed at Ancon on the Pacific side, which would be 100 feet wide by 1000 feet long, and construction of a massive dam at Gamboa to control the flow of the Chagres River.

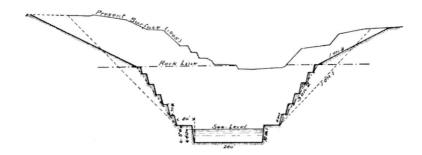

Figure 35. Maximum section at the Continental Divide at Culebra showing the extent of the French excavations and the intended excavation profile for a sea level canal in November 1905, which proved overly optimistic (Board of Consulting Engineers, 1906).

The foreign members of the board wanted to return to Europe no later than November 28th, but the hearings continued and they were not able to depart New York until December 7th. One more meeting was convened in New York on December 17th, after which the members of the three committees prepared their comments and submitted them for inclusion in the final majority and minority reports, which were passed to one another for comment and review prior to their being signed and submitted to President Roosevelt. General Davis traveled to Paris and to Brussels to secure the signatures of the five European engineers between January 7th and 15th, and the official date of the Majority Report was listed as January 10th although Davis didn't return to New York till the 25th. The board convened their last meeting on January 30th in New York, without their European members present. The Majority and Minority reports, along with 31 oversize plates were then submitted to President Roosevelt on early February 1906.

Testimony and reports by Wallace and Stevens. Upon their return to Washington, DC, they listened to testimony from former Chief Engineer John F. Wallace. Wallace was a staunch advocate of a sea level canal, and he followed these meetings up with extensive notes and written memoranda, titled *"Notes on the Panama Canal"* and *"Memorandum by Mr. John F. Wallace Accompanying Diagrams Illustrating Tentative Method of Culebra Excavation,"* and *"Statement of Mr. John F. Wallace, November 3, 1905,"* which were included in Appendix J.

Four of the five European members returned to their respective countries on November 5th (with Mr. Quellenac remaining until December 5th) and the Americans members of the board met briefly three times in December 1905, on the 5th, 17th, and 30th. General Davis, Alfred Noble, and Professor William Burr were appointed to a committee to supervise preparation of the board's final report, due around January 1st.

On December 19th Stevens sent a detailed report to the Isthmian Canal Commission and copied each member of the International Board of Consulting Engineers. His report contained extensive data on the fluctuating flow of the Chagres River and the other 37 water courses traversed by the canal, 18 of which were situated along the Culebra Cut. He recounted how he had been predisposed to constructing a sea level canal, but that upon observing the highly irregular flows of the Chagres River, he had changed his mind. He related how the base flow of the Chagres was only about 750 cfs, but that within a matter of 12 hours of storms, it could increase to as much as 150,000 cfs (Stevens, 1927). He asserted that the only plausible manner by which to control the Chagres would be to entrap it in an enormous reservoir with plenteous waste weirs (spillways). An enormous reservoir could absorb the sever fluctuations in flow and swallow the considerable volume of sediment that would surely play havoc with any alterative scheme, seeking to separate the Chagres from the canal.

Stevens asserted: *"A high-level canal, with an earth dam at Gatun to carry a head of 75 or 80 feet of water, this height to be overcome by a double flight of locks 37-1/2 or 40 feet in lift each; this level to continue to the vicinity of Pedro Miguel; there by similar double-lift locks to drop to sea level on the Pacific Ocean, with a tide regulating lock in the neighborhood of Miraflores."*

His concluding remarks: *"The sum of my conclusions is, therefore, that, all things considered, the lock type or high-level canal is preferable to the sea level type, so called, for the following reasons: It will provide a safer and quicker passage for ships, and therefore, will be of greater capacity. It will provide, beyond question, the best solution of the vital problem of how to safely to care for the floodwaters of the Chagres and other streams. Provision is made for enlarging its capacity to almost any extent as very less expense in time and money than can be provided by any sea-level plan. Its cost of operation, maintenance and fixed charges will be much less than any sea-level canal.*

The time and cost of its construction will not be more than one-half of that of a canal of the sea-level type.

The element of time might become, in case of war, actual or threatened, one of such importance that measured not by years, but by months or even days, the entire cost of the canal would seem trivial in comparison.

Finally, even as the same cost in time and money for each type, I would favor the adoption of the high-level lock plan, in preference to that of <u>the</u> proposed sea-level canal. I therefore recommend the adoption of the plan for an 85-ft summit-level, lock canal, as set forth in the minority report of the consulting board of engineers." (Stevens, 1935).

Majority report. The international board submitted its report to President Roosevelt in early February 1906. Predictably, the eight members of the Sea Level Canal Committee favored that option: General Davis, Professor Burr, Barclay Parsons, and the five foreign members of the board (Hunter, Tincauzer, Guerard, Quellennec, and Welcker). Their voluminous report and recommendations were referred to as the "Majority Report." The remaining members, Alfred Noble, General Abbot, Frederic P. Stearns, Joseph Ripley, and Isham Randolph favored a locked canal with a summit level of 85 feet above sea level (described below).

Roosevelt was disappointed that his international board had not reached a consensus, despite his admonitions to them. There was considerable technical disagreement about the safety and utility of locks for large ocean-going vessels, with the Majority Report citing every safety incident involving vessels damaging lock gates. The belief of the majority was that large locks could not be constructed with sufficient redundancy to "operate safely," and that whatever dimensions were chosen would within a decade or two, become hopelessly obsolete. The five European engineers had very little experience with locks, while those voicing the minority view (recommending a locked canal) had substantive experience with the design and operation of lock canals. The bottom line of the Majority Report was that it employed a very narrow navigation channel across the Continental Divide, just 200 feet wide with a depth of 40 feet (Figure 35). Like the old French plan, a tidal lock would be installed at Ancon on the Pacific side and damming the Chagres at Gamboa. The most troubling aspect was how the report dealt with the unpredictable flows of the Chagres River, which were to be handled using 19 lineal miles of diversion channels which would intercept the flows of natural watercourses and somehow convey it to a point where the canal could accept it, presumably near Bohio. The technical aspects of how such flows and their accompanying erosion and sedimentation would be handled were not described in any detail, a pivotal consideration that Stevens would make much about later, in the congressional hearings that followed. The majority group estimated that this scheme would take 12 to 13 years to construct, at an estimated cost of $247,021,200.

Authors of the Minority Report. The Minority Report was largely the work of Alfred Noble, Joseph Ripley, and Isham Randolph, with less assistance from Frederic T. Stearns and General Abbot. All three had substantive experience with the design, construction, and operation of lock canals. The appendices of the board's final report were filled with technical data on various locks and canals of the world. Most of these technical summaries were prepared by members of the Minority Report, including: James Ripley's description of the St. Mary's Falls (Soo) Canal and Locks (within Appendix D); General Abbot summarized the hydraulics of the Panama Canal (Appendix E); Alfred Noble and Joseph Ripley prepared a technical estimate of the "Traffic Capacity of a Lock Canal with Summit Level at Elevation 85" (Appendix L); Isham Randolph prepared a summary of the equipment recommended for canal excavation, with the capacity of each type (Appendix M); a historical sketch of the history of the Panama Canal by Alfred Noble (Appendix Q); a technical description of artificial waterways, their improvement and navigation by Joseph Ripley

(Appendix S); and the entire group prepared "Estimates of Cost of Lock Canal" (Appendix T). The experience and professional judgment of the authors are revealed in these detailed assessments, which likely lent credibility to their assessments. Brief bio sketches of the engineers who prepared the minority report are presented below.

Alfred Noble. Alfred Noble was born in Livonia, Michigan in 1844. After serving as a sergeant in Union Army during the Civil War he enrolled in the study of civil engineering at the University of Michigan, graduating in 1870. In 1885, Noble designed the first "movable" or "emergency dam" for the new Weitzel Lock on the Soo Canal at Sault Ste. Marie, Michigan. The unique and intricate structure was to be deployed in the event of a lock gate failure. Noble oversaw the installation and testing of his emergency dam on the St. Mary's River about 3,000 feet upstream of the lock. It operated like a swing bridge, but with a vertical axis. A similar swing bridge dam was constructed on the Canadian side of the Soo Locks around 1895. This precaution proved invaluable on June 19, 1909, when two freighters and a passenger ship were involved in a string of collisions that began when an up-bound freighter struck the lowest lock gate, breaking it. The outrushing water soon broke the hinges on the opposing gate and two locks of water and their respective vessels were swept down the mill race, initially, at a sensational speed (60 km/hr). In a matter of minutes the emergency dam was pivoted into place and deployed, much to everyone's relief (Goldmark, 1928). The Minority Report included a provision for emergency dams at the upstream end of the locks at Gatun, Pedro Miguel, and Miraflores for an estimated cost of $2 million. These were installed on the Panama Canal locks and replaced in the 1930s, but were never deployed because of the use of experienced pilots in the canal, so were removed in the mid-1950s.

Joseph Ripley. Joseph Ripley was born in St. Clair, Michigan in 1854. In 1877, Ripley was hired as an engineer for the St. Mary's Falls (Soo) Canal in Sault Ste. Marie, which had been established 1855 and operated by the State of Michigan. When the Weitzel Lock was opened in 1881 ownership of the canal passed to the Army Corps of Engineers. Ripley transferred to the Corps as a civilian Assistant Engineer, where he remained until 1906. In 1893-95, he supervised the detailed topographic and hydrographic surveys of the 75 miles along either side of the St. Mary's River between Whitefish Bay in Lake Superior and Lake Huron, while the Poe Lock was under construction. It was designed by Ripley's superior, Colonel Orlando M. Poe, Superintending Engineer of improvement of rivers and harbors on Lakes Superior and Huron. Its width of 100 feet with a length of 800 feet made it the largest lock in the world when it was completed in August 1895. In 1897, he was appointed general superintendent of the Soo Canal, by then the busiest ship canal in the United States and Canada. Ripley's surveys aided the construction of the approach walls to the Poe Lock. Ripley was a staunch advocate of employing lengthy approach walls with near-vertical faces, where ships could be checked and stopped at a safe distance from the lock, and against which they might moor using safety lines. He felt this technique was indispensable for the safe operation of a lock, and the Soo Locks on the American side had an impeccable record of safety. In May 1906 Ripley was hired by Chief Engineer Stevens as Principal Assistant Engineer in charge of designing locks, dams, and regulating works for the Panama Canal, based in Washington, DC. He was promoted to Assistant Chief Engineer in October. When Colonel Goethals assumed control of the project in April 1907, Ripley was supervising the structural design staff, but he resigned on June 18, 1907 to accept another position.

Isham Randolph. Isham Randolph was born in 1848 in New Market, Virginia where he was home schooled and attended private schools. Like John Stevens, he had acquired an extensive knowledge of civil engineering through hands-on experience. He began his professional career as a surveyor for the Baltimore & Ohio Railroad system, followed by work for the Lehigh Valley Railroad. In 1872, the B&O moved him to an extension they were making in Chicago, and the following year he was named Resident Engineer for construction of a roundhouse, engine shops, and 27 miles of track on the south side of Chicago. He then joined the Scioto Valley Railroad where he remained until 1880, when he became chief engineer of the Chicago & Western Indiana Belt Railroad, where he supervised construction of terminals and freight warehouses. From 1885-1893, he operated his own consultancy in Chicago, working on railroads and the Chicago Union Stockyards. In 1893, he was named Chief Engineer of the Sanitary District of Chicago, in which capacity he served until 1907. Between 1893-1900 he supervised construction of the 28-mile long Chicago Sanitary & Ship Canal, the largest civil engineering project in American history until the Panama Canal. The canal involved the excavation of 42,230,000 cubic yards of rock and soil and the placement of 460,000 cubic yards of concrete, creating the largest ship canal in the world until the Panama Canal's completion in 1914. His novel employment of Stoney Gates with sills 15 feet below normal operating pool were the model for the same gates employed at similar depths on the Gatun Dam spillway.

Frederic P. Stearns. Frederic Pike Stearns was born in Calais, Maine in 1851. In 1870, he began working for the Office of the City Surveyor in Boston, while studying engineering in the evenings, educating himself not only in civil engineering, but over the years continued to educate himself in chemistry, biology, bacteriology, geology, architecture, and landscape architecture. From 1872-80, he served as Leveler, Assistant, and the Division Engineer for the Boston Water-Works, working on developing water supplies from the Sudbury River. During this time he carried out a series of hydraulic experiments with Alphonse Fteley on water flow through the Sudbury Conduit and over weirs, summarized in an article that was awarded the Norman Medal of the American Society of Civil Engineers (ASCE) in 1882 (Fteley and Stearns, 1883). He also worked with the Massachusetts State Board of Health making exhaustive studies of water supply and water quality, and the means of controlling and improving both. His standards for water treatment and filtration were employed across much of New England. His crowning jewel was the Wachusett Reservoir constructed for the Metropolitan District of Boston Water Works between 1900-06, which included a mass concrete dam, a hydraulic fill embankment dam with the deepest cutoff trench ever constructed up to that time, and a 12 mile long aqueduct. On June 26, 1906, he delivered his presidential address to ASCE, titled

"The Development of Water Supplies and Water-Supply Engineering" at the society's annual meeting in Frontenac, New York. After serving on the International Board of Consulting Engineers in 1905-06, he was appointed to the Board of Consulting Engineers of the Isthmian Canal Commission, serving from 1907-09, during which time he took an active role in reviewing the design and construction of Gatun Dam.

Henry L. Abbot. General Henry Abbot was the most celebrated Army Engineer in America, and a member of the National Academy of Sciences. He was born in Beverly, Massachusetts in 1831, and graduated from West Point in 1854 as a topographical engineer. He served on the Pacific Railroad Surveys between California and Oregon, then a hydrographic survey of the Mississippi Delta, co-authoring "*Topographic and Hydrographic Survey of the delta of the Mississippi*" with Captain A.A. Humphreys in 1861. He saw service in the Civil War and rose to Brevet Major General of Volunteers during the Siege of Petersburg. After the war he reverted to the rank of major in Corps of Engineers. He continued studying the levees along the Mississippi River, then served as Post Engineer working out of Willet's Point, NY, which includes serving as Superintending Engineer of Construction for the building of Ft. Schuyler. In 1870-71, he was dispatched to Europe to observe an eclipse and examine torpedo defenses, and then assigned to the Board of Engineers for Fortifications. In 1872, he received appointment to the National Academy of Sciences, and was promoted to Lieutenant Colonel in March 1880. In 1886 he received his LL.D. from Harvard and was promoted to Colonel. He then served on the Board of Ordnance and Fortifications and as Division Engineer of the Northeastern United States, beginning in December 1888. He was stationed in New York City and also served as President of the permanent Board of Engineers, President of the Harbor Line of Boards of New York and Boston, and various other boards until retired from active service in August 1895. In March 1897, he was appointed to the French *Comite Technique de la Cie. Nouvelle du Canal de Panama* and lived in Paris in 1898-99 as part of these duties. In February 1899, he retired to Cambridge, Massachusetts. He was placed on the retired list as a brigadier general in April 1904. After his service on the International Board he authored a book titled *Problems of the Panama Canal,* published in 1907, which described the technical advantages of the lock canal scheme, and "*Hydrology of the Panama Canal*," published in the ASCE *Transactions* in 1913 (Abbot, 1913).

Recommendations of the Minority Report. The Minority Report cited five significant advantages of a lock canal over that of a sea level canal:

1. Greater capacity for traffic than the narrow waterway proposed by the Board.
2. Greater safety for ships and less danger of interruption to traffic by reason of the wider and deeper channels which the lock canal makes possible at smaller cost.
3. Quicker passage across the Isthmus for large ships or a large traffic.
4. Materially less time required for construction.

5. Materially less cost.

As mentioned previously the minority group considered a range of possibilities, with summit transit elevations between 30 and 90 feet above sea level. For the purposes of comparison to the Majority Report's sea level canal, the minority group selected "a project with summit level at elevation 85 maintained by a dam and duplicate flights of three locks at Gatun." This was recommended for adoption, with General Abbot preferring a lower dam with duplicate flights of two locks at Gatun, supplemented by a dam and duplicate single locks at Bohio, raising the summit level to elevation 85 feet, as before.

The most prominent component of their plan was a massive earthen dam at Gatun, eight miles downstream of the Bohio dam site. Here the world's largest earthen dam would retain a reservoir of 110 square miles (which later swelled to 164 mi^2), able to absorb anything the fickle Chagres might throw at it. The dam's foundations crossed two steeply incised bedrock channels, which extended as much as 258 feet below sea level before reaching bedrock (Figure 36). Based on 27 borings taken by September 1905, the principal difference between the Bohio and Gatun sites was that the valley fill was considerably *less pervious* at Gatun than the materials encountered at similar depth at the Bohio dam site, which could make underseepage and uplift pressure more troublesome. The Minority Report concluded that the thick cover of low permeability materials would make an excellent foundation for as earthen embankment dam. The authors also recommended that additional borings be undertaken at the site before finalizing the design of the dam, spillway, and adjoining lock structures.

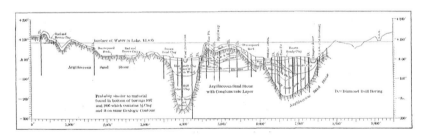

Figure 36. Geologic cross section compiled from borings in 1904-06 along the axis of the proposed earthen dam at Gatun. Note the deeply incised channels, the west diversion channel, the French Canal, and the channel of the Chagres River (Wegmann, 1911).

The Minority Report recommended a conservative cross section for the embankment, with the dam crest 100 feet wide lying 50 feet above the operating elevation of the reservoir, with a width of 374 feet at the operating level (elevation 85 feet), and a width of 2,625 feet at sea level. Foundation seepage and permeability comparisons were made with the Wachusett Reservoir, and the dam's cross section compared with the San Leandro (120 feet high) and Pilarcitos (95 feet high) dams in the San Francisco Bay area, constructed in the 1860s and '70s. The report estimated that the proposed Gatun Dam would employ approximately 21,200,000 cubic yards of material, making it the largest embankment dam ever contemplated up to that time (and was not exceeded until construction of the Fort Peck Dam 30 years later).

The spillway was to be a concrete gravity structure with its lower sill at an elevation of 69 feet, upon which would be fitted 30 x 30 foot Stoney Sluice Gates, patterned after those fitted to the Chicago Drainage Canal, supervised by Isham Randolph. The sills of these gates were to be fitted 16 feet below the normal operating pool at elevation 85 feet. If the lake were to rise one foot to elevation 86, the spillway could discharge 140,000 cfs. It was believed that the largest flood would not lift the reservoir pool more than two feet, and less than this if the spillway began releasing water before the crest of the flood wave reached the spillway. Randolph had successfully argued that this ability to draw down the reservoir in anticipation of significant inflows was the major advantage of installing the Stoney Gates with their sills 16 feet below the normal operating pool.

Cost and time to completion for a lock canal. The most significant finding contained in the Minority Report was the cost benefit of relocating the dam from Bohio eight miles downstream to Gatun. Due to the reduced need for tributary intercept canals and diversion channels, there would be a net savings of nearly $12 million, assuming the estimated cost of Gatun Dam to be about $7.8 million. This reduced the overall cost of the lock canal, which including a 20% contingency, was $139,705,200, more than $107 million less than the estimate for the sea level canal. In addition, more than 2/3 of the travel distance along the canal route would afford channel widths in excess of 500 feet, which was another advantage over the much narrower sea level plan, which would have required one-way traffic through the Culebra Cut and the mooring of vessels at either end.

The Majority Report estimated the time to completion for the sea level canal as 12 to 13 years, but the Minority Report makes a technical case for the time to completion of being "*not less than 15 years*," because the rate of excavation (cubic yards per day) would actually diminish as the prism of ground being excavated at Culebra approached its lowest levels, which was an accurate assertion. Assuming no slope failures would occur, the sea level canal required at least 103,795,000 cubic yards of excavation.

The minority group estimated that a lock canal with a summit level of 60 feet above sea level would require excavation of 72,800,000 cubic yards assuming no slope instability, and would take ten years to complete. They then estimated that the lock canal with a summit level of 85 feet above sea level would only require the excavation of 53,800,000 cubic yards assuming no slope instability, and would take seven and one-half years to complete. They unanimously recommended the latter scheme, utilizing a summit elevation of 85 feet.

Impact of the Board of Consulting Engineers reports. The majority and minority reports were bundled with the transcriptions of the Board of Consulting Engineer's meetings, with all of the supporting documents and reports, which altogether comprised 426 pages. These were forwarded to the ICC in mid-January 1906 for their review.

ICC Chairman Theodore Shonts lifted excerpts of Steven's report and paraphrased it, along with additional arguments in favor of a lock canal with a massive earthen dam at Gatun, and sent these to Secretary of War William Howard Taft and President Roosevelt. In January 1906, Taft and Roosevelt summoned Stevens to Washington and grilled him on his views and how he had reached his conclusions. Stevens was used to justifying good engineering before demanding and authoritative figures, having worked 25 years for Great Northern Railroad mogul James. J. Hill, who had an uncanny ability to ask probing questions that sought to *"differentiate facts from assumptions"* (Stevens, 1935).

The ICC commissioners reviewed the voluminous document and on February 5th submitted their recommendation to War Secretary Taft approving the Minority Report recommending a lock canal with a summit elevation of 85 feet. Stevens' written reports of December 19th and his written affirmation of the Minority Report dated January 26th seemed to tip the scales in favor of the lock canal scheme. These printed appeals were corroborated by personal appeals to Taft and Roosevelt during his visit to Washington, DC in mid-January.

The President decided to back the recommendations of the Minority Report, based on Stevens' persuasive arguments, which appeared to be based on cold, hard facts rather than optimistic desires. In this unexpected shift the Americans would avoid the colossal catastrophe that might have occurred, had they attempted a sea level canal across Continental Divide at Culebra, where some of the world's largest landslides had yet to occur (Figure 37).

Two weeks after the ICC vote the president handed the matter over to Congress, conveying the 426 page report of the International Board (which included Steven's report), the Isthmian Canal Commission's review and statement of support of a locked canal, and the Roosevelt Administration's view of the matter, which emphasized the practicality of supporting the plan that had the greatest likelihood of success without being cost prohibitory, pointing out *"the difficulties and dangers of navigation"* posed by a sea level canal that was only 200 feet wide.

The Senate Committee on Interoceanic Canals and House Committee on Interstate and Foreign Commerce began debating the subject of a sea-level versus lock canal. The Senate committee heard from Professor William Burr, Barclay, Parsons, and John Wallace, the staunch sea-level canal advocates. They denounced every aspect of the lock canal scheme, pointing out how all of the various elements, such as the proposed Gatun Dam, and the mighty ship locks would inevitably fail.

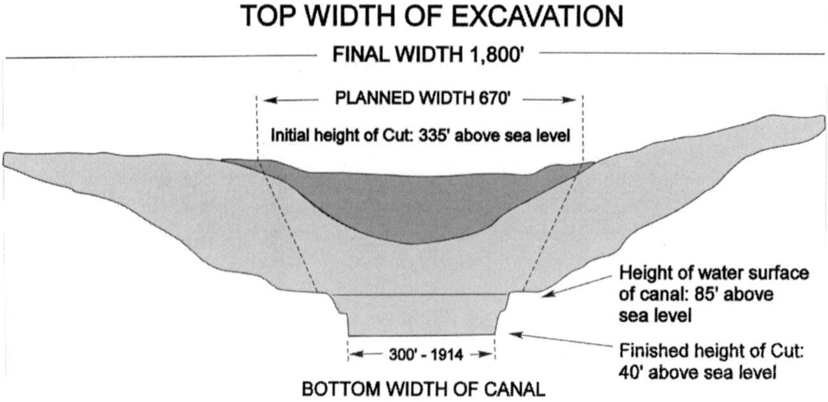

TOP WIDTH OF EXCAVATION
FINAL WIDTH 1,800'
PLANNED WIDTH 670'
Initial height of Cut: 335' above sea level

Height of water surface of canal: 85' above sea level

300' - 1914
BOTTOM WIDTH OF CANAL

Finished height of Cut: 40' above sea level

Figure 37. Comparison between planned width of the excavation at Culebra Cut. The portion excavated by the French (1882-1903) is shown as dark grey, while that removed by the Americans (1904-14) is shown in light grey. The additional excavation was hastened by slope failures that began to plague the project in 1911.

Given this unabashed criticism, in April Stevens was again summoned to Washington to testify before the Senate committee. He conceded a number of points, such as a sea level canal not being nearly as dangerous if its width was increased to something between 400 and 500 feet. His answers were truthful and unbiased by any predisposition or political agenda. He estimated that the time-to-completion of the proposed sea level canal was *"nothing less than 18 years,"* and that the cost would be prohibitory. But, he was confident that a lock canal with a summit elevation of 85 feet could be completed in nine years. By the time he was through testifying there was little doubt in anyone's mind that he was an expert on the construction of significant civil works, and not a starry eyed dreamer. Despite Stevens' credible testimony, on May 17th the Senate committee voted to recommend a sea-level canal by the margin of just one vote! The matter would now go to the entre Senate.

With things looking dour, Stevens was summoned once again from Panama to be subjected to more inquiry by the House of Representative's Committee on Interstate and Foreign Commerce. By now it was mid-June 1906, a year after Roosevelt had charged the international board with determining what type of canal should be constructed across the Isthmus. Stevens, a formidable and imposing expert witness, reciting facts and figures without referring to any notes for two entire days. Again, Stevens pointed out that *"the one great problem on the construction of any canal down there* [referring to Panama] *is the control of the Chagres River, which overshadows everything else."*

On June 15th, Congress in Committee as a whole voted 110 to 36 in favor of a high level lock canal (Figure 38). On June 19th, Senator Philander Knox of

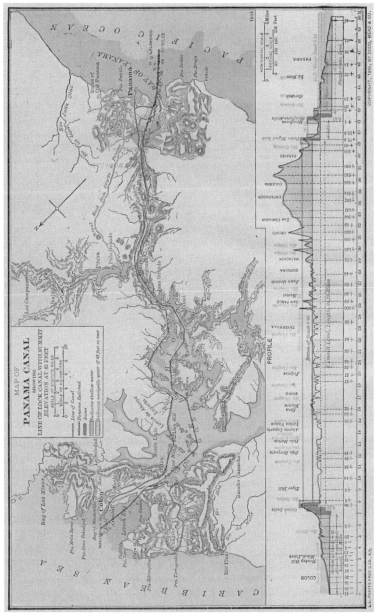

Figure 38. First plan and section view of the American scheme for a lock canal with a summit elevation of 85 feet above sea level, prepared in June 1906 (Dodd, Mead & Co.).

Pennsylvania, Roosevelt's former Attorney General, gave an eloquent speech about the merits of a lock canal, borrowing heavily on Steven's testimony before the Interoceanic Canal Committee. Two days later the full Senate voted 36 to 31 in favor of the high level lock canal advocated by Chief Engineer Stevens and the Administration. On June 29th, President Roosevelt signed the bill authorizing construction of the high level canal across the Panamanian Isthmus. It had taken the Americans seven years to select the path and the type of canal they would build to join the Atlantic and Pacific Oceans. It was now up to the engineers to make it happen.

JOHN FRANK STEVENS – SECOND CHIEF ENGINEER

Pathfinder and problem solver. Wallace was succeeded by another railroad engineer, John Frank Stevens (Figure 39). Stevens was everything that Wallace was not; he lacked any formal secondary education, having learned his craft through apprenticeship in surveying, engineering, construction, and management. He was born on April 25, 1853 in West Gardiner, Maine and was known by his first two names, John Frank, or just Frank (Foust, 2013). After graduating from Farmington State Normal School in Maine in June 1872, he accepted a position working on a field crew in Lewiston, performing surveys for mills and industrial canals. He had obliged himself to teach for two years, but sensing a natural inclination for this line of work, he began applying for positions out west, even applying to work for James Buchanan Eads on his famous bridge in St Louis. He decided to strike out for Minneapolis, where he had an uncle he could live with. He was hired as a timekeeper for restorative work on the St. Anthony Falls Dam, which had failed five years previous. This led to his applying for the position of Rodman for the City Engineer of Minneapolis in 1874. He was soon promoted to Instrument Man, then Assistant City Engineer.

In 1876, he began working for railroads, accepting a position with the Sabine Pass & Northwestern Railroad (SP & NW) in north Texas. This was followed in quick succession by a series of junior engineer positions on railroad extensions across Minnesota, then back to Texas as Roadmaster for the SP & NW in 1879. Then he took as position as Assistant Engineer for road relocation and construction of bridges for the famed Denver & Rio Grande line in Colorado under General William Jackson Palmer. In late 1880, he moved back to Iowa to work for the Chicago, Milwaukee & St Paul line, then being laid across the state.

In the spring of 1882, his career took a significant shift when he accepted a position with Langdon & Sheperd, a Minneapolis firm engaged in constructing segments of the Canadian Pacific Railway from Montreal westward to the Pacific. When Stevens arrived that spring the line was about 160 miles west of Winnipeg, heading across the prairies towards the Canadian Rockies. The construction season lasted just five or six months, but the days averaged had about 19 hours of sunlight.

Figure 39. Chief Engineer John Frank Stevens at his desk in his new office at Culebra in March 1907.

The contractors were paid for each mile of useable track they laid, and Stevens was responsible for overseeing simultaneous construction along a zone that usually stretched 125 miles beyond the railhead. He learned to sleep in the back of a buckboard wagon as his drivers took him from one location to another, moving 30 miles during the six hour nights. Despite the odds, Langdon & Shepard made a tidy profit, thanks to Stevens' penchant for efficiency. They laid more track each week until the winter shutdown on December 1st, at the Swift Current River. They had rough graded another 40 miles, to the banks of South Saskatchewan River. Years later, Stevens would write "*I did not get a very large salary that year, but I learned a great deal of the practical work from a contractor's standpoint, knowledge which was to be of great value to me in the years to come*" (Stevens, 1935).

In March 1883, Stevens accepted a position as Assistant Engineer of Location on the Canadian Pacific under Major Albert B. Rogers. They would be the pathfinders searching for viable rail routes over the Canadian Rockies. About 25 miles west of present-day Calgary they were divided into three survey parties, with local Indians providing the pack horses that would sustain them over the next 5-1/2 months. Stevens' party followed the valley of the Bow River to Kicking Horse Pass, from which his party continued surveying westerly, across the Selkirk Range via Rogers Pass, as far as the Columbia River. His party was detailed to remain in the field during much of the following winter, to examine an alternate route. They became snow-bound in the mountains, and Stevens struck out foot to seek help and provisions. By "sheer providence" he managed to make contact with a relief party five days later. Blazing the path of Canada's first transcontinental line over their two highest mountain ranges would have been sufficient accomplishment for most civil engineers, but John Frank Stevens was just getting started.

The next season (1883-84), Major Rogers promoted Stevens to Engineer in Charge of Location, working from British Columbia, on the West Coast of Canada. This season he would be supervising three survey parties, altogether about 25 men. They traveled up the Frasier River to its head of navigation at Yale, which had been reached by the Canadian Pacific, advancing eastward. From there, the survey parties moved east on foot to Kamloops, near the forks of the Thompson River, where they met Major Rogers. Stevens' party then began blazing a new route across the range, reaching the summit about seven miles from the Columbia River, without the aid of pack horses. Stevens later said that during that summer of privation he learned "*My cardinal principles in handling men have always been to pay them well, feed them well, and to work them hard, but no harder than myself and to ask them to take no chances that I did not take myself. The average man will do his best work when following a leader – but not, when under a coward or shirker*" (Stevens, 1935).

In 1884-85, Stevens continued working on the Canadian Pacific, bridging the mighty Columbia River at Revelstoke by employing an ingenious scheme for sinking rock-filled crib caissons during the winter, when the swift river's flow was at its lowest ebb. Stevens considered this feat his greatest engineering triumph, remarking: "*I felt very pleased at the outcome. It was merely one of the many instances in the life of an engineer which go to show that a purely technical education (I never had any) does not always fit him to grapple with and solve unusual problems that arise. "horse sense" and the nerve to try doubtful experiments often succeed in the face of difficulties*" (Stevens, 1935). He "wintered over" in a log cabin at Revelstoke along with his wife and two-year old son. The last two summers witnessed the completion of the principal bridges along the line and Stevens blazing a suitable route through the rugged Selkirk Mountains. He remained in Canada until the last spike was driven on November 7, 1885.

Stevens returned to his home in Minneapolis. The following spring of 1886, he was asked to supervise the construction of a 90-mile line from Sioux City to Manila, Iowa for the Chicago, Milwaukee, and St. Paul line. This was followed by laying out a new rail line between Sault Ste. Marie, Michigan and Duluth, Minnesota, a line of almost 400 miles which proved much more difficult than anticipated, working through the cold winter months. This line was then absorbed and completed by the Canadian Pacific in the fall of 1888, but fell into legal difficulties, requiring

Stevens courtroom testimony in Detroit for the better part if six weeks, another experience that paid handsome dividends in the years to come.

By 1889 the Great Northern Railway of James J. Hill had pushed westward from Minneapolis to the hamlet of Havre, in north central Montana. Stevens knew Hill from his early days with the Canadian Pacific near Winnipeg, when Hill was that line's managing director. Hill sought more than anything else to construct his own transcontinental line, but the path across the Rockies and the countless ranges of western Montana, Idaho and central Washington remained uncharted in regards to the relative elevations of all the lowest passes.

In November, Stevens was interviewed in Helena, Montana and requested to seek out the fabled Marias Pass across the northern Rocky Mountains, which Meriwether Lewis had first searched for in 1806. Neither Lewis nor those who had followed him had yet been successful, but the Blackfeet and Kalispell Indians all spoke of a great pass across the mountains at the head of the Marias River. Rising to the new challenge, Stevens immediately set out from Fort Assiniboine for the Blackfoot Indian Agency, 180 miles distant. There Stevens found a Kalispell Indian who had sought sanctuary for killing a tribesman, who was the only person willing to serve as a messenger in case Stevens met with disaster. Stevens succeeded in ascending the False Summit, a few miles east of the real pass. There the Indian refused to go any farther, it being wintertime and beginning to get dark. Stevens pushed on ahead and *"walked directly into the Pass, and was on the Continental Divide."* After 83 years of searching, Marias Pass had at last been located, along the eastern boundary of what is now Glacier National Park. The date was December 11, 1889. With darkness overtaking him, many feet of snow, and nothing to burn for a fire, Stevens was obliged to walk back and forth on his makeshift snowshoes all night long, in a battle to survive frostbite and freezing to death (the temperature was somewhere near 40 degrees F below zero). Stevens survived the ordeal and returned safely, with the realization that he had made a significant discovery. The crest elevation of the pass was surveyed the following spring and found to be 5,213 feet above sea level. The tracks of the Great Northern were laid over the summit and a 10 foot high bronze statue of Stevens was erected on the site in July 1925, commemorating his discovery.

Stevens' discovery of the mythical Marias Pass (so named by Meriwether Lewis) catapulted him into the permanent employment of James J. Hill. He went onto blaze a path through the Cascade Mountains in Washington along the Wenatchee River, which led to a new pass, which was named "Steven's Pass" by C. F. B. Haskell, who supervised the railway's survey party that followed him. With an altitude of 4,000 feet, this summit and the line's bridge over the Columbia River represented two critical elements in completing the Great Northern line to Seattle in October 1893.

Stevens was then named Assistant Chief Engineer of the Great Northern in 1893, elevated to the position of Chief Engineer, and later, General Manager of the

Great Northern System between 1895-1903. During this time he supervised the design of the 2.65 mile long Cascade Tunnel excavated beneath Stevens Pass, at an altitude 3348 feet, and the laying out of a most advantageous line into Portland along the Columbia River, the construction of numerous improvements such as the Seattle Tunnel, branch lines serving mines and ports, and a rail line following the Deschutes River in central Oregon, connecting to California. Part of the Stevens aura was the well-known fact that Hill was known as a difficult man to work for, who fired engineers for the smallest infractions (Budd, 1944).

In 1903 Stevens accepted the position of Chief Engineer and Vice President of the Chicago, Rock Island & Pacific Railroad in their Chicago headquarters, where he was working when President Roosevelt appointed him Chief Engineer of the Panama Canal project in the summer of 1905.

Appointment as Chief Engineer. When John F. Wallace suddenly resigned in late June 1905, it was an embarrassment to the Roosevelt Administration, in part because it was so unexpected. The crisis wasn't just Wallace's sudden departure, they were also battling the yellow fever epidemic that everyone was hoping wouldn't strike the Americans on the Isthmus. Panic-struck men were resigning their positions and purchasing steamer passage back to the states. Before the yellow jack scare subsided in late September 1905, almost three-quarters of the American workers had departed, leaving a management vacuum on the Isthmus and a public relations debacle of epic proportions back home. Many newspapers conjured up every negative thing they could about the infidelities and dangers of surviving in such a disease-ridden place as Panama. Something drastic needed to be done, and it needed to be done quickly.

War Secretary William Howard Taft was about to depart for the Philippines with legendary railroad engineer John Frank Stevens, lately of the Chicago & Rock Island Line in Chicago. Stevens had been tapped to go to the Philippines as a special advisor to the war Department on railroad construction. By 1905, Stevens had supervised the surveying, alignment, and construction of more miles of railroad than anyone else in the world (Budd, 1944). During a visit to Washington, DC in June 1905, railroad mogul James J. Hill had briefly met President Roosevelt, whom he didn't care for, but told him that the best man on the planet to construct the Panama Canal was John F. Stevens. When Roosevelt seemed interested, Hill offered to speak to Stevens when he passed through Chicago on his way back to Minneapolis. When later queried about why he should help Roosevelt after his administration had worked so feverishly to slap the Northern Securities Company launched by Hill, J. P. Morgan, and E.F. Harriman with antitrust suits for manipulating prices in the railroad industry, back in 1901-02, Hill was rumored to have replied "*I didn't do it to help Roosevelt, I did it to help the United States,*" referring to avoiding the embarrassment and national disgrace that would have accompanied an American failure in Panama, similar to that inflicted upon the French.

The day after William Howard Taft's confrontation with John Wallace in the New York hotel, he called Stevens at his office in Chicago and offered him the position of Chief Engineer of the Panama Canal, with an annual salary of $30,000, which was $5,000 more than Wallace had earned. Stevens' first reaction was to decline the offer, but he said he would consider it. He noted in the Chicago papers that Wallace was scheduled to return to Chicago that same day. Working in similar positions (Vice President and General Manager) for two of the nation's mightiest railroads in the same city, Stevens had known Wallace for many years and held him in high esteem, not only as a friend, but also admiring Wallace's abilities as an engineer. Stevens sent Wallace a note asking to meet at the Union League Club, downtown. Wallace likely refused because that same day (Tuesday June 30th) his professional reputation was being soiled in the nation's newspapers, who were printing the transcript of his interaction with Taft on Sunday afternoon in New York. Stevens decided to board the overnight train to New York where he was instructed to discuss the offer with attorney William Nelson Cromwell. Cromwell appealed to Stevens' sense of duty and patriotism, telling him it was "*the most important position ever offered to any engineer in the history of mankind.*"

Stevens returned to Chicago to talk it over with his wife Hattie, who encouraged him to accept the offer, reminding him that his amazing career had been in preparation for just such a moment as this, and that it was providential, a privilege that would allow him to bring credit to his profession, as well as his country. Stevens wired his acceptance to President Roosevelt that evening.

Four days later Stevens headed back to New York, where he met ICC Chairman Theodore Shonts and the two of them had lunch with President Roosevelt at his home on Sagamore Hill. Roosevelt seemed distracted; Secretary of State John Hay had died unexpectedly on July 1st, and the Japanese delegation had just arrived for discussions Roosevelt was brokering to negotiate a peaceful settlement of the Russo-Japanese War. Sensing the frustration that Wallace had apparently had with the ICC, Stevens was frank with Roosevelt about "*having a free hand in all matters, and would be hampered by no one in Authority*" (referring to Shonts and the other members of the ICC). To this demand he added the limitation that he "*would remain until the failure or success of the undertaking was assured according to his own judgment*" (Stevens, 1927). Roosevelt told Stevens that things in Panama were "*in a devil of a mess*" and that he would appreciate it if Stevens could "*get busy and buttle like hell*!"

After this meeting Shonts and Stevens were shuttled to the nearest Long Island train station to return to New York and were intercepted by the press. Shonts seemed a bit irritated with Stevens, and he answered all of the media inquiries, telling reporters that "*the Chief Engineer would have no one to blame if the work is not done right.*"

Stevens and Shonts returned to Washington, DC and spent several weeks lining up key people he sought to entice to join him in Panama for the greatest of all engineering undertakings. In this regard, he was able to glean some of the best and brightest engineers he had encountered in the railroad industry, such as Louis Rourke, W. G. Bierd, Ralph Budd, and Frederick Mears (Budd would later serve as President of the Great Northern Railway and the Chicago, Burlington & Quincy Railroad, while Mears would build the Alaska Railroad for the U.S. Government and after retiring from the Army, serve as Chief Engineer of the Great Northern Railway). Each would play a prominent role in the railroad operations on the project. Stevens made personnel decisions based on qualifications and experience, selecting the best possible individuals to head up each department. He believed that the most efficient administration occurs when he "*gave an official ample authority, and then to hold him responsible for results*" (Stevens, 1927).

Shonts and Stevens sailed on the same steamer from New York and arrived on the Isthmus on July 27, 1905, near the height of the yellow fever panic. When probed about what he "planned to do," Stevens responded that he viewed the canal project as "*just a big railroad job*," and that he "*didn't anticipate any significant problems, so long as he received adequate logistical and manpower support.*" The next day he began making a thorough reconnaissance of the project, which took an entire week. He traveled from place to place interviewing key personnel on the Isthmus, asking countless questions. Frank Maltby would later remark "*He could size up the abilities and experience of an engineer pretty quickly, sifting facts from presumptions*" (Maltby, 1945).

Morale, sanitation, and housing. What Stevens concluded during his initial inspection was that the project was leaderless and unfit for further work until a number of glaring problems could be wrestled under control, and either improved or altogether eliminated. Foremost among these was low worker morale, the result of disease and the absence of a clear sense of purpose and direction. One of the first senior people he met was Surgeon Colonel William Crawford Gorgas, who filled him in about the sanitation problems. He soon learned of all the struggles Gorgas had experienced trying to get Governor Davis and the old Isthmian Canal Commission to approve any of his requisitions because they belittled his theories on mosquitoes being the carrier of such diseases. Stevens had heard of Gorgas' work and that of Walter Reed in Havana, and saw no reason to doubt his abilities. Stevens would later state: "*But as should always be the case in such matters, an open mind was held on the unfamiliar subject, and it was well that the matter was so approached, for it seemed simply an act of Providence that just at that time such men as Gorgas and his very able assistants were on the job.* "*Gorgas was one of the finest characters that I ever knew*" (Stevens, 1927).

In an unexpected and unprecedented move on August 1, 1905, Stevens ordered a halt to excavation activities and shifted all of his worker's energies to preparatory work, which included "building the engineering plant" (Sibert and Stevens, 1915). Stevens showed that he was a seasoned construction superintendent when he asserted "*morale is more valuable than machines, because it's men who operate the machines.*" Another old axiom he often cited was "*the best way to restore morale is to keep workers clean and dry.*" As he saw it, he had three basic problems: 1) yellow fever, 2) malaria; and the worker's "cold feet."

The first order of business would be on improving every facet of the working conditions on the Isthmus, which took the first five to seven months of his residency. This included improvements to drainage and sanitation, followed by improvements to housing and dining facilities, such as canteens dispersed along the CZ to feed the burgeoning mass of workers. He also succeeded in revamping and expanding the commissaries along the line. Stevens would later tell audiences that *"the preparatory stage is the most important. If you plan things out carefully and think them through, only then will an efficient operation be realized."*

Not all of these activities were well received by ICC Chairman Theodore Shonts, whose displeasure with "wasting money trying to dry water pools, cutting grass and weeds, and installing miles of copper screens on every house and structure were not essential to the task at hand, which was to excavate a big ditch. Things came to something of a head when Shonts recommended Gorgas' replacement with an osteopath he knew in Florida, which was never carried out.

Stevens "preparatory work" included dramatic improvements in Colon and Panama City. Each city had its drainage revamped and was supplied with a potable water supply via a modern network of buried pipes. The streets were paved with brick from Illinois, and vitrified clay sewer pipe from St. Louis was laid beneath the new streets to convey sewage to a point of suitable discharge.

Stevens' next focused on worker housing and building the necessary structures to sustain a contented work force. In just over a year and a half these projects required a staggering 75 million board feet of lumber, most of which came from the Puget Sound area and was off-loaded on the Pacific side. This building program included the construction of 1,200 new structures and the renovation of 1,536 of the 2,148 buildings left by the French. Stevens placed Jackson Smith in charge of the Quarters and Labor Department. White workers were entitled to quarters whose floor space was a function of their rank: one square foot per dollar of monthly salary. Stevens felt that this policy *"proved a strong incentive to encourage individual ambition."*

Stevens also had the insight to appreciate the need for workers to commute efficiently between their work stations and homes without undue delay, and for their wives and children to be able to access the commissaries, bakeries, laundries, and schools. This included such novel innovations as the construction of several suspension bridges across the canal excavations to convey workers to the opposite side in a safe and efficient manner (Figure 40).

Stevens realized that the "creature comforts" questioned by Chairman Theodore Shonts were necessary in Panama because the job duration was going to be close to a decade. It wasn't like a railroad construction job, moving along at a brisk pace, laying miles of track each day. Stevens knew that worker morale and health depended on the workers and their families being fed and sheltered in a manner sufficient to "keep their feet dry." Mess houses were set up along the line of the

Figure 40. Stevens had this wire suspension bridge built across the canal excavation at Culebra to allow workers to cross easily over to the eastern side of the canal, to access the Panama Railroad.

canal, for both white and colored workers (in segregated facilities). Single white men who resided in dormitories were charged 30 cents per day for three meals served in one of the mess houses, while the married men were expected to eat at their homes.

PRR Commissaries. The Panama Railroad had always operated a diminutive commissary for its employees, serviced by their steamships that plied the run between New York and Colon. When Stevens arrived the only commissary was located in Colon, and residents of other areas did not have ready access to the commissary. He ordered a complete overhaul and re-organization of the commissary, building more commissaries along the line of the canal and dramatically increasing their inventory. Within a year of Stevens' arrival local commissaries were established up and down the rail line to make them more convenient to the workers, as well as laundries and bakeries.

The commissaries were serviced by a twelve car train that included five refrigerated cars loaded in the predawn hours at the cold storage warehouse in Colon. Two cars carried fresh baked bread, one with vegetables, and the remaining four carried staple commissary supplies (dry goods). This train would ramble through the Canal Zone each morning, reaching Panama City by nine o'clock in the morning. These outlets offered virtually any food item that could be purchased in the United States, including perishables, like vegetables, eggs, butter, milk, poultry, beef, pork, strawberries, or ice cream. Coupons were issued to each employee which were could be redeemed for merchandise at the commissaries. This practice angered many of the local Panamanian merchants, who had been gouging the canal workers for decades on the premise of supply and demand. In the spring of 1908, the ICC bakery started baking pies and pastries in large quantities, and a coffee grinding house and ice cream plant were established next to the refrigerated warehouse.

The Panama Railroad increased the wharfage and warehouse capacity of its port facilities at Colon and Cristobal. These improvements included construction of a new reinforced concrete refrigeration and ice-making plant. The refrigerated warehouses had three gradations of temperature, and being situated on the wharves, the means to pass refrigerated stores directly from ships holds to the warehouses without exposure to the warm tropical air. Block ice began flowing to the worker's homes, allowing them to store perishable items.

The railroad managed to purchase goods in bulk at substantial discounts because of the large lots, which were shipped via rail to ports in the states, then loaded on vessels owned by the Panama Railroad Steamship Co., which was under the authority of the Chief Engineer. By eliminating wholesalers and middle men, the commissaries were able to offer most products at about 70% of the typical retail cost in the United States. The collective inventory soon grew, exceeding a value of one million dollars (Stevens, 1927). The Commission began by paying the Panama Railroad $14,000 per month for the use of their trains and right-of-way in 1904. This figure had climbed to $90,000 per month by July 1909, and $2,400 per month to operate the telegraph and telephone system on the Isthmus.

Shifting the headquarters. Stevens also moved the headquarters and administration of the canal project from Ancon Hill at Panama City to the Continental Divide at Culebra, where the management would be more centrally located, along the railroad. He had a comfortable home built for himself and similar residences constructed for the senior engineering and administrative personnel (shown in Figure 25). Living next to one another allowed for easy staff meetings among the project's principal managers early or late each day, depending on where they were working.

Establishing the work force. One of the most tiresome problems was the tendency of many spirited individuals to overstate their experience and qualifications, which were supposed to be verified or vouched for prior to their seaborne transit to Panama. For instance, during that first year of operations men with supposed experience in "setting grade" and "laying" railroad track had been approved by the ICC commissioners in Washington and dispatched to Panama. Upon their arrival, it was soon determined that none of them had ever previously worked on a railroad!

The job also demanded all manner of personnel experienced in mining and heavy construction, including miners, quarrymen, blasters, muckers, laborers, and more than 900 drillers (Figure 41). The construction of innumerable structures required general and finish carpenters, roofers, "dirt jockeys" capable of excavating suitable foundations and constructing post-and-pier supports used everywhere to elevate floors several feet above the squalid jungle floor. They also needed blacksmiths, iron workers, longshoremen, quartermaster and supply men, construction foremen, mechanics of every type, sanitation workers, drainage tenders, fire and policemen, warehouse men, commissary managers, cooks, bakers, insect exterminators and "vector control specialists," domestics, and even teachers.

Figure 41. Drillers standing on their jackhammers, likely advancing small diameter blast holes 6 to 10 feet deep. The A-frames in the background were the production drilling rigs that could drill up to 50 feet deep, and were used to set the excavation benches critical to rail-mounted muck cars, which had to be close to wherever the shovels were working.

Of this sea of specialists, the most sought after were heavy steam shovel operators, mining men with drilling and blasting experience, and individuals with experience in dredging, which were almost impossible to procure. By the end of 1906, there were 3,243 American workers on the Isthmus, skilled in more than 40 different subspecialties. Their average monthly pay was $47.

Compressed air plant. When the Americans began working on the Isthmus in mid-1904, they employed steam powered drills, which required heavy wire-reinforced rubber hosing. In January 1905, Chief Engineer Wallace requested 25 pneumatic rock drills, followed by 75 more over the balance of 1905, along with six Rand Imperial air compressors. Soon after Stevens' arrival he conceived the establishment of an eight-mile long master air pipeline, pushed up the northern arm of the Culebra Cut, and fed from compressor plants at Las Cascades and Empire, four and a half miles apart (another compressor plant was subsequently constructed on the Pacific side, at Rio Grande). The compressed air powered countless rock drills, needed for loading sufficient dynamite to break the rock down so it could be excavated and transported by rail out of the cut.

Many of the drillers and drillers helpers were obliged to more or less train themselves with the operation and layout of compressed air lines, which lost pressure with increasing length. These had to be maintained and repaired on an almost daily basis, as they were being threaded across congested trails, tracks, and roads (Figure 42). The master air lines were eventually enclosed in rectangular wooden conduits to protect them from accidental decoupling and damage, which required several hundred carpenters and train loads of lumber.

Figure 42. The working areas were always crossed by flexible hoses carrying compressed air to the drills and jack hammers, and required constant maintenance.

Despite these precautions, the wood conduits were damaged by blasting, flooding, inadvertent burial, wood eating ants, and dry rot. Flooding dominated the wet season, while the termite ants dominated the dry season. The positions of the conduits had to be adjusted constantly, with special attention to keeping them dry or free of pests. Special mechanics had to be trained to maintain the compressors and repair the air lines on a daily basis, as the machines wore out.

Quarries. Crushed rock was needed in increasing quantities as aggregate for the various concrete structures, such as drainage ditches, smaller dams for the supply of drinking water, new supporting bents for the Panama Railway line relocations, all of the lock structures, and the Gatun Sam spillway and powerhouse. The French had established a rock quarry at Ancon Hill near Panama City in the mid-1880s, mining andesitic agglomerate of the Panama Formation. Soon after the Americans arrived in

1904, they set up a three 50-ton Allis Chalmers rock crushers at the Ancon Quarry that were capable of crushing about 1,000 cubic yards per eight hour shift, if all three were working.

By late 1906 the projects managers realized that the Ancon quarry was too far from Gatun Dam and Locks to efficiently pass the required volume of aggregate. Its transport would overtax the main rail line that was essential to the progress of excavation in the Culebra Cut. Surprisingly, no alternative sources of high quality aggregate were found along the line of the proposed canal. After making inquiries of the Panamanians, they suggested a site about 20 miles northeast of Colon, on a prominent peninsula, seaward of old Portobello. The rock here was pre-Tertiary basaltic and andesitic lavas intruded by diorite and dacite.

After inspection it was decided that the additional distance to this site would be justified by the quality of the derived aggregate, so a modern crushing and classification plant was erected at New Portobello. An Allis Chalmers No. 21 "Gates" Gyratory Rock Crusher, weighing 235 tons, was set up there. It employed two receiving openings measuring 3.5 by 9.5 feet, the largest in the world at that time. The rock was reduced to minus seven inch size. This primary crusher was capable of handling 5,000 cubic yards per 8-hour shift, but such a significant volume was never required. The maximum output of 3,500 yards per shift was attained in 1909, when the demand for concrete aggregate swelled with the construction of the locks at Gatun. The crushed rock was then run through a secondary crusher to bring it down to minus 3.5 inch size, which was loaded onto 70 ton barges for transport to Cristobal, about 35 miles by sea.

Organizing the system of supply. A vast array of hardware, construction materials, and merchandise had to be shipped to the Isthmus, offloaded, transported to various warehouses and holding facilities, and inventoried for disbursement. The majority of the structures were made of lumber, which came principally from the Pacific Northwest to the Panama City side of the isthmus. Timber piles were also delivered from the Pacific side, many of these redwoods from northern California. The eastern terminus at Cristobal received most of the mechanical goods, such as steam engines, railroad rolling stock, and railroad cars of almost every type, unwieldy steam shovels, and centrifugal pumps for the floating dredges.

Bulk goods shipped in massive quantities to the Canal Zone included millions of tons of coal and calcined Portland cement, lubricants, thousands of square feet of brass screens (the only screens that could resist corroding in the humid tropical weather), corrugated tin sheeting for roofs, whitewash, wires of all sizes and type, iron nails, screws, and spikes, and every imaginable size and type of pipe: steel, iron, and terra cotta.

One of the sore points of John Wallace's tenure as Chief Engineer was his inability to secure prodigious quantities of pipe that he requisitioned month after month, to no apparent avail. Pipe between 8 and 24 inches in diameter was needed for steam exhaust lines, compressed air, and water supply lines, almost everywhere on the project. Most of this was wrought iron and steel pipe. The 1904 ICC appointed a committee to examine the matter of pipe specifications, to ascertain whether the more expensive, but lighter steel pipe was warranted. Various tests were performed and the Youngstown Sheet and Tube Company submitted specifications for ductile iron pipe, which was coated with asphaltum tar inside and out, which was eventually selected (Bennett, 1915). $400,000 worth of ductile iron pipe began flowing out of Youngstown, Ohio for use on the canal, but delays such as these hampered efforts to get the project up and going for over a year.

Added to these heavy elements were thousands upon thousands of smaller items, such as crescent wrenches, hammers, picks, shovels, canvas sheeting, bleach, lye soap, starch, lime for privies, more than 12,000 tons of paper, and tens of thousands of citronella candles. An inexhaustible supply of copper water pans of every imaginable size were also imported to surround every post and pier supporting element of each building foundation and every bedpost to prevent the ascendance of wood eating termites and ants, which could destroy wood structures in a matter of days (Figure 27).

Railroad locomotives and rolling stock. Stevens ordered prodigious quantities of the more powerful steam locomotives, retiring 146 French and Belgian engines as the new American models rolled into the Isthmus. Most of these were Mogul engines manufactured by the American Locomotive Company (who purchased the Rogers Locomotive Works in 1900) of Patterson, New Jersey (Figure 43). These employed a pair (2) of leading wheels on one axle under the nose of the engine, six powered and couples driving wheels (6), and no trailing wheels under the cab (0). This arrangement increased the weight on the driving wheels, providing greater traction. This configuration as referred to as a 2-6-0. It was the most long-lived steam locomotive in American serving from about 1900 to the 1950s, mostly on light branch rail lines.

Another change made under Stevens' tenure was to switch to screw spikes for the new double-tracked mainline of the Panama Railroad. Screw spikes were beginning to be employed in the states on approaches to bridges, across bridges and trestles, and in vicinity of switches and frogs, and at all rail crossings. The screw spikes required special tie plates and took much longer to install, but gave far better service over the long haul. Stevens justified the cost based on the lower maintenance costs and dry rot of the ties. Loose spikes could be re-pounded, but they lost about a third of their skin friction each time they were pushed back into the tie, and the damage to the wood fibers of the tie around the spike almost always led to accelerated dry rot.

Streamlining the removal of muck. When Stevens arrived in the Isthmus it didn't take him long to see that the steam shovels were not operating at a high efficiency because there were too few tracks and far too few muck trains. The steam shovels could excavate prodigious quantities of soil and rock, the limitation in their utilization was the problem of transporting the material to suitable disposal sites on either coast.

Figure 43. Stevens ordered 144 Mogul 2-6-0 locomotives like those shown here in the engine shops at Las Cascades. These were manufactured by the American Locomotive Company of Patterson, New Jersey and served as the workhorses of the canal project. 246 locomotives were employed in the canal construction.

The rock and soil excavated by the shovels was referred to as "muck," while "spoil" referred to excavated material that needs to be "wasted" (in mining parlance) or disposed of at some suitable location. After solving the sanitation, housing, morale, and choice of the type of canal during his first six months, Stevens attention was increasingly drawn to developing an efficient "engineering plant" to perform the required excavations, conveyance of spoils, and placement of excavated spoil.

In early 1906, Stevens brought in William Grant Bierd as General Superintendent of the Panama Railroad. The two men had worked with one another at the Chicago, Rock Island and Pacific in Chicago. Bierd's first assignment was to complete the double tracking of the entire lines, from coast-to-coast, laying fill, extending culverts, and constructing new bridges and constructing wye turn-arounds for the second line. Then he was assigned the task of developing a system of trackage that would support up to seven levels of simultaneous excavation at either end of the nine-mile long Culebra Cut across the Continental Divide. By June 1906, Bierd had 350 total miles of track, almost four times what the French had operated. He devised many innovations to streamline the efficiency of the excavation, transport, and fill placement processes, described below. Within a few months he replaced the 69-year old Colonel James Shaler as General Manager of the Panama Railroad.

Bierd began by sloping all of the temporary excavation benches back, towards their respective coasts on either side of the Continental Divide. This aided drainage away from the working faces and helped keep the roadbed as dry as possible. It also aided the filled muck trains, which were always empty on the ascending grade, and always descending when the cars were loaded, taking advantage of gravity to reduce the demands on the locomotives. It was one of the "railroad innovations" seldom dealt with or considered by the Corps of Engineers.

The ultimate goal of the excavation operations was to keep an empty muck car next to the steam shovels every second of every work shift. This meant that a train of muck cars would slowly be pushed forward or backward as each car was filled by the shovels. When an entire train of muck cars was filled, it needed to vacate the line as quickly as possible and be replaced by empty cars as seamlessly as possible (Figure 44).

Figure 44. Muck trains had to be positioned close to the working face within easy reach of the steam shovel's bucket. This image shows a smaller steam shovel loading side-dumping rail cars in 1906.

If this process could be streamlined to the point where the muck trains were *always* in motion, the maximum yardage could be removed, increasing performance and reducing the unit costs for excavation and dumping of each cubic yard. The American muck cars had more capacity (12 to 35 cubic yards) than their older French equivalents (6 to 9 cubic yards). The 95-ton Bucyrus shovels could excavate five cubic yards in a single stroke, filling a dump car with just seven shovel-fulls of muck, which took about one and a half minutes. The car would then be edged forward to receive the next ~35 cubic yards of material. In this manner it took about 45 minutes to fill an entire muck train (Figure 45).

Figure 45. Stevens' new excavation scheme employed wider cut benches with dual tracks, with the shovel and the muck cars on the same level. This view shows the smaller side dumping muck cars being filled by a 70-ton steam shovel.

Stevens envisioned as many as 230 muck trains to keep 70 steam shovels busy. One muck train would be positioned next to a working shovel, which could fill an entire train in about 45 minutes, with the engine pushing or pulling empty cars into position every few minutes. Another muck train would be outbound to dump their loads in huge spoil piles at either end of the canal (filling Gatun Dam and locks on the Atlantic side, or La Boca Bay on the Pacific side), while an empty third muck train would be heading back to the shovel it was assigned to take the place of the one being loaded. When the excavations grew more distant, more muck trains had to be employed to maintain the string of empty muck cars with each shovel (sometimes as many as seven trains per shovel). If there were no complications with weather, rail traffic, or speed of dumping the round trip circuit took about 2-1/4 hours. As the distance of the active faces being excavated on the Culebra Cut grew farther from the spoil piles, the transit times of the muck trains would increase, and a fourth or fifth train might be assigned to a specific shovel.

In order to maintain high utilization of each steam shovel, real-time control of the muck trains was essential, to route them through the maze of switches that would keep them moving inbound or outbound. To this end, as soon as Bierd laid new lines he was obliged to install telegraph lines along the entire right of way, to control the flow of trains coming and going, crossing over from one side to another (they used wyes to turn the muck trains around at each dump site), working to keep their down time to a minimum, waiting on sidings. The train dispatchers were the critical cog in all of this, and the unpredictable weather on the Isthmus had to be dealt with almost every day, which irritated the shovel operators, who were paid bonuses based on productivity.

Much of the success the Americans enjoyed was due to more efficient means, techniques, and mechanical technology than was available twenty years previous. The work horses of the muck cars were wooden high-back flatcars manufactured by the Pressed Steel Car Company of Pittsburgh. These wooden cars were fitted with three-foot high side boards along one side, with sheet steel edges and aprons spanning the gap between adjacent cars (Figure 46). They had a rated capacity of 40 tons each, or about 33 cubic yards.

Figure 46. Wooden high-back flatcars were fashioned by setting timber side boards on one side, as shown here. When the side boards were on the left side (as shown here) they were referred to as "right-hand loading cars" (Bennett, 1915).

Unloaders and spreaders. One of the American's technological advantages was the Lindgerwood (Train) Unloader, built by the Lidgerwood Manufacturing Company of Brooklyn. Three of these devices were initially ordered by Wallace in December 1904, along with three ballast unloaders (commonly referred to as "unloading plows") manufactured by the Marion Steam Shovel Co. of Marion, Ohio.

After reaching the waste dumps the muck piled on the flat cars was pushed off in one continuous motion by a three-ton plow, as shown in Figure 47. In this manner, an entire train could be emptied of its contents in a matter of minutes instead of hours, dumping one car at a time.

Figure 47. Example of a right-handed Lidgerwood plow being used to empty excavation spoils from modified muck flat cars, using a cable winch powered by the locomotive. In this manner a 20-car muck train could be unloaded in about 10 minutes.

In order to work efficiently the rail cars needed to be positioned near a descending slope, to help the muck fall away from the tracks. As these depressions were infilled, it became necessary to spread the muck dumped on gently sloping ground. Either Lindgerwood or Jordan Bank Spreaders were used to push the loose fill away from the tracks, like a mechanical road grader (Figure 48). In this manner enormous tracts of land were filled with millions of cubic yards of excavated material, as shown in Figure 49.

Track shifters. W. G. Bierd's other significant contribution to the canal was his invention of the "track shifter," which increased the efficiency of moving soil and rock fill on the project. Bierd reasoned that he could employ lightweight rail-mounted cranes, which employed 10-ton stiff-leg booms with chain yokes to lift sections of panel track and shift them to the left or the right, as depicted in Figure 50. In this manner tracks could be quickly shifted and adjusted to meet the needs of steam shovels and their appurtenant muck trains at the point of excavation, or similarly adjusted at the spoil dumps near either coast. The 75 pound steel rails had sufficient flexibility to be shifted a few yards at a time, provided that the ties plates were securely attached. Stevens quickly ordered 70 additional wreck cranes, and these became an integral part of the "engineering plant" assigned the task of "making the dirt fly" on the Isthmus.

Figure 48. This shows a Lidgerwood Spreader, which employed mechanical arms to spread fresh muck and push it away from the railroad tracks, after it was dumped using a Lidgerwood plow. The steel plate blades of the spreader could be extended 13 feet from either side of the rail mounted car.

Figure 49. Some of the largest spoil piles were placed around La Boca and Balboa on the Pacific Coast, eventually connecting to the islands seen in the distance. Note the size of the locomotive, indicated this enormous expanse of fill is at least 30 feet deep.

Figure 50. This shows a track shifter lifting a section of panel track and swinging it to a new location, semi-parallel to the existing roadbed. In this manner the dump sidings could be shifted laterally to lay down an enormous pile of fill, sometimes filling hundreds of acres to depths of up to 35 feet.

Increasing rates of excavation. In the first three years the Americans only managed to excavate seven million cubic yards. Stevens stepped up the pace, removing a record 325,000 cubic yards in October 1906. By the close of 1906, there were almost 24,000 workers on the job, making it by far the largest construction project ever undertaken by the United States. In January 1907, the monthly figure had increased to half a million cubic yards, more than doubling the record set by the French. In February, it increased again, to 600,000 cubic yards, and by the spring was averaging 750,000 cubic yards per month, using 65 steam shovels. Stevens' preparatory work was paying some handsome dividends. In March 1909, a new record of 2,054,088 cubic yards was established, which would remain unbroken for the remainder of the project. That sum required 1,400,000 sticks of dynamite and 68 steam shovels feeding an average of 160 muck trains per day, along 76 miles of track. That figure testifies to the competent control and coordination of the rail traffic by the system of controls they established using telegraphs and telephones. It would forever after be referred to as "the railroad era" of the Panama Canal (McCullough, 1977).

Presidential visit. In November 1906, President Theodore Roosevelt was the first American president to ever leave the United States, to make a first-hand inspection of what many in the press referred to as "Teddy's big ditch in Panama." The ICC had known since July that a visit was being planned for the coming fall, and preparations had been underway for months in anticipation of his visit. The President and his entourage departed New York sailed on November 9th aboard the battleship *USS Louisiana*, which at 16,000 gross tons displacement, was the largest class of

American warship in an expanding navy seeking to exert itself as the dominant power in the Western Hemisphere (Morris, 2001).

As with just about everything else Roosevelt did, his visit was vigorous and full of surprises, dispensing of the modicum of protocol and schedule. He wanted to pay his respects to those involved in the great undertaking, but he also wanted to see first-hand how things were going, what the average worker thought of the management, and so forth. He arrived at the height of the rainy season, when things could be deplorable, and one day it rained three inches in two hours. None of this deterred the president, who at one point, simply hopped off the presidential train and seated himself in the operator's position of a giant Bucyrus steam shovel at Bas Obispo, while it was raining (Figure 51). There he began conversing with the operators asking a myriad of questions about how they were being treated, the quality of the food, what they thought of the management, and so forth. He told the workers that they were "*the pick of American manhood*" and to "*play their part like men among men.*"

Figure 51. Left - Theodore Roosevelt chatting with workers at Bas Obispo, during his visit in November 1906. Right image shows Roosevelt in white Panama suit climbing onto the ICC Shovel 114.

Roosevelt made so many impromptu stops (Figure 52), that it took him almost 10 hours to cross the Isthmus to Panama City that first day, a journey that usually took about an hour and 20 minutes. Stevens was five years older and an inch taller than the president, but after two days he confided to Frank Maltby that "*I have blisters on both feet and am worn out,*" then noting that "*Shonts is knocked out completely*"(Maltby, 1945).

Figure 52. Division Engineer Frank Maltby points to the Gatun Dam site as President Roosevelt looks on (in white suit), from the promontory forming the dam's right abutment.

Hattie Stevens hosted the First Lady, Edith Roosevelt, and the wives of the ICC engineers at their home in Culebra for lunch (Foust, 2013), and the presidential couple for dinner (Figure 53).

Roosevelt identified with Stevens because of his training on the frontier and *"reputation as a doer, and not so much as a talker."* A great gala was convened in the largest warehouse on the new wharf in Cristobal on the last evening of the Presidential visit. All of the engineers, administrators, managers, foremen, and medical staff on the canal were in attendance. Roosevelt made an impromptu motivational speech, telling his audience: *"Whoever you are, if you are doing your duty, the balance of the country is placed under obligation to you, just as it is to a soldier in a great war. As I have looked at you and seen your work, seen what you have done and what you are doing, I have felt just exactly as I would feel to see the big men of our country carrying on a great war."*

Roosevelt's seemingly distractive off-schedule forays did have an impact. On the return trip to New York he dictated a message to Theodore Shonts regarding his observations of the living conditions of the West Indian laborers. He noted the dirt floors and "shoddy nature" of their bungalows, which he felt were unhealthy. He recommended that their living quarters be upgraded and that they be fed in mess facilities set up by the ICC.

The battle over contracting. In July 1906, Stevens had been appointed a member of the Isthmian Canal Commission. That summer Shonts and Stevens began discussing by whom the construction work should actually be performed, by government force-account, or by private contractors. American railroads almost exclusively utilized

Figure 53. Lunch with the President at Culebra. From Left: John Frank Stevens, Harriet Shonts, Joseph Bucklin Bishop, Surgeon General Presley M. Rixey, Theodore Roosevelt, the president's stenographer M. C. Latta, Edith Roosevelt, Hattie Stevens, Theodore Shonts, and the 10-year old boy is Eugene Stevens, the Chief's youngest son, who became a close friend of the President's youngest son, Quentin Roosevelt (Hardy, 1939).

contractors because they maintained trained and experienced organizations. Stevens realized that he was training as much as managing a largely untried organization. Shonts asked Stevens to develop a suitable form of construction contract for its consideration.

Because of the sheer scale of the undertaking it soon became apparent that there were literally thousands of details that must be attended to, so many that the typical bid using unit quantities could not be employed down in Panama, at the end of a tenuous and lengthy tether. They decided to use a modified form of a percentage contract, which had been used in some of the largest railroad construction jobs back in the states.

Stevens took the draft contract with him to the White House for review by the president and his cabinet. The new Secretary of State, Elihu Root, insisted that a number of conditions be changed, such that Chief Engineer Stevens would not reign "*supreme over all matters and questions that might arise.*" With such changes and deletions, Stevens feared that such contracts would never realistically work, they would only serve to make the contractors wealthy.

Bids were advertised on October 9, 1906, which were due on January 12th. The ICC received bids from only four firms, each proposal for a percent of the gross profits. The lowest bidder was Oliver and Bangs of New York City, who bid 6.75 percent. Shonts cabled Stevens soliciting his advice. Stevens reply was that the lowest bid was too high, and that he wanted to see the record of work completed and capacity of each of the four firms before deciding. He was also concerned that there might develop "endless friction" over "conflicts of authority," based on the clauses Secretary Root had added.

The ICC allowed Oliver an additional ten days to form a $5 million corporation with new associates. During this interim Bangs was eliminated from the partnership when Oliver obtained "new partners." These last minute machinations rankled Stevens, who suspected that profiteering was likely their central motive, not completing the project on schedule and within the stated budget. These sorts of schemes had characterized the early transcontinental rail lines, which were built by syndicates that were bereft of construction expertise or experience, but managed to make record profits at the public expenses.

When Stevens learned of this he cabled Shonts informing him that he felt it would be a dreadful mistake to award the contract to Oliver, whom he did not believe to be qualified by "*nature, experience, or achievement*" (Duval, 1947). Stevens also argued that the additional time Shonts had granted Oliver was essentially allowing him to submit a new bid without any competitors.

Shonts departure and increasing frustration. What Stevens didn't know was that Theodore Shonts had received an attractive offer from the Interbourough Rapid Transit Company to lead a potentially lucrative rail transportation merger in New York City. On January 22nd, Shonts submitted his resignation, but not before informing Roosevelt and Taft that he was recommending the firm of William J. Oliver and Anson M. Bangs be awarded the canal excavation contract (Figure 54). This was done without having consulted Stevens or having chased down the additional information he requested. In the wake of Shonts resignation, Stevens was named Chairman of the Isthmian Canal Commission, consolidating his authority and, presumably, assuring him of Roosevelt and Taft's increasing confidence in his abilities as an able administrator.

Stevens had also cabled Secretary Taft with the same objections he dispatched to Shonts, and Taft requested a more detailed explanation of his disapproval. Taft and Shonts were unaware of the newspaper interviews with Oliver, a millionaire contractor from Knoxville, which had triggered considerable discomfort with the senior engineers on the Isthmus because of comments he had made about "*gathering together a new pool of specialists for each aspect of the canal work and to develop new engineering plans*" and his plan for using "*convict labor*" to carry out the most difficult work. The bottom line was that Oliver didn't have the qualifications or the track record to be overseeing the world's largest civil engineering project, all he had was a group of New York investors seeking to make some money. Stevens concluded with a statement that "*as a business proposition, the contract should never have been advertised*" (Duval, 1947).

This time Taft's response was much the same, requesting copies of Oliver's newspaper interviews. Stevens responded with a lengthy cable explaining that the interviews had appeared in just about every American newspaper, with Oliver bragging about how he and his associates were going to bring some real steam shovel men down to Panama and make the dirt fly, using thousands of Negro convicts from the deep South, who would stand up better to the heat and disease. Stevens concluded with the statement that *"a Napoleon was not needed here (Panama), but such an organization as outlined in my letter of July 27 (1906) to Chairman Shonts,"* which he asked Taft to retrieve from his own files (Baugh, 2005). Taft disagreed and decided to hold Oliver's bid in abeyance (Duval, 1947). This must have angered Stevens.

Figure 54. Newspaper article announcing the awarding of the Panama Canal construction contract to Knoxville millionaire William J. Oliver. He was given 10 days to find new partners if Anson M. Bangs' was unable to come up with $5 million in capital to secure the necessary bond (newpapers.com).

After some discussion with Taft, on February 8th, Roosevelt sent Stevens a letter in which he expressed his disbelief that he should work so hard to rescind the contract to Oliver, given the fact that he had developed the very plans from which

Oliver's forces were to be directed. He requested Stevens' assistance in seeing the matter through to completion. Stevens replied the next day that this was the first time he had been advised that this was a contractual matter with a reputable construction company, he had believed Oliver to be nothing more than a showman of sorts, of which there are many, bereft of any actual heavy construction experience. Stevens then listed all of the points that were different from Oliver and Bang's original bid. He reminded Roosevelt that he was opposed to paying the contractor on the basis of percentages, and to the contractor supplying all of the labor. During a meeting of the ICC in Washington in December he had opposed these same clauses, but had been overruled. He ended by telling Roosevelt that he would receive a personal letter which would clarify the matter.

Discouragement and resignation. Stevens (Figure 55) must have been discouraged by the chain of events during January 1907 battling with Shonts and Taft, and now Roosevelt, over a contract with someone who, in his mind, was clearly unqualified for the tasks at hand. In those days there was no qualifications-based selection process, the owner or his representative may or may not exercise due diligence to ascertain if bids were submitted by unqualified entities. If the contractors could present the required capital (one-fifth of the project cost) and secure a performance bond (for 50% of the project cost), the government was willing to award the contract to them.

Figure 55. John Frank Stevens served as Chief Engineer for 21 months, from July 2, 1905 to April 1, 1907. Stevens oversaw the preparatory work to establish sanitation, adequate housing, and establish schemes that were used to excavate the Culebra Cut and efficiently dispose of the fill spoils (Library of Congress).

During 1906, the average volume of excavation had been steadily climbing, reaching 457,000 cubic yards by December. The organization he had worked so hard to build over the previous year and a half was starting to jell. Reflecting of this progress, Stevens had changed his mind about hiring contractors. He now favored having the ICC carry out the construction work under the direction of its own engineering department.

It was under this pall of discouragement that on January 30th, Stevens wrote an ill-advised letter to Roosevelt, which the President received on the February 12th. In his letter he outlined in very plain terms how frustrated he was with the contractor selection process of over the previous two months. He said that he had been obliged to *"fight a continuous battle with enemies from the rear"* and that he was *"continually subject to attack by a lot of people, and they are not all in private life, that I would not wipe my boots on in the United States."* The reference to men "not all in private life" presumably referred to Shonts, Root, Taft, and even Roosevelt. At the end Stevens requested that Roosevelt relieve him of his duties in two or three months, so long as it would not *"embarrass in any way your plans"* (Duval, 1947).

It was not the sort of accusation a subordinate can level at their superiors, regardless of how frustrated they might feel. Roosevelt was shocked and angry. It was obvious to him that Stevens had cracked under the pressure, that he didn't realize that in public service one is obliged to numerous acts of compromise, as well as accountability. Roosevelt waited a few days for his anger to subside, and wrote to Stevens on February 14th, accepting his resignation effective April 1, 1907. Roosevelt never mentioned Stevens in his autobiography while describing the many challenges constructing the Panama Canal. Late in Stevens' life, he left to his eldest son Donald signed photographs of *"the two men I most admired."* They were of James Jerome Hill and Theodore Roosevelt.

When news of Stevens' resignation reached the Isthmus it created quite a stir, as he was fondly referred to as "The Chief," having established himself in the employee's minds as the champion of their welfare. The engineers appreciated Steven's prowess as a problem solver, who was tough, but fair. When they made inquiries he now responded *"don't talk, just dig."* The rank and file workers decided to appeal to Stevens' patriotism to stay on, circulating a petition that was signed by every one of the American engineers and employees then living in the Canal Zone. Stevens agreed to remain on the job until midnight on March 31, 1907, at which time he would officially turn over his position to Army Lieutenant George Washington Goethals, who had arrived on the Isthmus in early March, to meet with Stevens and familiarize himself with the project before the turn-over. Hattie and John Frank were feted in an enormous going away party held at the same warehouse in Cristobal where Roosevelt had been honored a few months previous. There Stevens was presented with a gold watch, a gold ring (to replace one that had been stolen from him), and silver coffee service by the employees. He was also presented with the official petition asking him to remain in Panama, a bound volume with nearly 4,000 signatures and salutations. His *"accomplishments [in Panama] were toasted with much remorse and not a few tears by all in attendance.*" For the next 36 years he would only state that he made the decision for "personal reasons," which he chose never to reveal (Stevens, 1927).

Epilogue on Stevens' resignation. The glaring difference between the demanding assignment in Panama and all of the previous engagements in Steven's career was the absence of absolute control. Everything Stevens did or didn't do was subjected to being overruled by politicians, a predicament that has always proved loathsome for engineers when it comes to their engineering judgment being questioned or overruled. In the beginning of his time in Panama he focused on problem solving, operating with more or less of a free hand, like he had for the Great Northern Railway. In the end, it was his perceived inability to control the project that brought on despair and discouragement.

Soon after accepting Stevens' resignation President Roosevelt met with Oliver and his partners regarding their revised bid. It appears that he was persuaded to agree with Stevens about the suitability of their proposal. On February 26th, Roosevelt sent a letter to the Isthmian Canal Commission summarizing his findings and decision to reject all bids, in large measure because of Stevens' unexpected resignation. Some of its highlights included:

1) *"There were two bids worthy of consideration...The Oliver and Bangs bid, at 6 percent, was rejected as not satisfying the specifications of the invitation."*

2) *"Mr. Oliver was allowed to perfect his bid with new associates and new financial responsibility; but this permission did not in any way change the situation from what it would have been had Mr. Oliver's bid in its present form been presented on January 12th, the day for receiving fixed bids."*

3) *"One of the chief reasons for adopting the contract as proposed was that in its main features it was formulated by Mr. Stevens, who was expected to supervise the work as Chief Engineer. He had had experience with contracts of this character and he had had eighteen months' actual experience with the work on the Isthmus. Less than ten days ago I received a letter from Mr. Stevens in which he asked to be entirely relieved from work on the canal as soon as he could be replaced by a competent person and that person could become familiar with the work. I have accepted his resignation. The withdrawal of Mr. Stevens takes away the special reason mentioned for proceeding under the present form of the contract."*

4) *"In order to secure continuity in engineering control and management in the future, I have decided to request you to assign the office of Chief Engineer, Major Goethals, a member of the Corps of Army Engineers."*

The President's February 26th announcement decreed that the project would, henceforth, be designed, and its construction managed by, the Isthmian Canal

Commission, a duly designated arm of the U.S. Government, which would be dissolved whenever the canal project was declared completed. In doing so, a significant precedent was set, which would be revisited with the construction of the Boulder Canyon Project in 1931-35 and the Tennessee Valley Authority in 1933-58, among others.

Stevens and his family departed Colon on the steamer *SS Panama* on Sunday April 7th. He accepted a position as Vice President of the New Haven & Hartford Railroad in New York City, where they would maintain an apartment on Park Avenue until Hattie's death in 1917. A number of senior engineers decided to resign rather than to work for Army officers. These included Frank Maltby and William G. Bierd, the latter whom took a new position working for Stevens on the New Haven lines. Others, like D.W. Bolich, departed in mid-1908 after Colonel Goethals re-structured the organization of the project, placing all of the work under three division engineers (Sibert, Gaillard, and Williamson).

Those closest to Stevens felt that one of his prominent characteristics was his sensitivity to criticism (Budd, 1944). This may have been ascribable in some measure to his lack of formal engineering education, although this was more than compensated by his wealth of practical construction experience. Others have pointed out that as a railroad engineer, he had little experience in the design or construction of harbor or port facilities, such as concrete locks, their steel gates, and dams. Despite any ignominy associated with his departure, Stevens flourished as a consulting engineer to railroads, as Chairman of the US Railroad Commission to Russia appointed by President Wilson in 1917 (and where he remained until 1923), and served as ASCE President in 1927. In 1935, *Engineering News Record* ran a series of articles about his professional career, which were then compiled into a special autobiography published by ENR (Stevens, 1935). In 1946 the Canal Zone issued a regular five-cent stamp featuring Stevens' portrait (Figure 2). Two new biographies of Stevens have recently appeared in print; one by Baugh (2005) and another by Foust (2013).

THE FOURTH ISTHMIAN CANAL COMMISSION

On March 4, 1907, President Roosevelt issued an executive order that consolidated the offices of Chairman and Chief Engineer in John F. Stevens. On March 16th the remainder of the Commission with the exception of Colonel Gorgas, were asked to resign, followed by the resignation of Stevens, to be effective March 31st.

George W. Goethals was promoted to Lieutenant Colonel on March 2nd and appointed to the Commission on March 5th. He would assume the role of Chairman and Chief Engineer on April 1st. Unlike the previous commissions, Goethals authority would be supreme. President Roosevelt was through with bickering and resignations. From the outset he informed everyone: "*Colonel Goethals here is to be chairman. He is to have complete authority. If at any time you do not agree with his policies, do not bother to tell me about it – your disagreement with him will constitute your resignation*" (Bishop and Bishop, 1930).

Joseph Bucklin Bishop would continue serving as the Commission's Secretary in Washington, DC. Between March 5th and 27th, new commissioners were appointed, which included: Major David D. Gaillard, Major William L. Sibert, Surgeon Colonel W. C. Gorgas, Rear Admiral Harry H. Rousseau, Senator J.C.S. Blackburn, and Jackson Smith. Goethals, Gaillard, and Sibert were graduates of West Point serving as officers in the Army Corps of Engineers, while Rousseau was Chief of the Navy's Bureau of Yards & Docks.

Goethals presided over his first ICC meeting on April 8, 1907, in Culebra. Sibert, Gaillard, and Rousseau were appointed Supervisory Engineers. Sibert would have charge of design and construction of locks and dams; Gaillard was placed in charge of all excavation work; and Rousseau of all municipal engineering in the Canal Zone, Colon, and Panama City, and the Division of Motive Power and Machinery. Blackburn would supervise the Commission of Governmental Affairs; Colonel Gorgas would remain in charge of the Sanitary Department, and Smith was to continue serving as Chief of the Department of Labor & Quarters.

Goethals then appointed three-man committees for consideration of specific matters, such as Engineering, Governmental Affairs, Sanitary Matters, and Labor, Quarters and Subsistence. The Engineering Committee was comprised of Sibert, Gaillard, and Rousseau. After Bishop moved to Panama in August 1907, all of the commissioners resided on the Isthmus year-round, and were granted up to one month's leave each calendar year.

After Joseph Ripley resigned in June 1907 as Assistant Chief Engineer overseeing the various aspects of design in Washington, DC, General Mackenzie was persuaded by Taft and Roosevelt to release Major Harry F. Hodges, who was temporarily assigned the role of General Purchasing Officer in Washington, DC, while the engineering design section was reorganized into the Division of Design of Locks, Dams, Regulating Works, and Accessories. Hodges was promoted to Lieutenant Colonel on August 27th, named Assistant Chief Engineer of the canal project, and given charge of the new division. This division was relocated to the Isthmus shortly thereafter, and worked out of the administration building in Ancon.

In mid-1908, the management of the project was re-organized, according to geographic sectors, while retaining some of the technical expertise departments, such as structural design. Sibert was given charge of the Atlantic Division, which included Cristobal Harbor, its breakwaters and port facilities, Gatun Locks and Dam, and the spillway structure and outfall channel. Gaillard was placed in charge of the Central Division between Gatun and Pedro Miguel, which included most of the excavations along the Culebra Cut through the Continental Divide. Rousseau became Assistant to the Chief Engineer, supervising the construction and operation of shops and terminal facilities, including dry docks. Sydney B. Williamson was given charge of the

Pacific Division, which included construction of the Pedro Miguel Locks, the Miraflores Locks and Dam, and excavation of the channel connecting Miraflores to the Pacific.

In July 1908, Jackson Smith resigned and Lieutenant Colonel Harry F. Hodges of the Corps of Engineers was appointed to replace him. Senator Blackburn resigned in December 1909 and Maurice H. Thatcher of Kentucky was appointed by President Taft to replace him in April 1910 (Figure 56). Mr. Thatcher resigned in August 1913 and was replaced by Richard L. Metcalf, who served until the Isthmian Canal Commission was abolished by Congress effective April 1, 1914.

Figure 56. Members of the Isthmian Canal Commission in mid-1910, from left: William L. Sibert, Joseph Bucklin Bishop, Maurice H. Thatcher, Harry H. Rousseau, George W. Goethals, David D. Gaillard, Harry F. Hodges, and William C. Gorgas. Five of the eight members were civil engineers (Linda Hall Library).

GEORGE W. GOETHALS – THE THIRD CHIEF ENGINEER

Career Army Engineer. George Washington Goethals was born in Brooklyn in June 1858, to a Belgian immigrant family, a few miles from where Theodore Roosevelt was born the following October. When he was 11, the family moved across the East River to New York, where he completed his secondary education. In

1873, he enrolled at the City College of New York with the intent of becoming a physician. In January 1876, he learned of a vacancy for an appointment to West Point, so he applied, passed the examination, and joined the plebe class at the Academy in June 1876. Like Robert E. Lee before him, Goethals managed to matriculate through the Military Academy without any demerits, holding the rank of Cadet Captain and graduating Second in the Class of 1880 (Figure 57). He accepted a commission in the Corps of Engineers and spent two years at the Engineering School of Application at Willets Point and in Washington, DC. His next duty station was at Vancouver Barracks in Washington Territory, where he managed to replace a bridge that had been washed out over the Spokane River, which he referred to as *"The hardest job I ever tackled"* (Bishop and Bishop, 1930).

In the summer of 1883, he met Effie Rodman, daughter of Captain Thomas R. Rodman of New Bedford and sister of Lieutenant Samuel Rodman, West Point Class of 1882. They courted and corresponded with one another and in September 1884, Goethals was posted to Cincinnati to work on Ohio River improvements. He and Effie were married in New Bedford the following December. From 1889, the Goethals would vacation in New England close to Effie's childhood haunts in Martha's Vineyard. In 1893 they purchased a lot in Vineyard Haven, where they built their permanent home the following year.

In August 1889, Goethals was transferred to Nashville to work on the design and construction of locks and dams on the Cumberland and Tennessee Rivers. In January 1891, he was made engineer-in-charge of Tennessee River improvements working out of Florence, Alabama. This work entailed completing the Muscle Shoals Canal and building the Colbert Shoals Lock, which were on a fast track because of legal proceedings on freight rates to be heard the following year in Chattanooga. Goethals organized his forces accordingly, forming day and night shifts, placing an assistant in charge of the day shift and himself of the night shift. In December 1891, he was promoted to Captain, and he remained at Muscle Shoals for three years. This work included the construction of 14 miles of railroad that was needed for the construction and for pulling ships through the canal.

The most difficult lock was at Colbert Shoals on the Tennessee River, where Goethals designed the Riverton Lock, a flight of two locks each with a lift of 13 feet. The 26 foot lift was the largest in the United States at that time. When the contractor began excavating the lock trough he encountered quicksand. He installed sheetpiling, but these failed miserably, and the project became stalled. The contract was annulled and Goethals hired a new Assistant Engineer named Sydney B. Williamson, who was an 1884 graduate of the Virginian Military Institute. Williamson managed to sink test pits down to bedrock, then led the workmen into the dangerous excavations to lend confidence to the scheme he devised for excavating the running sands. Goethals was transferred to Washington, DC before the Riverton Locks were completed by Williamson. No one came between the two men after that, Goethals sending for Williamson wherever he was tackling difficult problems.

Goethals' work on the Tennessee River was under the supervision of Thomas L. Casey, who served as Chief of Engineers of the Army from 1888-1895. Casey summoned Goethals to the Capitol to serve as his assistant, reviewing engineering specifications, contracts, surveys, and financial accounts for the Corps' work. He quickly established himself as an able trouble shooter and remained in Washington four years, working for Generals Casey, Craighill, and Wilson.

When the Spanish American War erupted in April 1898, Goethals services were sought by Generals John R. Brooke and William Ludlow. He accepted the position as Chief Engineer of the 1st Army Corps with the temporary rank of Lieutenant Colonel, and he embarked with that force sailing from Newport News to Puerto Rico in July. Upon landing, Goethals met innumerable challenges in scavenging materials, equipment, and the necessary docks and wharves to unload the expedition's vessels or their horses, artillery, firearms, and stores. Just as the Corps was beginning to undertake offensive operations against Spanish forces, the war was concluded on August 13th.

Figure 57. Left – Cadet Captain George Washington Goethals graduated #2 in the Class of 1880, which had 52 cadets (USMA). Right – Colonel Goethals at the desk of his office in Culebra in 1908.

In February 1903, the General Staff Act was enacted, which created a Chief of Staff for the U.S. Army who would be assisted by two subordinate generals and a corps of 42 officers as representatives of the various branches, such as infantry, cavalry, artillery, engineers, quartermasters, ordnance, and so forth. This Staff Corps was selected by a panel of generals and were established in the Army-Navy Building in Washington, DC in August 1903. That first group included Majors Goethals and Gaillard from the Corps of Engineers, as well as Major Peyton C. March of the infantry and Captain John J. Pershing of the cavalry (25 of the 42 officers selected in 1903 went into become generals).

Selection as Chief Engineer and Chairman. On February 1, 1904, President Roosevelt named William Howard Taft as his new Secretary of War, bringing him back from the Philippines where he had served as Governor-General. Major Goethals (Figure 56) began working with Secretary Taft when he assumed the position of Secretary of the National Coast Defense Board (known as the "Taft Fortification Board"), established in January 1905.

Taft was so impressed with Goethals that on June 30th, 1905 (two days after Taft accepted John F. Wallace's resignation), he encouraged Isthmian Canal Commission Chairman Theodore Shonts to find a place for the Army Engineer in administering the work on the canal in Panama. On July 2, 1905, Shonts sent a note to John Frank Stevens, within hours of his verbal acceptance of the Chief Engineer's position. He suggested that Stevens consider bringing Major Goethals onto to his staff when he arrived in Panama. He even mentioned Secretary Taft placing a letter in Goethals' personnel file detaching him to Panama as soon as the Chief Engineer made the request. Stevens declined the suggestion, not feeling that he could be assisted by a military engineer.

Undeterred, Taft took Major Goethals with him on his next inspection tour of the canal project in early November 1905 (Figure 58). Goethals was accompanied by Lt. Colonel William M. Black, who oversaw the ICC's engineering department in Washington, because he was intimately familiar with Panama. Goethals' initial impression of the Canal Zone was that it was "chaotic." This was soon after the yellow fever scare, while Stevens was focused on "preparatory work." Taft's pretense for bringing Goethals was to look at the matter of fortifications to protect the American interests there, but that may have been a pretense for introducing Goethals to Stevens. Whatever his motive, Taft seemed to value Goethals' opinions, and upon their return Goethals returned to his General Staff job and teaching at the War College.

Wary of appointing a third civilian engineer who might also resign, Theodore Roosevelt sought advice from his Secretary of War William Howard Taft and from the Army's Chief of Engineers, Brigadier General Alexander Mackenzie. Both men recommended Major George W. Goethals of the Corps of Engineers.

On February 18, 1907, Goethals was officially notified that he was to be assigned duty in the Canal Zone. That evening, while he and his wife were hosting a dinner party at their quarters for an old West Point friend named Colonel Fieberger, he was unexpectedly summoned to the White House by President Roosevelt. He hurried upstairs to don his dress uniform and dashed to the White House, while Fieberger and Effie impatiently awaited his return. Roosevelt told him he was being appointed Chief Engineer of the canal project and that it was "*essential for the successful prosecution of the work*" that there not be "*frequent changes of leadership*" (Bishop and Bishop, 1930). Roosevelt then described how the composition of the Isthmian Canal Commission was being altered once again, and that Goethals would serve as Chief Engineer and Chairman simultaneously, to avoid the conflict that had often arisen between these authorities. He then apologized for the unwieldy nature of the commission, which was required by the Spooner Act.

Figure 58. Secretary of War William Howard Taft and his entourage of Army and Navy aides and Marine guards making an inspection of the area around Empire in April 1907. It was on a similar tour in 1905 that Major Goethals first viewed the canal project.

Their discussion then turned to the matter of reviewing the bids that had been submitted to the government on January 12th, which Stevens had rejected. Goethals first task was to review these and report back to the President as soon as possible. He penned a report five days later, pointing out most of the same problems Stevens had raised. He began by stating that only two of the bids were worthy of consideration, but that the two remaining bids should be rejected because the contractors proposed to operate on borrowed capital, which would significantly reduce their profit margins, especially if they sought out qualified engineers and foremen to administer the job. The proposal also assumed that Stevens would be supervising the work, and he had tendered his resignation. Goethals didn't feel comfortable "*jumping into Stevens shoes,"* whom he knew to be a very capable and experienced railroad man. Goethals suggested that he be allowed to supervise government-paid forces to complete the Herculean task, *"using the organization that Mr. Stevens had built up."* The President concurred and the announcement was made on February 26, 1907 that all bids were being rejected.

Division Engineers. Goethals had only a few days to select his new management team, knowing that Roosevelt preferred Army or Navy officers *"who would see the job through till completion."* The first officer he chose was Major Harry F. Hodges, because of his experience working on the design and construction of the Soo Locks. This choice was initially opposed by General Mackenzie because Hodges served as one of his principal assistants in the Office of the Chief Engineer, and he could ill afford to lose him. The next officer he selected was Major David D. Gaillard (West Point Class of 1884), a fellow Army Engineer who was also serving on the General Staff in Washington. The third candidate was Gaillard's former West Point roommate, Major William L. Sibert (West Point Class of 1884), who had more construction experience with canals and locks than any other officer at the time. Navy Lieutenant Harry Rousseau was also selected to be the Assistant to the Chief Engineer. In January he had been named Chief of the Bureau of Yards & Docks for the Navy, with the rank of Rear Admiral. His rank would revert to senior Lieutenant. Rousseau was an 1891 graduate of Rensselaer Polytechnic Institute, the nation's foremost engineering school in the late 19th Century. The selections of Majors Gaillard and Sibert and Rear Admiral Rousseau were among those announced by President Roosevelt on February 26th.

When Goethals assumed the role of Chairman and Chief Engineer of the ICC, he couldn't have imagined how many details would need attending to over the next 7-1/2 years. But Goethals was not just the Chief Engineer of the project, he also served as the head of the Panama Railroad, the Commissary, the Panama Railroad Steamship Company, the tugs and dredges operated by the Canal Commission, 100 steam shovels, 246 locomotives, 2,200 rail cars, hospitals, dispensaries, and even a sanitarium on Taboga Island. He also served as the leader of the civil government, courts, schools, post offices, police, a penitentiary, and a battalion of Marines. Goethals was ably assisted by a group of about one hundred engineers and managers, who, by the end of 1907 he was overseeing the activities of about 4,400 Americans and roughly 42,000 laborers.

After moving to Panama, Goethals summoned a few engineers that would play prominent roles in the canal's construction. Foremost among these were his old colleague Sydney B. Williamson and Lieutenant Colonel Harry F. Hodges. Williamson was an 1884 graduate of Virginia Military Institute, while Hodges graduated from West Point Class in 1881, a year behind Goethals. After General Mackenzie retired, Hodges was transferred to Panama in July 1908 as Assistant Chief Engineer and appointed to the Isthmian Canal Commission.

Goethals establishes a presence. On March 2nd, Goethals was promoted to Lieutenant Colonel and officially apprised of his new responsibilities on March 4th, to take effect on April 1, 1907. His new salary would be an astounding $15,000 per year, five times his major's pay of $3,000 per annum, but only half of what Stevens had received. Goethals would not be required to wear his Army uniform and he was allowed just a few days to prepare for the voyage to Panama so he could spend a few weeks with John Frank Stevens to effect a smooth transition of responsibilities.

On the evening of March 17th, Goethals was formally introduced to the Americans working on the canal at a "smoker" (welcoming party) at Corozal. Stevens was not attending, but every time his name was mentioned the assembled crowd cheered. When Goethals was introduced there was stone silence. Many in the audience believed the Army's assumption of the project's management would usher in an era of uniforms and salutes. In Goethals' inaugural address to the senior engineers, managers, and foremen he promised *"that I would look after your interests as they would be my own; that every man would have the right of audience. I will say that I expect to be the chief of the division of engineers, while the heads of the various departments are going to be colonels, the foremen are going to be the captains, and the men who do the labor are going to be the privates. There will be no more militarism in the future than there has been in the past. I am no longer a commander in the United States Army. I now consider that I am commanding the Army of Panama, and that the enemy we are going to combat is the Culebra Cut and the locks and dams at both ends of the canal. Every man who does his duty will never have any cause to complain on account of militarism"* (Bishop, 1915).

Retaining Stevens' engineering plant. Stevens' system of moving muck to spoil piles by rail and issuing cash bonuses to steam shovel operators when they exceeded their monthly quotas was beginning to have an impact, as were the numerous rail linkage improvements and the sustained flow of newer, larger, and more capable equipment. Each month new excavation records were being set: in March 1906 the figure was 800,000 cubic yards, in April it rose to almost 900,000, a staggering figure compared to the French efforts. There were now 500 muck trains per day entering and leaving the confines of the Culebra Cut, heading either (towards Gatun) or south (towards La Boca).

Daily routine. Goethals began by spending his mornings in the field during the cool of the day and his afternoons at his headquarters in Culebra. He initially hitched rides on the regularly scheduled trains running north and south, across the Isthmus. In 1905, the ICC had purchased a gasoline-powered railroad "motor car" (Figure 59), manufactured by the Sheffield Car Co. of Three Rivers, Michigan. This allowed the Chief Engineer more freedom of movement and make "surprise inspections" along the Panama Railroad's right-of-way. The workers quickly dubbed it the "brain wagon" or the "Yellow Peril" because of its yellow hood (Keller, 1983). Traveling in the Motor Car, Goethals regularly viewed the progress being made at the various construction sites. These included two dams, six sets of locks, two artificial lakes, regulating works, entrance channels, breakwaters, telephone and telegraph systems, a hydroelectric station, a rebuilt railroad, and the excavation of the challenging Culebra Cut.

Figure 59. Colonel and Mrs. Effie Goethals pose in front of the gasoline-powered motor car that Goethals used to travel up and down the rail line, visiting various parts of the canal project.

Goethals employed a sizable group of "spotters," men who would circulate amongst the workers in different parts of the Canal Zone, listening for any rumblings of discontent, or noting loafers who didn't appear to be working (Parker, 2007). Being out in the sun each afternoon Goethals soon became very tanned, which contrasted noticeably with his snow white hair. He smoked cigarettes during his waking hours and his white moustache was stained brown by the nicotine.

Family life. One of the great disappointments for Goethals must have been his wife's extended absence from the Isthmus. She remained in Washington, DC for many months after the Colonel was transferred to the Isthmus "*doing society at a great rate*" according to McCullough (1977), who also described her as "*tall and vane.*" Mrs. Effie Rodman Goethals (Figure 58) was from a respected New Bedford family, who seldom felt comfortable sharing the spotlight with her famous husband. Her position obliged her to entertain an astonishing array of politicos who frequented the Isthmus to see "the big ditch" being dug. Many of those, such as President William Howard Taft, stayed with the Goethals in their home at Culebra during his visits to the canal in 1910 and 1912 (Figure 60). Effie was never comfortable in her role as "First Lady of the Canal Zone," even with the array of servants that came with the Chief Engineer's residence. This was in considerable contrast with Hattie

Stevens, who had found the Isthmus "charming" and its residents "full of zeal," often accompanying the couple's youngest son Eugene on his forays into the hinterlands. Most residents felt that the Colonel's wife preferred coastal New England society to that of Culebra.

Figure 60. Goethals pointing out some feature to President William Howard Taft. Taft made five trips to the Canal Zone as Secretary of War and two as the nation's chief executive, in 1910 and 1912. The President and his wife Nellie always stayed at the Goethals' home when they visited the Isthmus (American Press Association).

In the spring of 1912 the Colonel and Mrs. Goethals were accompanied by their younger son Tom, who received a leave of absence from his last semester at Harvard for a vacation tour of Europe, the only one Goethals ever made to the old countries. Goethals accepted invitations to visit some of the locks of the Kiel Canal in Germany, the harbor at Hamburg, and the Teltow Canal in Berlin, and a military engineers' demonstration in Potsdam. The Colonel even had breakfast with Kaiser Wilhelm on March 10th. Their discussion centered on military fortifications of canals and harbors, which the German press subsequently denied (Bishop and Bishop, 1930).

The Goethals' older son, George Rodman, graduated fourth in the Class of 1908 at West Point, receiving a commission in the Corps of Engineers. In early November 1910, Second Lieutenant George R. Goethals and his new bride, the former Priscilla Jewett Howes of Watertown, New York, arrived in the Canal Zone, bringing some semblance of family to the Colonel, who spent most of his time alone, working long hours and avoiding most social functions. Lieutenant Goethals was initially posted to the transportation and operations department of the Panama Railroad. Priscilla Goethals was lively and beautiful, "*cutting an attractive figure*" wherever she went. Her presence at any function was coveted, and she was capable of coercing her stately father-in-law to attend more social functions than he had previously. After his promotion to First Lieutenant in February 1911, the younger Goethals was transferred to the Pacific Division to work as an Assistant Engineer under Sydney Williamson. He and Priscilla were assigned new quarters at Balboa, on the Pacific Coast. There he supervised construction and materials handling at Balboa and at Miraflores until January 1912, when he was placed in charge of construction of fortifications on Naos, Perico, and Flemenco Islands, off Balboa. Here he remained so engaged until August 1914, when he was posted to West Point as an instructor of engineering.

Labor challenges. In early May 1907, the steam shovel operators, already the highest paid laborers at $210/month, decided the test Goethals by requesting a pay increase to $300/month and threatening to strike if their demands were not met. They figured they were the most important operators on the project, who "made the dirt fly." Wary of how he should act to avoid unnecessary confrontation, but exhibit an upper hand, Goethals acted cautiously. He decided to refer the matter to War Secretary Taft, who was staying with Goethals on an inspection tour of the Isthmus. Taft acted cautiously as well, not desiring any bad publicity. He passed the operator's demands onto President Roosevelt for comment. The President told Taft to give an appearance of reasonableness by countering their demand with a 5% pay increase, to $220.50 per month.

The operators rejected Goethals' counteroffer and walked off the job, abandoning 55 of the project's 68 steam shovels. This reduced excavation to about 25% of the previous month's level. Goethals quietly hired strike-breakers and ever-do-gradually filled the empty operator's seats, so that by early July he had replaced all of the striking operators. The strikers gave up and asked for their jobs back, but were told to either depart for the states or start over again as shovel helpers, at the lowest rate of pay.

In November, the boilermakers at two of the largest machine shops struck. This time Goethals acted more swiftly, bring in replacement workers, and the strikers were rounded up by the Canal Police Department under Colonel James Perry Fyffe (formerly of the Tennessee National Guard). This was the manner by which Goethals handled all of the threatened strikes henceforth, informing the workers that they would be expelled if they did not show up for work the following day.

Hearing grievances and dispensing justice. As John Frank Stevens had initiated, Goethals continued the practice of the Chief Engineer hearing complaints from "Gold Rolls (white) employees on Sunday mornings. This gave people a chance to air their grievances and receive a swift judgment, one way or another. Goethals' had a secretary and stenographer take notes of the proceedings, and an administrative staff that were responsible for following up on everything that was promised by the Colonel. Goethals had a dry sense of humor, and years later would describe "most of these proceedings" as *"Mrs. So and So's husband gets paid the same salary as my husband, so is entitled to the same size house, and so forth, but they have one more rocking chair than we have, so we would like an additional rocking chair"*(Bishop and Bishop, 1930).

When the Commission's Secretary Joseph Bucklin Bishop moved to the Isthmus in late August 1907, he suggested hiring an intermediary to deal with the grievances of the Spanish and Italian workers, which had become more frequent because of harsh treatment and the high cost of goods on the Isthmus was making it difficult for them to save much of their earnings. Many had become discouraged realizing that they wouldn't be able to save sufficient funds to purchase a return ticket by steamer back to their homelands. Recognizing the negative impact of such stories on public relations, Bishop hired Joseph Garibaldi, the grandson of Giuseppe Garibaldi, who spoke fluent Italian and Spanish. Garibaldi prepared a report that pointed out how the ICC's crackdown on "strikers," essentially, anyone who complained or stopped working, were often beaten, when in point of fact, some of the "stoppages" were simply problems in translation and understanding what the American bosses wanted. The number of Spanish worker continued to diminish, from 6,000 to about 2,500 by the time the canal was completed in 1914.

Being an investigative journalist, Bishop appreciated the potential disaster that stories of unnecessary cruelty could do to the Roosevelt Administration if they appeared in American newspapers. The strikes subsided, but so did the influx of European workers. When a liberal government assumed power in Madrid in February 1909, they passed a law forbidding the ICC to recruit workers in Spain. The Italian parliament followed suit a few months later.

One positive aspect of these problems was Bishop's volunteering his services to serve as a similar role as Goethals hearing grievances on Sunday mornings, from the "Silver Roll" employees, those foreign and minority workers that were paid in silver, dominated by West Indians from British and French colonies in the Caribbean. This did not necessarily improve their lot, other than to possibly retard unnecessary beatings by Canal Zone foremen or policemen. Like the European workers, the West Indian workers' plight continued to spiral downward as the cost of living gradually escalated on the Isthmus, without receiving any salary increases. Mess houses had been set up to feed them a "square diet" after Roosevelt's visit in November 1906, but the quality of the food was never what it was for the Gold Roll employees, and the West Indian attendance at meal time gradually diminished, as they found alternative sources of food they preferred to the white man's diet.

Work force and increasing production. All of the Canal Zone workers received seven paid holidays per year: New Year's Day, Washington's Birthday, Memorial Day, the 4th of July, Labor Day, Thanksgiving, and Christmas. In the wake of Roosevelt's visit in November 1906, the ICC sought some form of recognition that might motivate employees to "remain on the job," instead or rotating back home after only a year or so. In April 1908, the ICC decided to begin issuing "Roosevelt Medals" to American citizens who had worked at least two years in the Canal Zone. Additional "Two Years" bars were awarded for additional service at the four, and six year anniversaries (Figure 61).

Figure 61. In April 1908 the Isthmian Canal Commission began awarding the "Roosevelt Medal" to canal employees with at least two consecutive years of service. This shows both sides of a medal awarded to an employee who served more than four years, entitling them to a second "two year bar" (Collection of Robert Karrer).

During 1907, the canal work force grew from 24,000 to 32,000 and the Americans managed to excavate an astounding 16 million cubic yards of soil and rock, more than the combined total between 1904 and 1906! In 1908, the Americans excavated 37 million cubic yards, more than double that of the previous record year. By 1910, the number or workers was nearly 40,000. During the last three years of construction this figure would hover between 40,000 and 50,000. The total number of Americans working on the Isthmus reached a peak figure of 5,362 people in 1913, of which, about 300 were women. In addition there were approximately 2,500 dependents. Their average wage was $150 per month. The highest paid workers were steam shovel operators, who received $310 per month during 1913-14.

The pivotal motivation for this accelerating efficiency was "bragging rights" between the crews of the 77 steam shovels now working on the project. In early September 1907, ICC Secretary Joseph Bucklin Bishop began publishing the *The*

Canal Record, a weekly newspaper distributed free of charge to the Americans living in Canal Zone (Figure 62).

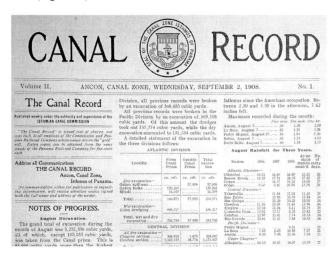

Figure 62. Typical cover page of The Canal Record, the weekly newspaper that heralded the accomplishments and construction records being recorded for posterity (author's collection).

Bishop was a veteran newspaperman with a wealth of journalism experience, having worked for three of the largest papers in New York City. Bishop's editorial policy forbade him from praising any American official, but to simply provide an accurate assessment of the progress being made on the construction project, which was part-and-parcel of Bishop's charge as the Commission Secretary (to record the work's progress). *The Canal Record* included general information on ship sailing schedules, and all manner of extracurricular activities, such as church picnics, school program, sporting events, dances, new items for purchase in the commissary, brief personnel notes about new arrivals and departures, and some letters to the editor. Copies of *The Canal Record* were also distributed in Washington, DC and to the major newspapers in the states. It was a public relations success of unimaginable magnitude.

Bishop also began printing the official figures for the cubic yards of excavation and spoil placement by each of the ICC's steam shovel crews (Figure 63), all of the muck trains, and all of the dredges. The citing of these figures in print for everyone to see quickly set in motion a sort of competition between the various crews, each one seeking to better the other's productivity, and to set new records. Records were sure to garner headlines and news stories mentioning the foremen and worker's names, which could be clipped out and sent home. The articles in *The Canal Record* had enormous impact on the morale and productivity, elevating efficiency in a manner that little else could.

Figure 63. Steam shovel crews like that pictured here, hoisting a 34-ton boulder, worked in friendly competition with one another for the honor of being singled out in *The Canal Record*, the project's weekly American newspaper.

Pacific locks moved from Sosa to Miraflores. In July 1907, excavation began on the Sosa-La Boca Locks near Ancon, on the Pacific side. Over the next five months two million cubic yards were excavated, almost exclusively by dredges. Wood trestles were then constructed along the intended toes of the earthen dike of the Sosa-Corozal dam, from which spoils from the Culebra Cut were dumped. This fill material began to slide laterally and dropped 8 to 10 feet, carrying the incomplete embankment with it, resulting in a slope of just 1 on 12 (vertical to horizontal), far too flat to complete a dam at that location. The massive landslide was likely the result of a foundation bearing failure, in the soft sediments underlying the trestle embankment. The west trestle experienced a similar landslide failure, bringing the project to an immediate halt.

Borings and test pits were made to ascertain the condition of the underlying foundation materials. The material overlying the bedrock and upon which they were attempting to lay fill for the embankment was found to be *"unctuous blue clay without grit, possessing but little supporting power, instead of the stiff clay shown on existing profiles."* They determined that the blue clay would have to be removed down to bedrock, but the depth of such excavations varied between 10 and 70 feet below sea level, and all of this excavated material would need to be replaced by fill spoil from the Culebra Cut. The Commission concluded that construction of an earthen dam at the Sosa Locks site would cost about $7.26 million more than had been allotted for the structure in the Minority Report of the *Report of the Board of Consulting Engineers for the Panama Canal* in February 1906.

In December 1907, four proposals for locks at the Pacific end of the canal were considered. The Third Project proposed to move the lower Pacific locks three miles inland to Miraflores, where excavation to bedrock appeared more feasible, and for the third lock to remain at Pedro Miguel, where the Minority Report had tentatively placed it in February 1906. It was felt that the bedrock could be reached with the existing dredges for no more than two locks at Miraflores, but this was subject to further examination with exploratory borings and soundings. In the end, this was the option that was selected, based on the borings (Williamson, 1931). Sibert was critical of this decision, favoring a grouping of all three locks at Miraflores, similar to the configuration he constructed at Gatun on the Atlantic side.

The site at Miraflores would also increase their areal distance from potentially hostile warships in the Gulf of Panama. In 1906, the U.S. Navy began producing 12-inch/45 caliber Mk. 5 naval rifles, capable of lofting 870 pound projectiles up to 20,000 yards (10 nautical miles). This rapid increase in firepower and range would soon make existing coastal fortifications obsolete (although it would take another ten years before effective sighting could direct fire at ranges in excess of 10,000 yards).

On October 29, 1907, the Navy's General Board requested that the width of the locks be increased from 100 to 110 feet in width, because the beams of the future battleships would likely "be in excess of 100 feet." This was indeed fortunate, as the largest class of battleship then being designed was the Florida Class, with a maximum width of 88 feet. The Navy's recommendation was approved by the ICC and implemented.

Mid-1908 reorganization. On June 30, 1908, the management and structure of the canal project was re-organized according to geographic sectors and divisions of responsibility, where the Division Engineers and their assistants would live. The headquarters with Colonel Goethals and his staff would remain at Culebra, alongside the central Division offices.

The Atlantic Division, under Sibert, was responsible for constructing the approach channels 41 feet deep and not less than 500 feet wide the Atlantic side, a seven mile-long channel leading to three massive locks at Gatun, as well as Gatun Dam, spillway, and powerhouse. The spillway discharged into an enormous concrete lined mill race that emptied into the ocean.

Gaillard would oversee the expansive Central Division from Culebra Heights. This encompassed 40 miles of work between Gatun and Pedro Miguel Dams, which included an extensive network of drainage channels and the landslides that made the project so problematic, extending its completion by more than two years. There would also be a significant amount of dredging of the navigational channel across Gatun Lake all of the way to Gamboa.

The work of the Pacific Division would be put under the charge of Sydney Williamson, who had an extensive experience working with Goethals as an Assistant Engineer and Construction Engineer on Corps of Engineers projects. His work would include the approach channels across La Boca Bay, up the Rio Grande to the twin locks, dam, and spillway structure at Miraflores, and continuing three miles north to the lock and dam at Pedro Miguel. A great naval base and the world's largest dry-dock would be constructed later at Balboa.

Standard of living. By mid-1908, there were over 1,000 families living in the Canal Zone, a dramatic increase over the previous four years. Marriage was encouraged as a means to keep the men out of trouble with the free flow of alcohol and brothels on both coasts. The Panama Railroad even ran extra trains on Saturday nights to take men to the bars and fleshpots of Colon (which had 140 saloons) or Panama City (which sported 220 saloons). These Saturday evening junkets received ample critique from the "married folks" and the ministers who arrived to tend to people's spiritual welfare. In May 1908, the Isthmian Ministers Association sent a resolution to the ICC asking them to forbid the making and sale of "*spirituous liquors*" because they represented "*one of the greatest curses of the present age*" and that they did not feel that the U.S. Government controlled territory should knowingly condone its sale and distribution "*where thousands of young men, without the benefits of home influence, are employed.*"

There is no record of whether this resolution was accepted or approved by the ICC, but the Commission's Department of Civil Administration prescribed those areas within which saloons and drinking places could be located in the various towns and villages of the Canal Zone. The Commission issued "*licenses for the sale of intoxicating liquors at retail*" to a number of individuals operating businesses in specific areas or lots of various hamlets along the line, such as Bas Obispo, Empire, Gorgona, Las Cascades, Gatun, Tabernilla, Matachin, Bokio, Rio Grande, West Culebra, or the 'French Dump' at Culebra. These licenses could not be transferred to others without approval of the Commission. Much of the Commission's time was taken up granting and approving transfer of liquor licenses, as well as movement of said businesses from one place to another.

The Isthmian Canal Commission provided married couple with "quarters" similar to the military, and these couples soon learned that virtually everything was provided free of charge, including piped water and utilities (after 1908), ice, fuel, distilled water, janitorial and landscaping services, schooling, medical and dental care, and hospitalization. Family quarters were held for employees who took leave for no more than eight weeks. Employees were also granted 14 days of paid sick leave per six months, and 30 days leave each calendar year. The benefits led to many of the younger men getting married, and single women (mostly nurses) found themselves being courted within days of their arrival, because there were numerous social events (including chaperoned tours) and church services scheduled on the weekends.

During John Frank Stevens' tenure, the carpenter force began constructing dining rooms for "guests with and without coats," bringing a sense of respect and proper decorum to the project. Stevens battled the stodgy ICC Commissioners over the funding of hotels, optional dining-out facilities, clubhouses, and recreational facilities. In November 1905, President Roosevelt overruled them and construction gangs began constructing attractive recreational facilities, such as men's clubs, pistol clubs, and several YMCA clubs, which included libraries, reading rooms, card rooms, billiard rooms, exercise rooms, dancing halls, and even bowling alleys. Numerous bands were formed and music lessons made available by those who could teach. Underground sewers were installed and streets were paved in not only the American zone, but in Colon and Panama City as well. The Chief Engineer's overarching goal in establishing this infrastructure was to create the most efficient environment for the tasks that they knew lay ahead.

By 1908, $2.5 million was being allotted per year for entertainment and recreation, which averaged about $750 per Gold Roll employee. Baseball diamonds, skeet, pistol, and rifle ranges, golf courses, and polo fields were constructed, along with clubhouses. Churches were built free of charge for those mainline denominations promising to send ministers to staff them. YMCA Clubs were constructed at Cristobal, Culebra, Empire, and Gorgona, and a Rest House and Food Station operated by the Salvation Army.

Drilling and blasting. The construction of the American canal consumed 61,000,000 pounds of dynamite, more than the aggregate total explosives used by the United States in all of the wars it had fought in! Most of the dynamite was shipped in wood tongue-and-groove boxes weighing 50 pounds, much of it coming from the Giant and Hercules Power Companies in the San Francisco Bay. A steamer could carry as much as 20,000 boxes, or one million pounds, in a single shipload. All of this was unloaded by hand at the wharves in La Boca or Critsobal. The 50-pound boxes were usually carried one at a time by West Indian laborers, who were referred to as "*powder monkeys*" (Figure 64), a friendly sobriquet that has been used ever since to describe gangs of men that handle explosives, regardless of race. After running a series of experiments, it was concluded that most of the miss-fires were ascribable to the holes being wired "in series." When the fuses were connected "in parallel" and fired using an ordinary electric current, the problems with miss-fires ceased.

The drilling and blasting activity required nearly half of the project's laborers to operate the mobile drills, operated using compressed air. The maximum number of drills employed in the Culebra Cut simultaneously was 377, of which 221 were of the smaller tripod type (Figure 41), and 156 were well drills (Figure 65). There were three principal drill sizes employed on the project. The largest were five inches in diameter and 100 feet deep (Figure 65), which were used for the deepest straight-sided excavations, for the lock structures, and the Gatun Dam spillway and millrace channel, at Gold Hill, the Gaillard Cut, and Hodges Hill. The medium size was three inches in diameter and 30 feet deep, which was used for most of the cut benches in the Culebra Cut. The smallest were 1-3/4 inches in diameter and up to 12 feet deep (Figure 40). These were used for production blasting whenever more dynamite was required to break up the rock between the deeper three-inch diameter production holes. They were also employed for "soft shots" at the back of finish cut slopes, where care was taken not to over-blast the country rock that would be left intact. Most of the production blast holes in the Culebra Cut were of the medium size, used to break up the rock into small enough pieces to enable easy excavation by steam shovels. The cuts were deepened and widened using multiple retreating benches, sloped one percent longitudinally, towards their respective coasts. An aggregate total of up to 475,000 lineal feet of boreholes could be drilled in a single month, if all the drills were working.

Figure 64. "Powder monkeys" loading drill holes with dynamite. Each wooden box contained 100 sticks of dynamite.

All of the smaller diameter drills were fed by a master compressed air system established by Stevens, and enlarged by Goethals, which included compressor plants at Las Cascades, Empire, and Rio Grande. The plethora of drill holes were each filled with sticks of dynamite, each weighing half a pound and eight inches long. These were loaded end-to-end, or separated by stemming, depending on the hardness of the rock. The workmen consumed an average of 800,000 sticks of dynamite per month, which required nearly half of the labor force to be involved in some aspect of drilling and blasting, or the support, thereof.

At the peak of excavation (March 1909), there were 300 rock drills working in the Culebra Cut, drilling 600 blast holes each day, and that month, expended 700,000 pounds of explosives. An aggregate total of 345,223 lineal feet of blast holes were drilled in the Culebra Cut. The role of the explosives was to fracture and disaggregate the rock sufficiently for the 75-ton and 95-ton steam shovels to excavate the resulting muck with their 3.75 and 5-cubic yard buckets. The dynamite loads in the holes were estimated by the rate of advance while drilling the holes, and the upper three to six feet of the holes were usually stemmed and tamped. The results were usually adequate for easy mucking, but there were occasional oversize blocks, as show in Figure 62. The largest of these oversize blocks that was successfully dumped onto a muck car was 34 tons. If the shovels couldn't handle the oversize blocks (known as "knockers"), they were pushed aside, where trailing teams of drillers would drill a few small diameter holes into the boulders using jackhammers, and a team of blasters would load the holes and shoot them at the end of a shift.

Figure 65. A group of six mobile well drill rigs with upright masts are advancing production blast holes to widen the bench in the foreground. These were steam powered and could drill 5-inch diameter holes to depths of up to 100 feet. Increasing blasting efficiency.

As the project droned on, the amount of dynamite used per cubic yard of excavation continued to decrease. For August-September-October 1908, 2,181,760 pounds of dynamite was used to blast 2,977,415 cubic yards of rock, or about 0.73 pounds of dynamite per cubic yard excavated. This figure dropped to 0.64 pounds/cubic yard in Aug-Sept-Oct 1910, and to just 0.36 pounds/cubic yard in Aug-Sept-Oct 1910. These figures suggest that as the blasting crews gained more experience, they were able to more intelligently load their holes and design their production load blasts, accordingly. This would have resulted in much lower unit costs for excavation.

Tragedy strikes. There were occasional premature blasts, the worst being on December 12, 1908, at Bas Obispo. In those days the standard stick of dynamite was 1-1/2 inches in diameter. The blast gangs probably bundled the sticks together in groups large enough to easily slide down the five-inch diameter boreholes, 100 feet deep. Then they would have slit the wax paper diagonally wrapped around each stick, so that when tamped, the dynamite material would flow out of the sticks to make the best contact (for acoustic coupling) with the rocky wall of the boreholes. That day they loaded 44,000 pounds of dynamite into just 53 boreholes, which suggests they were trying to fracture a fairly hard and resistant rock (all of the rock rip rap for the Gatun Dam came from Bas Obispo).

Detonations were always scheduled to occur during lunchtime or shift changes, when the fewest men would be nearby. That day the Bas Obispo blast was set for the afternoon shift change. As the blasting foreman and his assistant were tamping one of the last holes around 12:30 pm, the entire array detonated prematurely, killing 23 and injuring 60 workers. It was the worst accident suffered on the canal project during the 10 years of construction. The blast also destroyed Shovel No. 261, one of the 90-ton Bucyrus models, which after being deemed unsalvageable (the only steam shovel lost by the ICC), was scavenged for parts. The cause of the detonation was never determined for certain.

Moving mountains of earth. By March of 1909, there were 100 steam shovels working the entire length of the project. 68 of these behemoths were excavating spoil in the Culebra Cut, a staggering figure. This required the coordination of 160 muck trains servicing the shovels, which were tracked from two towers built at either end of the Culebra Cut, and about a dozen smaller sheds and signal platforms strategically located along the cuts, reporting train positions and "directing traffic." Their observations were relayed via telegraph to a centralized train control office at Empire, which issued dispatches directing each train's movements. There were 209 lineal miles of track in the Central Division, exclusive of the double-tracked Panama Railroad. Track supervisors were kept occupied coordinating track shifters, repositioning 76 miles of connecting track in the Culebra Cut, solely for excavation and haulage of muck. Most of the muck trains were grouped together "in sequence" to serve a particular shovel. The number of muck trains serving a shovel was based on the distance to the spoil dumps; longer distances required more trains to keep the shovels busy. The dispatcher's job was to keeping the trains in the correct line-up, so they would replace one another, as seamlessly as possible.

The spoils dumps were located up to 23 miles from the point of excavation (at Gatun Dam) and 60 different locations were used to accept fill. The largest dumps on the Atlantic side were 14 miles north of the Culebra Cut, at Tabernilla. Here 16 million yards of spoil was wasted, most of which was later covered by the rising waters of Gatun Lake. There were two enormous fills constructed on the Pacific side; the first was in La Boca Bay opposite Balboa (Figure 66), and the other was the 676 acre prism upon which the town of Balboa was subsequently built, on the eastern side of the Rio Grande Estuary. Each of these fill dumps contained approximately 22 million yards of material.

A breakwater was then extended 1.5 miles seaward of the balboa fill to create a continuous breakwater connecting Naos, Perico, and Flamenco Islands to Balboa, to protect the Canal's dredged approach channel from being silted in. The first stretch of breakwater was 1.5 miles long, but it ultimately required 2.5 million cubic yards of fill (about ten times what was originally allocated) because the soft estuarine clay off Balboa had very little bearing capacity, and the embankments kept slumping off to either side of the railroad fill trestle. In the coming decades these three islands would be fitted with an impressive array of coastal fortifications, interconnected by tunnels and connected to the mainland by a rail spur.

Figure 66. 22,000,000 cubic yards of fill was placed at Balboa, shown here, and a 2,500,000 yard causeway/breakwater was constructed connecting the island seen in the distance, where protective fortifications were installed.

In March 1909, more than two million cubic yards of material was removed, ten times the volume achieved by the French in a single month. The monthly record for a single steam shovel was set by ICC Shovel No. 123, a 95-ton Bucyrus model, working 26 of 31 days in March 1910, when it managed to excavate 70,000 cubic yards.

Flooding the canal and continuing to dredge. Goethals left the details of how to excavate the canal most efficiently to his subordinates, principally, Lieutenant Colonel Gaillard. Between May and September 1913, a series of enormous landslides involving millions of cubic yards of material began moving into the main excavation, shutting the project down just when it appeared that the end was within a few months sight. The worst of these were at East and then West Culebra, which took out about a third of the town of Culebra, 1000 feet west of the channel. Serious sliding was also occurring along the southern side of the highest cut at Gold Hill, in what came to be known as the Cucaracha Landslide. After due consideration with Gaillard and the ICC geologist Donald MacDonald, it was decided to open up the outlets and blow the cofferdam dike at Gamboa and allow the Culebra Cut to be inundated.

On October 10th, the dike was blown and water entered the channel for the first time. Within a few days every floating dredge that could be positioned in the Culebra Cut began working away at the slide debris, using dipper and suction dredges (Figure 67). Goethals determined that the rest of the canal excavation would occur beneath water, using dredges and shovels to aid in the work (Figure 68), and hoping that the water would have some stabilizing influence on the lower slopes, serving to buttress them.

Figure 67. Goethals ordered the Culebra Cut flooded on October 10, 1913 to allow his fleet of dredges to excavate millions of cubic yards of slide debris. This shows one of the largest dipper dredges working the plugged channel below Culebra.

Maximum effort was expended to continue subaqueous excavation with shovel, ladder, dipper, and hydraulic (suction) dredges, whose effective depths of excavation were limited to something between 45 and 50 feet. The dredging continued unabated for the next 10 months, until mid-August 1914, when the first steamer officially passed through. Approximately 46% of the 262,000,000 yards excavated for the project was accomplished using subaqueous excavation. Expansive dredge tailings dumps were set up in parts of Gatun Lake and near Balboa, to accept millions of cubic yards of material (Figure 69).

Figure 68. Six dredges working the channel adjacent to the infamous Cucaracha Slide on February 8, 1914. The slide lies along the southern flank of Gold Hill, the highest cut along the canal at the Continental Divide.

Figure 69. Outfall from the suction dredges at Balboa, where 22 million yards of fill and dredge tailings were dumped into rock-lined dikes, covering an area of several square miles.

The plan envisioned in the Minority Report of the International Board of Consulting Engineers in February 1906, estimated that 72,800,000 cubic yards needed to be excavated. Because of landslides, the final volume excavated by the Americans was 232,440,945 yards, an increase of 319%. Added to this was about 30,000,000 yards of excavations made by the French that were incorporated into the final scheme, making for a grand total of 262,000,000 cubic yards. It is estimated that the cost of excavating the nine-mile long Culebra Cut was about $90 million, or $10 million per mile (McCullough, 1977).

Tourists and politicians. After Theodore Roosevelt's well publicized visit in November 1906 and copies of The Canal Record began circulating among influential people back in the states, viewing the massive construction activities in Panama became chic, and anyone with money or established eastern society made the voyage south, so they could say they had seen the "8th Wonder of the World."

For Goethals it meant a never-ending stream of politicos and dignitaries descending upon him, every one of them wanting him to shake their hand and hear some sort of advice that had to offer. It was the sort of mindless obligation that a construction engineer like Stevens could never have endured. Goethals sought relief by seldom being in his office during those hours when visitors were most likely to drop in (Figure 70).

Figure 70. Goethals with his ever-present umbrella in a rare pose by the Miraflores Dam and spillway gates during the last year of the project.

In 1908, one of the tourists who came down to Panama was 18-year old Rose Fitzgerald, the lively daughter of Boston Mayor John Francis "Honey Fitz" Fitzgerald, who a few years later married a Boston banker named Joseph P. Kennedy, eventually becoming the matron of the famous Kennedy family. By 1911, the number of tourist had swelled to 15,000, a staggering number for that time. Everyone wanted their picture taken in front of the massive Culebra Cut (Figure 71). In 1912, the number reached nearly 20,000 tourists, and hotel rooms were difficult to find the entire year. Altogether, about 100,000 people visited the project during its construction.

Figure 71. Thousands of gawkers traveled to Panama to see the world's largest excavation and the Cucaracha Slide, shown here in March 1914.
The Canal Zone Navy.

Goethals was probably one of the few Army officers to command a sizable merchant marine force. An unusual aspect of the Panama Railroad was its secure tie with the Panama Railroad Steamship Company, which was owned and managed by the railroad, whose headquarters were always in New York City. Their steamers ran between Colon and New York, from which virtually all of the food and supplied had flowed to Panama since 1850. When the Americans began construction, the Panama Steamship Company was operating five seagoing vessels on the New York to Colon route, which took seven days each way (they added one more during the canal construction). In 1905-06, all of these vessels were modified to support refrigerated holds so that perishable foodstuffs could be transported to the Isthmus in bulk.

The Americans inherited two tug boats from the French when they took over in mid-1904. Each year the ICC's fleet of tugboats grew, increasing to a fleet of 36 tugs by 1922 (Figure 72). These were used to help guide the larger vessel through the narrow confines of the Culebra Cut at Gold Hill, next to the vexsome Cucaracha Landslide, which has shut the canal down on at least 17 occasions.

Figure 72. An ICC tug helps a Hog Islander streamer pass by the East Culebra Slide in 1920, when landslides were still plaguing the canal.

Five of the old French dredges were refloated and repaired, providing solid duty during the first few years. Using design specifications of Frank Maltby, the ICC contracted for an array of modern and more capable dredges, which were delivered in 1905-06. These new dredges were "knocked down" and shipped in pieces to be re-assembled on the Isthmus. Several seagoing ladder dredges were built on the East Coast and steered under their own power to Panama, the dredge *Corozal* having to make the perilous trip around Cape Horn to get to the Pacific side of the Isthmus.

These vessels were crewed by men who worked for the Panama Steamship Company, from whose ranks were trained those individuals needed to operate and maintain the flotilla of dredges that worked the line of the canal seaward of the Miraflores and Gatun Locks.

In 1910, the Panama Railroad Steamship Company purchased seagoing cargo ships from the Boston Steamship Line, the *SS Ancon* and *SS Cristobal*, which were used to transport passengers and freight between the United States and Panama. They eventually hauled almost 5,000,000 barrels of cement from the United Building Material Company and the Atlas Portland Cement Company (Bennett, 1915) to build the foundations, retaining walls, slope facing, locks, dams, and spillways from New York on bi-weekly runs.

End of the Isthmian Canal Commission. With completion of the canal seeming imminent, the Isthmian Canal Commission was abolished by an executive order signed by President Woodrow Wilson on January 27, 1914, to become effective on April 1st. Wilson then nominated Goethals to be the first Governor of the Panama Canal, which was quickly affirmed by Congress. In accepting this new post, Goethals soon discovered that his salary would be decreased by 33%, from $15,000 down to $10,000 annually, which was still more than a regular Colonel's pay of $6,000 per annum. The City of New York offered him the post of Police Commissioner, but this seemed premature, since the canal was not yet completed, and he remained on active duty, though eligible for retirement.

On March 4th, the U.S. Senate confirmed his promotion to the rank of Major General, and extraordinary leap meant to convey the thanks of a grateful nation, Congress, and Executive Branch. That same day they also approved similar elevations of Hodges and Sibert to Brigadier General, and a few days later, for Rousseau to Rear Admiral.

The canal is completed. By May 1914, the cut at Cucaracha had been opened sufficiently through dredging to allow passage of towed barges for the American-Hawaiian Steamship Line. By early August, the dredges removed another 2,750,000 yards of debris from the Cucaracha slide. On August 3rd, the *S.S. Cristobal*, a 9,300 ton steamer operated by the Panama Railroad Steamship Line (Figure 73), made the first coast-to-coast transit of the new canal, but not without complications in the locks at either end, which experienced problems with the new electric tow locomotives. The next day Great Britain declared war on Germany, and most of Europe became embroiled in a global conflict. All of the pomp and circumstance expected with the canal's official opening in mid-August more or less vanished, as the Western Europe became embroiled in the largest war in history, up to that time.

The canal's official christening occurred 12 days later, on August 15th, when the *Cristobal's* sister ship, the steamer *S.S. Ancon* (Figure 74) made the first official ocean-to-ocean transit, carry media and dignitaries from the Atlantic to the Pacific. On October 10th, the *USS Jupiter*, a Navy collier of 19,300 tons, built at Mare Island on the West Coast, made the first west-to-east transit of the Canal. She was the Navy's first surface vessel to employ turbo-electric propulsion, and became the Navy's first aircraft carrier, the *USS Langley*, in 1922.

The Panama Canal, dreamed of for so many centuries, was finally a reality. The final price was $23,000,000 less than what the Minority Report had predicted back in February 1906 (very close to the 20% contingency). The World War erupting in Europe would greatly retard the envisioned flow of merchant marine traffic, although 1,258 vessels passed through the canal during its inaugural year of operations, carrying 5,675,261 tons of cargo and generating tolls of $4,909,151. It was well below the pre-canal projections, but was compensated to some extent by the utilization of the canal by the U.S. Navy, which enlarged its facilities in the Canal Zone and of the Pacific Fleet in general in the years following the First World War.

Figure 73. The S.S. Cristobal actually made the first transit of the Panama Canal in an unheralded test run on August 3, 1914, shown here.

Figure 74. The S.S. Ancon makes the official first passage through the canal, from the Atlantic to the Pacific on August 15, 1914. Note dredges working along the toe of the Cucaracha Slide.

Theodore Roosevelt died at the age of 60 in January 1919, but five months later his dream of a powerful and mobile United States Navy was realized when 33 American warships, including nine of the largest battleships, transited the Panama Canal in just a bit more than two days. Ten years later the canal was transiting 5,000 ships per year and was rapidly opening up Asian markets in the Pacific.

Postscript on Goethals' career. The canal continued to be plagued by landslides, shutting it down repeatedly. In 1914-15, he wrote up the story of the canal's construction and the American administration of the new Canal Zone, released in 1915 as *Government of the Canal Zone* (Princeton University Press). In this volume he makes no mention of Dr. Gorgas' name, and demeans any significant contributions that Gorgas made to the canal building effort.

Goethals was unable to get away from the Isthmus until September 1916, when he visited his home in Vineyard Haven on Martha's Vineyard. He decided he had served his country long enough, and longed to return home. The previous July he had applied for retirement from the Army, and on November 15, 1916, he was placed on the retired list of the Army, after 40 years of service.

In January 1917 he announced his intention to open a consulting engineering company in New York with George M. Wells and Sydney B. Williamson, his "right arm" from days past, who returned from an assignment in Chile to accept his new position. Goethals was unexpectedly named General Manager of the Emergency Fleet Corporation of the new U.S. Shipping Board by President Wilson. He ended up in a highly publicized controversy over the merits of constructing wooden merchant ships, which the chairman of the Shipping Board opposed.

On April 6, 1917 the United States declared war on Germany. In early August, Goethals wrote to General John J. Pershing, Commanding the American Expeditionary Forces in France, and offered his services to supervise the military engineering work there. Pershing declined his offer, but in December, he was unexpectedly recalled to active duty to serve as Acting Quartermaster General of the Army in Washington, DC. His service had been requested by the Secretary of War, Newton D. Baker, and his West Point classmate, Peyton C. March, then serving as the Army's Chief of Staff. Goethals took up his duties on December 26th in the uniform of a major general. He made some progress in grouping the purchases made by the Army, so that the Quartermaster Corps would solicit competitive bids instead of each department of the Army individually soliciting bids for similar or identical lots. In late January 1918, he selected Brigadier General Robert E. Wood, his Chief Quartermaster on the canal project, as his replacement. Goethals was then assigned Assistant Chief of Staff and Director of Purchase, Storage and Traffic, in charge of all purchasing for the War Department, railroad transportation and shipment overseas of troops and supplies, between April 1918 and his second retirement on March 4, 1919, at age 60.

He continued to work as a consulting civil engineer through his firm George W. Goethals & Co., and was joined his son, Colonel George R. Goethals in August 1919, who retired from the Army. They worked together on an array of projects, including construction of the lock for the Inner Harbor Navigation Channel in New Orleans, which proved most troublesome. He was also appointed a Commissioner for the Port of New York. Throughout the 1920s, the senior Goethals was asked to serve on numerous review boards and panels for a variety of projects, from jetties at Palm Beach to the Boulder Dam in Nevada, and numerous water supply projects in California. Along the way he received honorary degrees from Columbia, Harvard, Yale, Princeton, University of Chicago, and Rutgers.

In the late summer of 1927, his health began to fail. In November, he went to his younger son Tom, who was a doctor. X-rays conformed that he had cancer. He died peacefully in New York on January 21, 1928, in his 69th year, and was laid to rest in the cemetery at West Point. Goethals was the only engineer honored by three postage stamps. The Canal Zone issued a red two-cent regular issue of him shortly after he died, in 1928, followed by a violet 3-cent stamp in mid-1934 (Figure 2), and the 25th anniversary commemorative issue of 1939 by the U.S. Post Office (Figure 1). Effie received personal letters of condolence from President Coolidge, Secretary of War Dwight F. Davis, Governor Alfred E. Smith of New York, Army Chief of Staff General Charles P. Summerall, Chief of Engineers General Edgar Jadwin, and host of others. Most came from the men who had worked with him in Panama, between 1907-16. Effie Goethals returned to Vineyard Haven and died in Wellesley, Massachusetts on December 31, 1941.

HARRY H. ROUSSEAU

Harry Harwood Rousseau (Figure 75) was born in Troy, New York on April 19, 1870. After graduation from Troy High School as the valedictorian of his class in 1887, he entered Rensselaer Polytechnic Institute (RPI) that same fall, and worked for the Troy Public Improvement Commission during his summer breaks. At that time RPI enjoyed a reputation as one of the nation's most coveted engineering schools, having sired a greater many bridge and railroad engineers of considerable distinction. He received his bachelor's degree in civil engineering in 1891 and promptly accepted a position with Charles F. Stowell, Structural Engineer in Albany, where he remained two years. In 1893-94 he worked for the Brooklyn Elevated Railroad Co., then joined the Pittsburgh Bridge Company, as Superintendent of their Structural Department. A year later he joined their Design and Estimating Department, and, in 1897 was named the firm's Principal Assistant Engineer.

In 1898, he sat for the competitive examination for the Navy's Civil Engineering Corps, which in those days required a degree from a technical institution and five years of engineering practice, of which, at least two years were to have been in responsible charge. Of the 43 individuals who sat for the exam that year, Rousseau was ranked number one. On October 12, 1898, he was commissioned a Lieutenant Junior Grade and began working for the Navy's Bureau of Yards & Docks in

Washington, DC. The agency had a very small staff, but was responsible for the design, construction, and maintenance of all port facilities, as well as dry docks and ship building slips and graving docks. During his first four years Rousseau's responsibilities were in the general charge of designing dry docks at Portsmouth and Mare Island. From 1903-06 he had the good fortune to "follow his designs" to the Mare Island Naval Shipyard near San Francisco, where he supervised the construction of the massive dry dock he had designed in Washington. While at Mare Island he also supervised channel and harbor dredging, the construction of masonry quay walls, and cofferdams, providing excellent background for the sorts of structures and activities with which he would subsequently be engaged in Panama.

Figure 75. Harry H. Rousseau in the uniform of a Navy Rear Admiral, in March 1915, when he was 45. He remained on active duty until his untimely death at age 60, while traveling on a steamer bound for Panama (U.S. Navy).

In an unprecedented move, on November 26, 1906, Rousseau was named Chief of the Bureau of Yards and Docks upon the recommendation of the retiring Chief, Rear Admiral Mordecai T. Endicott, who felt that despite, despite being just 36 years old, he was the most qualified engineer available to succeed him. President Roosevelt agreed, confirming Rousseau's appointment and promotion to Rear Admiral (the highest regular rank in the Navy at the time) for a term of four years, beginning January 6, 1907. A lot of heads were turned, but Rousseau had hardly settled into his new office, when his star suddenly moved in a new direction.

Almost coincident with Rousseau's appointment was the unexpected resignation of John Frank Stevens as Chief Engineer and Chairman of the Isthmian

Canal Commission. Frustrated with the frequent changes in leadership of his coveted Panamanian canal, President Roosevelt decided to turn the canal project over to the War Department and its Corps of Engineers. In addition, Roosevelt sought younger engineers who would establish themselves in Panama.

Rousseau's technical expertise was needed on the canal project because the design and construction of massive locks was nearly identical to that for masonry dry docks. After less than three months in his new position, he reverted to his rank of Navy Lieutenant and was dispatched to Panama, where he joined the Isthmian Canal Commission on March 16, 1907. He did receive the higher compensation allowed by Congress for members of the Commission ($7500/year).

Rousseau's new title was Head of Departments of Building Construction, Motive Power and Machinery, and Municipal Engineering, where he oversaw the activities of more than 10,000 employees. When Goethals reorganized the project in mid-1908, Rousseau's title became Assistant to the Chief Engineer, and he oversaw the Department of Construction and Engineering until March 31, 1914, when the Commission was dissolved. He was then named Engineer of Terminal Construction. In 1909, he was promoted to Lieutenant Commander, and in late 1914 to Commander.

Port improvments. In 1911, Rousseau was given charge of the design and construction of the canal's new marine terminals, which included ship repair facilities, such as shops, piers, coaling stations, fuel oil plants, breakwaters, floating cranes, and dry docks. The sheer volume of work required that Rousseau remain on the Isthmus for several years after the canal was completed, supervising $20 million in port facilities (Figure 76).

Rousseau's department expended considerable effort improving the government harbor facilities at Cristobal, constructing new wharves, piers, warehouses, and expansive coal loading facilities located at this important terminus for the majority of the freight coming from the United States to sustain operations in the Canal Zone.

A new navigational channel was dredged three miles across Limon Bay to a minmum depth of 30 feet to provide adequate depth for seagoing vessels. The project that ate up far more time than anticipated were the two massive breakwaters, which took five and a half years to complete (1910-15) and cost $11 million. These involved more than 8,000 lineal feet of breakwater, requiring 4,700,000 cubic yards of yards of riprap to protect the anchorages in Limon Bay at Cristobal. The eastern breakwater was not completed until late 1915.

The rock for the breakwater came from Bas Obispo and from the quarry down the coast near Portobello. These improvements were supervised by Major Edwin Jadwin, who became Chief Engineer of the Corps in the late 1920s. The two breakwaters took almost four years to complete because the engineers were limited

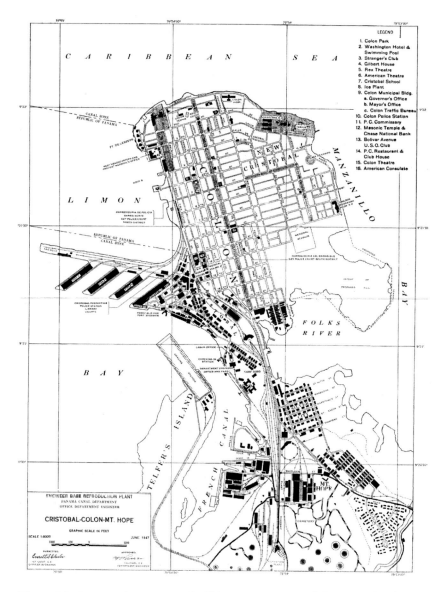

Figure 76. Map showing the port facilities constructed at Cristobal (left middle) in Limon Bay, which was part of the Canal Zone. Colon lies within the Republic of Panama (Canal Zone, 1947).

by how many trains per day they could run along the narrow strips (Figure 77). The rock from Portobello came by ship and was placed using barges, while that from Bas Obispo came by train.

Figure 77. View from Toro Point of the breakwater extending 2.4 miles into Limon Bay, to provide sheltered anchorage at Cristobal, at a cost of $7 million. Much of the interior fill for the breakwaters came from the excavations for the Gatun Locks.

Impacts on career and personal life. Rousseau had an "*unusually reserved*" demeanor, he possessed an "*extremely kind nature*," and was "*ever ready to help*" (Chambers, 1931). He was socially responsive with a "keen sense of humor," which led gayety to any party he attended. Not long after arriving on the Isthmus Rousseau met Gladys Fargo Squires, the vivacious daughter of Herbert G. Squires, the American Minister to Panama, and previous to that, Cuba. Harry was smitten by her charm and she by his "eminently good breeding" and "fear of the almighty," for the two met at the Episcopal Church, of which, Harry was its most faithful male attender. Despite their age difference (she was 12 years younger), the two were married in New York in 1908 and made their home in Culebra shortly thereafter. In Panama Gladys bore him three sons: Harry H., Jr., William P, and Bard S. Rousseau.

On March 4, 1915, Congress recognized those who had "*rendered distinguished service in constructing the Panama Canal,*" and authorized President Wilson to advance Rousseau to the Rank of Rear Admiral, in which he would have been eligible to retire with a comfortable pension, but as described by Admiral F. T. Chambers: "*He felt that his government had treated him well; and that he owed it his active assistance so long as his efforts were found of particular value.*" (Chambers, 1931). Harry Rousseau chose to remain on active duty until his death in 1930.

Post Panama duty. Rousseau's duties in Panama were essentially completed by late 1915 with completion of the eastern breakwater, opposite the Toro Point Breakwater. In early 1916 and he returned to Washington and was appointed to the Commission on Navy Yards and Naval Stations. This body was charged with making surveys of potential sites for naval bases, navy yards, submarine and aviation facilities in the continental United States. Particular efforts were expended on the West Coast, in San Francisco and San Diego Bays.

When war was declared in Germany in the spring of 1917 the national mobilization included massive shipbuilding acts, under the aegis of a newly established U.S. Shipping Board. The Emergency Fleet Corporation of the Board was made responsible for the design and construction of the needed ships, but the paucity of existing shipyards greatly hindered any plans for rapid expansion of America's merchant marine. Rear Admiral Rousseau was promptly named Assistant General Manager and Head of the Shipyards Plant Division of the Emergency Fleet Corporation. In addition to selecting suitable sites for, and then, hastily constructing new shipyards, his organization also had to construct marine railways and dry docks to support the shipbuilding efforts.

During the World War Rousseau also served as Associate Director of the United States Housing Corporation, which controlled naval housing matters. He was also a member for boards that over saw the expenditure of funds for plants and shipyards, and on the Government Munitions Board. At the war's conclusion he was awarded the Navy Cross "for exceptionally meritorious service in a duty of great responsibility." During the war he also served as Vice Chairman of the Port Facilities Commission of the national Shipping Board.

The decade of the 1920s witnessed his involvement in a broad array of issues and problems facing the U.S. Navy, in addition to his duties on the Commission on Navy Yards and Naval Stations. Rousseau served on numerous Examining Boards for promotion of Corps of Civil Engineers officers, the Interdepartmental Board on Contracts and Adjustments, dispute resolution boards for severed contracts and commandeering of property during the war, Chairman of the Metals Committee of the Federal Specifications Board in the Bureau of Standards, as Government Receiver for the controversial Naval Petroleum Reserve No. 1 at Elk Hills, which led to his later selection as Director of Naval Petroleum and Oil Shale Reserves. In 1923 he was awarded an honorary doctorate by his alma mater, Rensselaer Polytechnic Institute.

Rousseau also had a keen understanding of engineering economics, honed by years of managing large budgets. In 1930 he was named Chief Coordinator of the Bureau of the Budget, for all Federal coordinating agencies, a position of significant national responsibility, which many of his intimates credited with his untimely demise not long afterward (Chambers, 1931).

Since his days on the Isthmian Canal Commission Admiral Rousseau served as a Director of the government-owned Panama Railroad Company and the Panama Railroad Steamship Company. In July 1930 he was aboard the Steamship Company's vessel *S.S. Cristobal* sailing for Panama to make an inspection when he suffered an attack of apoplexy on July 24th, from which he died at sea. He was 60 years old. His funeral was held at the St. Thomas Protestant Episcopal Church in Washington, where he served as a Vestryman. The Canal Zone issued a 20 cent stamp featuring Admiral Rousseau in July 1939 (Figure 2). He was survived by Gladys and their three sons, and buried with full military honors in Arlington National Cemetery. Gladys lived another 44 years, and was the last matron of the large mansions adorning Culebra Heights when she died in 1974.

HARRY F. HODGES

Brief sketch of military service. Harry Foote Hodges (Figure 78) was born in Boston in February 1860, and received an appointment to West Point in 1877. Just 5 feet, 5 inches tall, he was respected for his academic abilities, graduating 4th in a class of 53 cadets in 1881, a year behind George Goethals. He accepted a commission in the Corps of Engineers and initially served in Washington as a member of, and purchasing agent for, the Canal Board. In this role he oversaw the contracts for the acquisition of hardware, equipment, and supplies.

This experience with canals led to his next assignment in 1885, working under Colonel Orlando M. Poe of the Corps of Engineers (West Point Class of 1856), Superintending Engineer for improvement of rivers and harbors on Lakes Superior and Huron. This was shortly after Poe had completed construction of the Weitzel Lock on the St. Marys Falls Canal at Sault Ste. Marie, Michigan (known as the Soo Canal). Hodges assignment was to assist Poe on the design of a much larger lock, 800 feet long and 100 feet wide. This was later named the Poe Lock, and it was the largest lock in the world when it was opened in 1896.

Hodges next tour was as an instructor of engineering at West Point, between 1888-92. During this time he continued working on the design of miter gates for locks, and wrote a book titled "*Notes on Mitering Lock Gates*," published as Corps of Engineers Professional Paper No. 26 in 1892, when Hodges was a First Lieutenant. During the summer breaks he assisted Colonel Amos Stickney in construction of locks and dams.

Figure 78. Left – Lt. Colonel Harry F. Hodges as he appeared while working on the Panama Canal (Jackson & Son, 1911). Middle view shows him in the dress uniform of a major general during the First World War (USACE). Right image was taken in the 1920s, after he had retired from the Army.

In April 1892, he was given charge of surveys for river improvements on the upper Missouri River and its tributaries, as well as the upper Mississippi River and its tributaries, working out of Sioux City, Iowa. In May 1893, he was promoted to Captain and appointed to the Board of Engineers in New York, where his principal responsibility was the design of seacoast defenses. After the declaration of War with Spain, in June 1898, he was promoted to Lieutenant Colonel of the 1st U.S. Volunteer Engineers Regiments, forming up at Peekskill, NY. From early August to late November, he was in Puerto Rico, engaged in surveys, construction of roads and defensive works, timber and masonry bridges in the district of Ponce. He assumed command of the regiment in September and was promoted to Colonel in January 1899. Between February and April, he was detached to inspect the Spanish fortifications of Puerto Rico.

Upon his return to the States in late April 1899 he was mustered out of the Volunteers and restored to the rank of Captain the Regular Army. He was then given charge of certain river improvements of tributaries of the Ohio River in Ohio, West Virginia, and Kentucky. In May 1901, he was promoted to Major and named Chief Engineer, Department of Cuba, under General Leonard Wood.

In August 1902, he was transferred to Washington and placed in charge of the Rivers and Harbors Division in the Office of the Chief Engineer, and appointed a Member of the Board of Engineers for Rivers and Harbors from 1902-03. This placed him in close proximity to the Army's Chief Engineer, including Generals George L. Gillespie (1901-04), Alexander Mackenzie (1904-08), and William M. Marshall (1908-10).

Everyone in the Corps of Engineers must have known about Hodges's work on locks and dams with Colonels Poe and Stickney, and his book on mitering lock gates was the only such text then in existence. Then there was his role in coordinating the Board of Engineers for Rivers and Harbors. So when George Goethals was tapped by President Roosevelt to become the new Chief Engineer and Chairman of the Isthmian Canal Commission in February 1907, Hodges was his first choice for an assistant. But, the Chief Engineer, General Mackenzie, rejected his request, telling him that Hodges was too valuable to spare because he depended upon him for too many things in the Office of the Chief Engineer.

Hodges' selection to the Isthmian Canal Commission. In August 1906, Joseph Ripley, a member of the International Board of Consulting Engineers and former Engineer in Charge of the Soo Locks at Sault Ste. Marie, Michigan, was named Assistant Chief Engineer in charge of designing locks, dams and regulating works for the Isthmian Canal Commission in Washington, DC. By March 1907, all of the essential features of the locks and dams had been conceptually worked out, although many of the construction details and foundation conditions required verification. In June 1907, Ripley resigned to accept a new position, leaving the "design bureau" without an experienced chief.

War Secretary Taft put pressure on General Mackenzie to allow Hodges to be detached from his position in the Office of the Chief Engineer to assume Ripley's responsibilities on the Isthmian Canal Commission. In late July 1907, Hodges was transferred to the ICC's office a few blocks away from the Army-Navy Building. Ripley and Hodges must have known one another from their association with Colonel Poe and the Weitzel Lock back in the 1880s, because Ripley had been a civilian engineer employed by the Corps at the Soo Locks for 28 years, before being tapped for the International Board of Consulting Engineers in June 1905.

Hodges was promoted to the rank of Lieutenant Colonel, Corps of Engineers on August 27th 1907. His new official title was Purchasing Officer, but this was later was changed to Chief Engineer, Purchasing Department, Isthmian Canal Commission. He was promoted to Lieutenant Colonel on August 27th, and was placed in charge of purchasing, which involved a web of contracts for the purchase of equipment and supplies. Some of the items being requested, like five million barrels of Portland cement, 100 Bucyrus and Marion steam shovels, 2,000 rail muck cars, and 120 tons of pyrethrum insecticide, exceeded the combined yearly output of these items in the United States at that time.

Hodges was also responsible for supervising the technical design and drafting of engineering plans for the canal project. This responsibility included a myriad of details that needed to be worked out, involving the design of culverts, outlet works, Stoney gates, valves, lock walls, piers, recessed miter walls, floor sills, rolling gates, miter gates, interlocking pin devices, operating bridges and machinery, and the reaction of Portland cement concrete with brackish water.

Much of this work involved research and the coordination or correspondence with a wide array of individuals with specialized expertise. Gradually, a library of technical data was acquired for ready reference and comparison, and the design engineers became intimately familiar with the sorts of structural details that had been employed on other locks around the world. Fortunately, the Soo Locks were, by far, the largest, most technical sophisticated, and busiest locks in the world, so Ripley had built a sound technical base for the American effort in Panama, and all of the technical concepts that were eventually employed in Panama had been determined by March of 1907, when Stevens was passing off his responsibilities to Goethals (Sibert and Stevens, 1915). This was the group that Harry Hodges would lead for the next seven years.

Move to Panama. As part of Colonel Goethals' massive reorganization of mid-1908, Hodges was ordered to the Isthmus on July 3, 1908, along with the engineering department designing locks, dams and regulating works. His responsibilities as Purchasing Officer were assumed by Major Frank C. Boggs of the Corps of Engineers. During the previous 15 months Major William L. Sibert had been in responsible charge of the Department of Lock and Dam Construction in Panama, which oversaw the final details on all construction plans and drawings, from which material and supply requisitions were promulgated and construction activities scheduled.

Hodges arrived in late August 1908 and was given the title Engineer of Maintenance. He began attending the Isthmian Canal Commission meetings in Culebra on Septmber 5th. For the next six years he also served as the Acting Chief Engineer whenever Goethals was off the Isthmus. Over the next six years Hodges would sign thousands of engineering plans as "H. F. Hodges." He also spoke fluent French and German, so was frequently called upon to serve as a translator when visitors to the Isthmus spoke those languages. His hobby appears to have been reading, and he spent a great deal of his leisure time reading not only technical articles, but contemporary English, French, and German literature, which impressed many of the foreign dignitaries that visited Panama.

Engineering design. Hodges brought his principal assistants in the design group with him from Washington, DC. These included four department heads: Henry H. Goldmark, responsible for the design and installation of lock gates; Edward Schildhauer, resposible for electrical and mechanical design and installation; L.D. Cornish, overseeing general lock and design of the mass concrete used to form the locks; E.C. Sherman had general charge of the hydraulic design of the various spillways; and, T.B. Monnicke was given charge of the design of the emergency dams.

Harry Goldmark (Figure 78) was a native of Brooklyn. In 1874, he graduated from the Polytechnic Institute in Brooklyn, with a second degree in physical science from Harvard in 1878. He then received additional engineering training at the Royal Polytechnic University in Hanover, Germany in 1880. That same year he joined the

Erie Railway as an design engineer working on projects in the United States and Canada. He joined the Isthmian Canal Commission in October 1906, where he had charge of the Engineer Office in Washington, DC. He moved to the Isthmus in October 1908 where he was given charge of the design of the steel lock gates and the protective apparati attached thereto. After the canal project wrapped up, Goldmark established himself as a consulting engineer in New York City. In the early 1920s he worked with George Goethals on the Inner Habor Navigation Canal lock in New Orleans, designing the temporary dam constructed there.

Edward Schildhauer (Figure 79) was from New Holstein, Wisconsin. He received his B.S. in electrical engineering from the University of Wisconsin in 1897.

Figure 79. Three of Harry Hodges' principal design engineers were Harry Goldmark (left), Edward Schildhauer (middle), and Edward C. Sherman (right). Images from Jackson & Son (1911).

He worked in electrical railway construction in Chicago for the Edison Company. In November 1906, he was appointed Elecrical and Mechanical Engineer for the Isthmian Canal Commission in Washington, DC. In 1908, he toured England, France, Belgium, Holland, and Germany to see their principal canals and locks. He moved to the Isthmus in September 1908. He designed the machines that operate the sluicegates of the large spillways, the valves of the culverts controlling the flow of water into and out of the locks, the balance vales for the cross sulverts beneath the locks, a mchine to operate the large miter gates allowing the gates to fully retract into recesses of the lock walls, and the "electric mules" that carefully pull the ships through the locks. All of these were innovative solutions that have since been employed elsewhere.

Edward C. Sherman (Figure 78) was from Kingston, Massachusetts, descending from three of the original passengers on the *Mayflower* in 1620. He received his bachelors in engineering from the Massachustts Institute of Technology in

1898. For three years thereafter he worked on the Cambridge Bridge acrooss the Charles River, then on river improvements along the channels surrounding Cambridge and Boston. In 1909, he came to Panama and was placd in charge of designing spillways for the Gatun Dam and Miraflores Dam, as well as other technical aspects of the mass concrete dams. He was one of the youngest engineers on the project in responsible charge of a deisgn team.

Fabrication and assembly. The great quantities of steel needed for the project required considerable lead time to allow for transportation to the Isthmus, and the preparation of suitable warehouse or stockpile areas. On-site fabrication yards were able to use steel plate stock to make flat concrete forms, using common structural steel shapes (dowels, channels, I-beams, and H-beams), as well as iron and steel bolts, rivets, nuts, washers, picks, mattocks, and claw bars. In addition, miles of wire, welded wire fabric, fencing, and nails were used to assemble jigs, forms, dollies, and lifting cradles. Most of the heaviest items were cradled to allow their lifting with conventional yokes. These items could then be transferred from the fabrication yard by cableways and set into their proper place using the berm or chamber cranes, which were all rail-mounted.

Mass concrete. The first mass concrete was laid at Gatun on August 24, 1909. More than five million barrels of concrete were consumed on the locks and the Gatun Dam spillway, with an average cost of $6 per cubic yard. About 2,000,000 cubic yards of concrete was placed on the project, using about 100 concrete mixers if various types, ranging in size from one-third to two cubic yards. These were mostly old fashioned "cube mixers" and "improved cube mixers," a type which the Corps of Engineers officers and civilian engineers directing the work were familiar with using because they produced a homogeneous mixture. 1,209,506 yards of concrete was batched using the small cube mixers, while the balance was produced using the newer rotating drum mixers, which gained popularity on the project because they were far more efficient.

Even with such small mixers the average rate of output was one cubic yard every 75 seconds. Two mixers working at the Gatun Locks produced 80,544 cubic yards in just 1,175 hours, for an average production of 68.54 yards per mixer per hour. The Gatun Locks formed a single mass of 1,945,457 cubic yards, at that time the largest concrete structure in the world. The total cost of mass concrete employed on the project was $30 million, which gave an average in-place unit price of $15 per yard. By later standards, this was considered quite high. Most of the concrete mixers were mounted in pairs on railcars which were run into the lock chambers, where chamber cranes could manipulate two-yard hopper buckets (Figure 80). These were dropped in front of the mixers, and two batches would generally fill a bucket, which was then hoisted to the location where the pour was occurring that day, and the crane operator would the release hopper flaps to drop the concrete. The concrete volumes were so low that nobody could get "buried" in the wet mix, but they could easily be injured by the buckets. "Spotters" were stationed on both axes to provide hand signals to the crane operator surrounding any pour, as seen in Figure 80.

Figure 80. Mass concrete being placed in the floor of the Miraflores lock using a two-yard hopper bucket, hoisted by a chamber crane. Note spotters signalling the crane operator (USACE).

Rail-mounted "auxiliary cranes" were also used to place concrete on smaller pours, using open mix bins, as shown in Figure 81.

Figure 81. Rail-mounted auxiliary crane being used to pour mass concrete at the Pedro Miguel Lock. All of the concrete was placed using mix batches of just one half to two cubic yards. This shows placement using a swivel bucket.

The concrete used on the Panama Canal utilized a six-sack mix (6 sacks x 94 pounds of cement per sack, per cubic yard) using volume measurement of the cement, sand, aggregate and water to produce high quality concrete. The high cement and low water ratios gave compressive strengths of as much as 7,000 psi, which was a record for compressive strength at that time. It was fortunate that such a low slump mix was used, to reduce shrinkage, but the heat generated by hydration of the cement was very high by later standards. Proper curing was aided by never pouring lifts more than three feet thick and allowing 10 to 21 days between pours (most of the time). The quality of the concrete placed on the original canal job was remarkable for the era in which it was placed, when there was little technical understanding of concrete heat of hydration and shrinkage. The supervising engineers knew that a well graded, homogenous mix with low porosity and minimal laitance would always perform better in the long run. That was the type of concrete they poured, and it has worn very well over the past 100 years (Figure 82).

Figure 82. The center walls of the Pedro Miguel Locks were poured as a monolithic block.

Collapsible forms. For the Miraflores and Pedro Miguel Locks the Commission purchased telescoping Blaw Collapsible Forms for the tunnel culverts in those locks (Figure 83). Collapsible forms of 10, 18, 20, and 22 feet in diameter were used to construct most of the culverts, and were delivered to Panama in 1909 and 1910.

The office engineers also designed the steel forms that were used for the concrete locks, and developed assembly details for the gantry, berm, and chamber cranes, as well as scaffold works. Brownhoist Cantilever Cranes built by the Brown Hoisting Machine Company of Cleveland, OH, were widely employed on all of the locks, borrowing on structural concepts that proved successful in the construction of the Chicago Sanitary and Ship Canal.

Figure 83. Left - The 22-foot diameter horseshoe culverts within the locks were formed using collapsible steel forms, manufactured in five foot sections by the Blaw Steel Construction Company of Pittsburgh. The right view shows a retracted form for an 18-foot diameter circular culvert, the most common employed on the locks (Bennett, 1915).

Traveling cranes. The primary cranes selected for the terminal port facilities were manufactured by the Browning Engineering Co. of Cleveland, Ohio. Known at the time as "locomotive cranes" they were self-powered and could rotate and hoist up to 100 tons. They were fitted with booms 100 feet long and placed on trolley tracks. Most of these were employed along the wharves on the Pacific and Atlantic costs to unload ships.

The cranes used in construction of the Panama Canal locks were an array of traveling chamber cranes (Figure 84) and berm cranes (Figure 85). These were used to hoist steel forms, hardware such as culverts, and transfer mass concrete from rail-mounted mobile mixers to where it needed to be placed. The towers rose 115 feet above the floor of the locks and could hoist concrete or steel to a height of 98 feet. The chamber cranes were able to move freely up and down the axis of the locks. The berm cranes were used in much the same fashion, and could work both sides of a lock simultaneously.

Lock gates. The locks employed miter gates, so called because they close in a wide V. They are the canal's most storied moving parts, and the only vital elements that were actually fabricated by a private contractor, McClintic-Marshall of Pittsburgh. They began assembly at Gatun in May 1911, at Pedro Miguel in August, and at Miraflores in September 1912.

The miter gates swing open like double doors. They were carefully fashioned to be hollow, and of watertight construction in their lower halves, to create buoyancy in the water. This greatly reduces the working loads on their hinges. All of the lock

Figure 84. Traveling "chamber cranes" were employed in all of the locks, to hoist heavy items like the collapsible forms for the 22-foot diameter culverts, shown here at the Pedro Miguel Locks.

Figure 85. Traveling berm cranes at Pedro Miguel Locks, as well as steel forms used for pouring mass concrete for the locks in foreground.

gate leaves are about 64 feet wide by seven feet thick; but their height varies from 47 to 82 feet, depending on their position in the locks. For example, the lower chamber gates closest to the Pacific Ocean (at Miraflores) are the highest because of the 20 foot Pacific tides.

The operating mechanisms employed in the lock gates were designed by Edward Schildhauer. His challenge was to design a hollow gate that could easily swing open while withstanding considerable hydrostatic pressure from an imbalanced load of water on either side of the gate. The leaves of the lock gates are attached to steel arms, or struts, which are connected to huge "bull wheels" sitting inside the lock walls. Each of these wheels is 20 feet in diameter, laying horizontally (dead level) and geared to an electric motor. The bull wheel and strut work like the connecting rod attachment to a driving wheel on a railroad locomotive to close and open the gates.

As a safety precaution, each lock chamber (except for the lower locks at Miraflores) is equipped with a set of "intermediate gates." Their purpose is to shorten the locks (reducing their volume) whenever possible so as to conserve water, if the ship transiting the canal is not of near maximum dimensions and can be accommodated by lock chamber just 600 feet long (instead of 1100 feet).

Since locks are a type of dam with moving parts that are subject to accidents or collisions, precautions were taken to protect them from being damaged in such a manner that might allow the water of Gatun Lake to escape into the sea. One precaution is to employ double gates ahead of a vessel, which are referred to as the "operating gate" and the "guard gate." These are strategically located where collisions between vessels and gates could compromise the water levels in adjacent lock chambers (e.g. this could easily occur at Pedro Miguel because there is only a single lift).

The locks on the Pacific end of the canal were completed in May 1913. Figures 86, 87, and 88 show how the massive steel miter gates were assembled in the locks by McClintic-Marshall. There were 46 miter gates, each weighing between 354 and 662 tons. When they were designed the physical chemistry of corrosion was not as well understood, so the only cathodic protection they were afforded was through the employment of sacrificial zinc anodes. The lock gates suffer a great deal of corrosion in the brackish water at either end of the canal. During the first 75 years of operation they had to be removed and overhauled to battle corrosion. That maintenance necessitated the loss of one gate for four months every two years (17% downtime), which hindered ship transits. In the past several decades vast improvements have been made in equipping the gates with modern cathodic protection, using adjusted live-current anodes, which sense water salinity.

Control of the locks. The canal was one of the earliest megaprojects that utilized electricity to power many of the elements essential to its construction, such as: traveling cranes, cableways, cement mixers, and rock crushing and classification.

Figure 86. Steel frames for two of the 46 miter gates fabricated by McClintic-Marshall from drawings prepared by Hodges' Maintenance Division.

Figure 87. Braces used to prop up the gates while steel plating is affixed to the strucural frames. The gates had to be precisely balanced so they would swings back and forth properly. Those shown here are at the Pedro Miguel Locks.

Figure 88. Six million rivets were used in the 46 lock gates. The steel gates had to be removed and overhauled to battle corrosion for four months every two years. Note men for scale.

After the canal was completed the locks were operated using more than 1,500 electric motors, and all of the controls were electrically powered. About half of the electrical motors and equipment used in the canal's construction was purchased from the General Electric Co., and all of motors, switches, buss bars, relays, switches, fuses, wiring and generators used in lock operations were also manufactured by GE.

The mechanical mule towing engines in the locks designed by Edward Schildhauer were also built by GE. These mules controlled the speed and movement of the ships as they were towed through the locks at two miles per hour. They were built in Schenectady, New York, with a cost of $13,000 per unit.

Schildhauer collaborated with engineers at General Electric to come up with the lock control systems. All of lock operations are made by operators looking down on the locks from control towers situated between the two upper lock chambers. This vantage offers an unobstructed view of all the locks, so that the operators can make keep each vessel clearly in view while operating the switches controlling water levels and opening or closing of the gates. The only activity the tower operators do not control are the mechanical mules.

The boards in each control tower are carefully laid out as a working model of the locks. Everything that occurs in the locks, such as draining or filling of the lock chambers, occurs in miniature form on the control board, at the same time it occurs outside. The control switches for the lock's miter gates are located beside the model representations of the gates on the control board. To fill a lock chamber the operator simply turns a small handle next to that chamber.

In order to prevent an operator from turning the requisite switches out-of-sequence, Schildhauer and General Electric devised a system of interlocking buss bars beneath the control panel to mechanically interlock the control switches. The controls for each operation must be switched on or off in the proper sequence, or the operation will be nulled. This feature was installed to reduce the possibility of operator error.

By employing electrical controls the locks are easily managed from one central position, much like the navigation bridge on a ship. The electric motors doing the work can be as much as half a mile distant from the control tower. This system of operational control has resulted in a remarkable record of safety over the past century, which is a real testament to the practical ingenuity of the American design team (Figure 89).

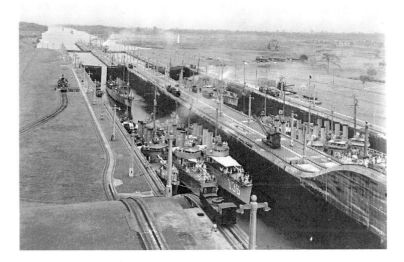

Figure 89. During two days in July 1919 thirty-three warships transited the canal from east to west in just over 48 hours. This shows two squadrons of destroyers in the second lift of the Gatun Locks, just as the miter doors are closing behind them.

Postscript on Hodges' career. Harry Hodges continued serving as the Assistant Chief Engineer in charge of design of locks, dams and regulating works. In July 1911, he was promoted to Colonel in the Corps of Engineers, and his titled changed to Engineer of Maintenance for the Panama Canal, where he remained until January 1, 1915. On March 12, 1915, he received the thanks of Congress for services rendered in constructing the Panama Canal, and was promoted to Brigadier General in the Line of the Army, effective March 4, 1915.

After departing Panama he took two months leave and then reported to the Office of the Chief Engineer in Washington, DC, where he was given charge of re-designing and supervising the renovation of the water supply system for the nation's capital. Soon after war was declared on Germany in April 1917, Hodges was given command of the 76th Infantry Division at Camp Devans, Massachusetts. The 76th sailed for France in February 1918, where Hodges remained until January 1919, when he and the division returned to the states. He was then given command of the 20th Infantry Division at Camp Seiver, South Carolina, which he retained until the end of February. He returned to the Office of the Chief Engineer and worked on special projects until his retirement from the Army on December 21, 1921, at the age of 61. He was promoted to Major General upon retirement. Hodges died at age 69 on September 24, 1929 while living in Lake Forest, Illinois. "Hodges Hill," a steep cut slope just north of the "Gaillard Cut," on the western side of the canal, was named in his honor.

DAVID D. GAILLARD

Brief sketch of military service. David DuBose Gaillard (pronounced *"gee-yard"*) was born in September 1859 in Fulton, South Carolina, of Huguenot ancestry. He was raised in Clarendon until age 13, when moved to Winnsboro, SC, to live with his grandmother Gaillard while attendeding Mount Zion Institute, where the schoolmaster encouraged him to take the competitive examination for West Point. He used his own funds to attend a prepartory school for West Point at Highland Falls, and was one of just two Southerners to be admitted to West Point in 1880 (just after George Goethals graduated). His roomate was the "other Southerner," W.L. Sibert of Alabama, who was much taller than Gailard, so they became known at the Academy as "David and Goliath." Both cadets did very well at West Point, Gaillard ranking fourth and Sibert seventh out of a cadre of 37 graduates in the Class of 1884. Both men sought and received coveted commissions in the Army's Corps of Engineers (along with Cassius Gillette, who ranked 3rd in the same class).

Second Lieutenant Gaillard was sent to the Army's Engineering School of Application at Willet's Point, graduating in March 1887. This was where the valuable engineering courses were taught, with much emphasis on construction tecniques and management thereof. In April, he was sent to Florida to serve as an assistant engineer to Engineer Captain William L. Black (would would be the first Army Engineer stationed in Panama, and Chief of Engineers during the First World War), working on rivers and harbor fortifications in Flordia. This apprenticeship would prove invaluable in the years to come because Black was one of the Corps' rising stars. The following October, he was promoted to First Lieutenant and married Katharine Ross Davis of Columbia, South Carolina, sister of Professor R. Means Davis, his proctor at Mt. Zion Institute, who had encouraged him to apply toWest Point years earlier.

In November 1891, he was assigned to the International Boundary Commission between the United States and Mexico, and served as an assistant in local charge of defensive works being constructed at Fort Monroe, Virginia. From February to October 1895, he was given local charge of construction of a portion of the Washington Aqueduct, raising the old dam at the Great Falls of the Potomac. He was promoted to Captain in October 1895. Between August and November 1896, he surveyed the Portland Channel in Alaska, the boundary between Alaska and British Columbia. After completing this he returned to Washington DC to work on the aqueduct, with which he was engaged when war broke out with Spain in April 1898.

Gaillard was assigned as Engineering Officer on the staff of Major General James F. Wade, U.S. Volunteers for a month and a half before being given command of the Third U.S. Volunteer Engineers with the rank of Colonel of Volunteers. He spent time reciuting men in the deep South, working out of Chickamauga Park, Georgia, then formed the regiment up at Jefferson Barracks, Missouri. They undewent engineer training near Lexington, Kentucky before continuing on to Macon, Georgia, from where the regiment sailed for Cuba on February 1, 1899. They were too late to see any combat, but were split into three battalions, stationed at Cienfuegos, Pinar del Rio, and Matanzas, where they peformed civil, mechanical, sanitary, and hydraulics duty. One of Gaillard's brightest officers and best engineers was a fellow Southerner named Captain Sydney B. Williamson, an 1884 grdauate of the Virginia Military Institute, whom he would work with again in Panama.

In mid April, the Third Engineers were ordered back to the United States and were mustered out at Fort McPherson, Georgia in mid-May 1899. The regiment was praised for its orderly conduct and maintenance of good order and mutual respect towards and from the Cubans amongst whom they "*performed many works of progress*." Gaillard returned to his former rank of Captain in the Regular Army. Years later, the men of his regiment prepared a book memorializing his life constributions and accomplishments (US Army, 1916).

After the dissolution of his engineer regiment, Gaillard was detailed to design the substructures for the gatehouses on the new tunnel for the Washington Aqueduct, which required four months. He was then assigned as an assistant to the Engineer Commissioner of the District of Columbia until February 1901. He was then dispatched to take charge of river and harbor work on Lake Superior, out of the Corps of Engineers office in Duluth. While there he began investigating the effects of wave action on engineering structures. After coming up short searching for any theoretical treatments of shallw wave action like that along the margins of the Great Lakes, he decided to undertake a more through study of the subject, and constructed models using diaphram dynamometer with gagues and a clockwork mechanism that enabled him to make precise observations of wave action in all simulations. He completed a professional paper [book] while stationed at Vancouver Barracks in 1903, and in early 1905, his manuscript was approved by the Chief of Engineers and printed for the use of all engineering officers with the title *"Wave Action Upon Engineering Structures."*

In April 1904, he was promoted to Major of Engineers and was one of just 43 officers appointed to the new General Staff of the Army, along with Major George Goethals. In this position he was occassionally dispatched to perform miscellaneous tasks. These forays usually lasted a few months and he would return to Washington. The first was to serve as Engineer Officer of the Northern Division in St. Louis between January to November 1904. He was the posted to the General Staff at the Army War College (along with Goethals) from November 1904 to October 1906. From October 1906 to February 1907, he served as Chief of the Military Information [intelligence] Division in the Army of Cuban Pacification in Havana, while disturbances were occuring there.

Appointment to the Isthmian Canal Commission. In mid-February 1907, his colleague and the senior Army Engineer on the General Staff, Major George Goethals, was selected by War Secretary Taft and President Roosevelt to serves as the next Chief Engineer and Chairman of the Isthmian Canal Commission. Goethals was given just a few days to suggest names of other Army Engineers he would like to nominate for his staff. He immdiately thought of Gaillard, who was still in Cuba on his intelligence mission. Having taught with Gaillard at the War College, Goethals knew he was from the South, spoke some Spanish (from his time with the Boundary Commission), and knew something of the tropics, serving his second tour in Cuba.

He sent a telegram to Gaillard in Havanna and Gaillard responded that his old West Point classmate, Major William L. Sibert, was an engineering officer who had a lot of experience with the construction of locks and dams on the upper Ohio River and many of its tributaries. Unable to get Harry Hodges, Goethals decided to nominate Gaillard and Sibert to War Secretary Taft. These selections were approved and both majors were appointed to the Isthmian Canal Commission on March 16, 1907.

The biggest surprise was that their Army salaraies as majors would be elevated from $3,000 to $4,500 per year, a 50% increase. Goethals was four years senior to both men. He was promoted to Lieutenant Colonel on March 4th and given a salary of $15,000, which was more than the Chief Engineer of the Army made. All three men realized that the assignment was a "plum," but that it also meant being uprooted from the creature comforts of their homeland and being detailed to remain in Panama until the job was completed, which could be "8 or 10 years."

Settling into Culebra. David Gaillard (Figure 90) arrived on the Isthmus on March 12, 1907. He was assigned quarters in the hamlet of Culebra, adjacent to where the future canal would cross the Continental Divide. Culebra was more like a military base than an actual town, but was comforably situated on a hill whose crest elevation was a bit over 380 feet above sea level, just enough to catch evening breezes across the sweltering Isthmus. One of the first things he learned was that the proposed canal didn't run east-to-west, but north-northwest-to-south-southeast. Culebra was, therefore, on the "west" side of the canal excavations, and the Atlantic Ocean lay 31 miles to the north, with the Pacific just 16 miles to the south.

Figure 90. Portraits of David DuBose Gaillard in his Army whites after he was promoted to Lieutenant Colonel of Engineers in April 1909 (USACE).

It's doubtful that upon his arrival Gaillard knew much about railroad engineering, steam shovels, production blasting, mining, or the excavation of millions of cubic yards of soil and rock, but he was about to become a world expert on those subjects. He was briefed about the situation, told that the French had excavated 23,000,000 cubic yards in the Culebra Cut and that 85,000,000 remained to be excavated. The highest cut, at Gold Hill, was within easy view of his new quarters, and when the channel was completed this chasm would be 494 feet deep, a world record.

Culebra overlooked a spectacle of sorts, with thousands of men, hundreds of steam engines powering locomotives and rock drills, dozens of flat car trains carrying muck criss-crossing their way northward, towards the spoil dumps at Tabernilla and the Gatun Dam. He couldn't see the work moving in the opposite direction, towards the Pacific, because his view was eclipsed by Zion Hill and Contractor's Hill, which were both higher than Culebra. The Pacific Ocean was much closer than the Atlantic, and the travel times of muck trains on that side of the great divide were much less than those heading north.

Keeping things moving. Major Gaillard's new job title was "Chief of the Department of Excavation and Dredging," which was comprised of two dredging divisions and one dry excavation division. The Pacific Dredging Division handled the area between Miraflores and the Pacific Ocean, while the Colon Division handled the are between the site of the Gatun Locks and the Atlantic Ocean. When he arrived nothing was being excavated along the 23-mile stretch between Gatun and Gamboa, where the canal left the Chagres River. Gaillard immediately began planning for the execution of work along the stretch between Gamboa and Gatun. The only equipment readily available (until dredges ordered from the states arrived) were those left by the French. He assigned six steam shovels to this zone and 46 old French and Beligian locomotives, pulling the smaller side-dumping muck cars. With this "secondary force", work began on the 23-mile segment that had hitherto been more or less ignored.

Louis K. Rourke. In the Culebra Cut, Gaillard kept the "engineering plant" established by John Frank Stevens running as well as he could, ably assisted by a young red-haired engineer named Louis K. Rourke (Figure 91), who had been working on the Culbra Cut for almost two years and seemd to know what he was doing. Louis Rourke was born in Abington, Massachusetts in 1873 and after matriculating through Abington High School, enrolled at the Massachusetts Institute of Technology in Cambridge, receiving his bachelors degree in civil engineering in 1895. He worked several years in the Maintenace of Way Department for the Boston & Maine Railroad. Then he decdied to strike out for Panama, serving as the Panama Railroad's Supervisor of Track from 1897-99. He then moved to Ecuador to become Superintendent of Construction and Contractor for the Guayaquil & Quito Railroad until 1904, when he returned to Massachusetts as a contractor for the Massachusetts Highway Commission.

Rourke was enticed by former associates with the Panama Railroad to make application to the ICC and return to to the Isthmus, where they were desperately in need of "good railroad men" to work on the canal. He arrived at Colon in November 1905, when things were at a pretty low ebb and work on sanitation and housing was just winding up. His first meeting with John Frank Stevens had been a *"real grilling,"* but he passed muster and Stevens made him Superintendent of Construction in the Culebra Division. He was subsequently promoted to Superintendent of Tracks, then Superintendent of Tracks and Dumps, and finally as Division Engineer of the Culebra Division when Goethals, Gaillard, and Sibert arrived in mid-March 1907. His crews had been making spectacular progress on excavating the big cut over the previous four or five months, and Gaillard was fortunate to hang onto him. Rourke took a month long vacation to marry Teresa Ryan in May 1907, returning with his new bride to Culebra, where they were given new quarters. He took the title "Assistant Division Engineer" when the project was reorganized in mid-1908, and served as Gaillard's principal deputy until he and his wife departed for the states in the summer of 1910, to accept the position of Commissioner of Public Works for the City of Boston, the youngest man to ever hold the title.

Figure 91. MIT-trained civil engineer Louis K. Rourke was a pivotal figure in keeping the excavations along the Culebra Cut moving forward between March 1907 and July 1910, when he became Commissioner of Public Works for Boston (Jackson & Son, 1911).

Reorganization of 1908. On July 1, 1908, Colonel Goethals restructured the organization constructing the canal (described earlier), abolishing the specialty departments and establishing three geographic divisions. Gaillard was named Division Engineer of the expansive Central Division, shown in Figure 92. This was the largest division, stretching 33 miles, from Gatun to Pedro Miguel, including the nine treacherous miles of the Culebra Cut, of which, seven and a half miles would only be excavated to a width of 300 feet (500 feet elsewhere) with a depth of 45 feet.

When Gaillard assumed control of the Culebra Cut in April March 1907, the maximum monthly excavation reached 815,270 cubic yards. By March of 1909 that figure had risen to 2,054,088 yards, an increase of 152%. That month there were 68 steam shovels working in the Culebra Cut, while the total number of shovels on the canal eventually reached 100. The muck trains were pulled by 115 locomotives and 2,000 rail cars, allowing up to 160 train loads per day to come and go out of the cut. The average distance to dumping sites was about 12 miles, and the longest haul was 33 miles. Between 1907-09, Gaillard and Rourke increased the miles of trackage in the cut from 76 to 100, and averaged nine parallel pairs of tracks working the various faces to widen and deepen the cut.

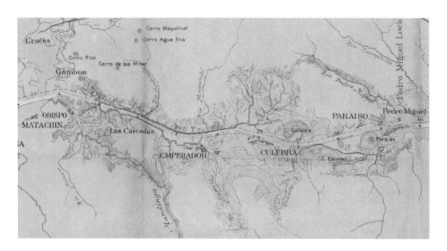

Figure 92. The Central Division encompassed the area between Gatun and Pedro Miguel, which included nine miles of excavation across the Continental Divide. Note the diversion channels that were excavated along either side of the canal to intercept and convey runoff from tributary channels, east of Culebra (USACE).

Serious drainage. In 1908 and 1909, more than 120 inches or rain were recorded at Culebra. Maintaining a workable sytems of drainage was a daily battle, as was keeping the worker's feet dry. There were 18 tributaries spilling their discharge into the Culebra Cut. Numerous diversions had been attempted, but each of these had been overwhlemed and breached at some point. It was finally decided to excavate capacious channels along either side of the Cut, east of the Continental Divide. The eastern (or upper in Figure 88) ditch was dubbed the Obispo Diversion. It stretched five and a half miles east, to Obispo, and had a minumum width of 50 feet. The companion ditch running along the western side of this same reach was named the Camacho Diversion, which required 1,000,000 cubic yards of excavation.

An earthen cofferdam extending to elevation 78 feet above sea level was constructed across the canal channel at Gamboa, to prevent the ranging Chagres River from flooding the channel prism. The only problem with this was that every drop of water that fell into the channel had to be pumped over the dike, into the Chagres. Whenever it rained hard, a sizable lake would form behind the "Gamboa Dike" and its backwater could extend several thousand feet up the canal floor. Sometimes it took several days to pump it down again. An outlet valve and pipe was extended through the dike to provde additional gravity flow whenever the level of the Charges had dropped to a normal flow level.

Improving efficiency. Like any railroad man, Rourke was always seeking greater efficiency, while Gaillard considered improvements that would reduce unit costs. By improving the tracks and decreasing grades they found that they could increase the number of muck cars pulled by each locomotive. The muck trains usually waited by their assigned shovels until each car was loaded. This system was shifted to allow the muck trains to slowly move along their assigned benches without stopping, allowing different shovels to contribute spoil onto the same train.

By holding the shovel steady while the train was moving the shovel operators were able to better spread the muck along the high-back flatcars. In the old system the slowest shovels would determine the rate that muck trains in succession could move out with their loads. This new technique removed the "lowest common denominator" that tended to bottleneck the muck trains, keeping things moving all of the time.

By early 1913, the shinning marvel of the entire project was the yawning chasm sliced through the Continental Divide at Gold Hill, just shy of 500 feet deep (Figure 93). Morale was running high and there was much expectation that the canal would be completed by August 1913. Man had, at last, conquered nature, or so it seemed.

Figure 93. The deepest excavation at Gold Hill on the Continental Divide, before landslides began really plauging the project (the Cucaracha Slide toe can be seen at lower right). Side slopes were excavated at an inclination of 56 degrees.

Punching through. On May 20, 1913, ICC shovels No. 222 and No. 230 slowly narrowed the gap in the floor of the Culebra Cut, as seen in Figure 94. They met on the "40 foot bench" (40 feet above sea level), the floor elevation of a 45 foot deep channel when the waters of Lake Gatun eventually rose to 85 feet above sea level. The last concrete had been placed on the Gatun Dam spillway just eleven days earlier. Above-ground excavation ceased just three months later.

Figure 94. Steam shovels from the Atlantic and Pacific sides meeting in the floor of Culebra Cut on May 20, 1913.

The Cucaracha Slide. In the fall of 1907, the toes of the slope abutting the southern flank of Gold Hill began to move into the canal excavation (Figure 95). Unlike other slumps, this one just kept on moving, at a near-constant speed of 14 feet per day, until it had ripped out every track and moved all the way across the cut, and 30 feet up the opposite side, blocking the canal excavation. It was an early harbinger of things to come.

In 1910, Cucaracha reactivated twice, blocking the floor of the canal cut prism once again, this time for several months. As the cut fell ever deeper into the earth (Figure 89), things began to appear increasingly ominous. Cucaracha would keep moving, almost every year. It became one of the most studied landslides in the world, and for good reason. No matter what the engineers tried, the problem just seemed to worsen with time, and the slide kept enlarging itself. Canadian geologist Donald F. MacDonald (described later) was dispatched by the U.S. Geological Survey to become the first project geologist on any civil works project in the United States, living on the Isthmus for the next five years (MacDonald, 1915). He soon discovered that the materials comprising the slides were radically different from the volcanic andesite flows forming the massif of Gold Hill. The slides were occurring in a slippery material called the Cucaracha Shale.

The unit was not really a true shale, although it behaves and looks like shale. It is actually of volcanic origin, something geologists refer to as a "volcanic agglomerate." These materials are deposited by ancient volcanic mudflows streaming down the sides of volcanoes and filling in low lying areas, similar to landslide debris, but much warmer. The bits of volcanic ash eventually hydrate and weather into smectite clays, capable of swelling 300 to 800% of their dry volume through cationic absorption of moisture (the same material when sold commercially for drilling mud is called "bentonite").

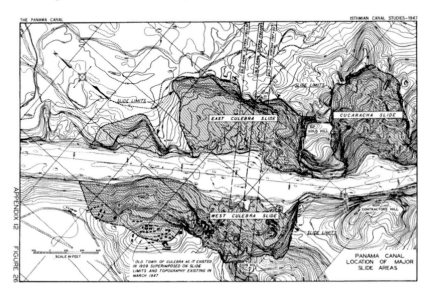

Figure 95. Topographic map prepared in 1947 showing the locations of the canal; the Cucaracha, East Culebra, and West Culebra Landslides; Gold Hill; and Contractor's Hill. The old town of Culebra has been destroyed by land sliding. Note the narrow passages caused by the slides (Canal Zone, 1947).

The Cucaracha, Culebra, and LaBoca Formations contain smectite clays, which are subject to significant strength loss upon shearing (Lutton et al., 1975). This loss of shear strength occurs with increasing movement, as shown in Figure 96. Geotechnical engineers have come to refer to this dramatic loss of strength as "residual strength loss"(Canal Zone, 1947). It seems to occur most often in over consolidated shales with swelling clay, which have previously been subjected to high loads, either from deep burial in the earth or, as in the case of Panama, by tectonic action. As tectonic plates collide with one another, which is how the Central American massif has been sustained for millions of years (tectonic forces keep pushing the ground upward and the area is perturbed by a system of active earthquake faults).

During the summer of 1912, the Cucaracha Slide deposited another 3,000,000 yards of debris into the channel cut. In January 1913, a new slide at Cucaracha spilled 2,000,000 cubic yards of earth into the channel, severing all the rail lines, taking muck to the Pacific-side dumps, and further complicating plans to complete the project on time.

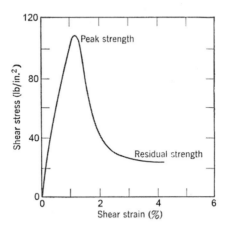

Figure 96. Drained direct shear test on Cucaracha Shale, illustrating how the material loses about 80% of its peak strength (y axis) as it undergoes slippage, shown here as "shear strain" (x axis) (Canal Zone, 1947).

West and East Culebra slides. In 1911, cut slope failures/landslides occurred up an down the line at various times, at Las Cascades, La Pita, Empire, Lirio, and East Culebra. There were twenty-two major slides tabulated that year by geologist MacDonald. It took Gaillard's crews three entire months of round-the-clock digging just to get back down to where they had been before the slides kicked off.

In the summer of 1912, the West Culebra Slide (Figure 97) began moving into the cut, shutting down canal excavation for four and a half months while removing additional slide debris. Thirty structures in the town of Culebra had to hastily moved to prevent their loss. The slippage eventually swalled up 75 acres of land (Figure 94). By October 1st, over 7,500,000 yards of additional debris had been excavated from the West Culebra slide alone, leaving 2,000,000 more yards of debris in-place.

By late 1912, the side slopes around Culebra had laid themselves back to an inclination of 5:1 (horizontal to vertical), or about 11.5 degrees, far less than their original inclination of 56 degrees. It was rapidly appearing that steep cut slopes couldn't be sustained unless they were made in massive volcanic flow rock, like basalt or andesite, like Gold Hill.

By mid-1913, the East Culebra Slide (Figure 98) had contributed 10,000,000 yards and the West Culebra Slide and additional 7,000,000 yards, completely blocking the cut channel. The Americans were expending an enormous effort, but seemingly getting nowhere.

Figure 97. Shovels begin attacking the toe of an new slide on the oposites side of the canal excavation, which they name the West Culebra Slide on September 12, 1912.

Figure 98. Crown scarp of the East Culebra Slide on September 19, 1912, looking south.

About a dozen steam shovels were buried at one time or another in the various slides, and it took weeks to painstakenly salvage them, as shown in Figures 99 and 100.

Figure 99. Steam shovels and railroad tracks were swalled up by the enormous landslides, which often lifted up the floor of the canal excavation through the process of "toe thrusting." This view is at Cucaracha.

Figure 100. Work crews salvaging one of the buried Bucyrus 75-ton rail-mounted shovels caught by movement of the East Culebra Slide on March 13, 1913.

Flooding the channel. It was decided to flood the channel and finish channel excavating the channel prism using dredges. There were several sound arguments for doing this. For one, the Cucaracha Slide debris was "flowing" into the cut floor, like so much molasses. It seemed like the kind of material that should be easily dredged. And, it was getting more and more difficult to excavate with shovels because it was so gooey and viscous, sticking to the buckets of the big shovels. The second sound reason was that filling the channel with 40 feet of water should provide some lateral support of the lowest slopes on each side of the channel, which were feeling the greatest stress. The hydrostatic pressure of the water column would exert an "outward force" against the submerged slopes, proportional to the square of the depth of water (this was all in the days before the fundamental theorems of soil mechanics had been developed).

The last steam shovel lifted the last rock in the cut on the morning of September 10, 1913, to be hauled out on the last dirt train by locomotive No. 260. On October 10, 1913, President Wilson pressed a button inWashington, DC and the Gamboa Dike was blown up, creating a chasm 100 feet wide through which the waters of Gatun Reservoir rushed through, allowing inundation of the canal channel as far as the East and West Culebra Slides, which, along with the Cucaracha Slide, were blocking the channel.

The day after the Gamboa Dike was blown workmen were told to excavate a "pilot channel," around the toe of the Cucaracha Slide, shown in Figure 101. This was because the water rushing in from the Gatun Reservoir was now 30 feet higher on the north side. Engineers reasoned that if they could open up even a small outlet that the outpouring waters would rapidly excavate a much larger, deeper channel with very little effort. Six days later they set off a series of charges buried in this pilot channel, as seen in Figure 102. The blasts were impressive, but without the desired results. The sticky clay closed whatever gaps were formed by the explosions and no meaningful amount of water made it through the slide debris.

At this juncture the engineers received some heavenly assistance in the form of intense precipitation in the highlands, which caused the Chagres River to flood, rasing the level of Gatun Lake by three feet in just three days (the lake had been rising at a rate of two inches per day). The rising waters spilled through the hand-excavted trench, and began filling the channel prism south of the Cucaracha Slide. This allowed dredges from the Pacific side to begin knawing at the far side of the slide, while all of the dreges that could be brought to bear on the northern side were passed through the Gatun Locks and dispatched with no little urgency to the same scene, working from the northwern side. The engineers then resolved to dredge the slide debris for as long as it would take until the slopes stabilized themselves and came to equilibrium with the water level in the canal (Figure 103). The ladder dredge *Corozal* succeeded in making a pioneer cut through the debris dam of the Cucaracha slide on December 10, 1913, two months after the channel was flooded. By April 1914 the channel was of sufficient width to allow tungs and lighters to pass through.

Figure 101. Laborers hand excavating a trench around the toe of the Cucaracha Slide on October 11, 1913, in hopes of getting water on the north side (from Gamboa) to cut a channel as it flowed towards the Pacific.

Figure 102. Unsuccessful blast to excavate a channel around the toe of the Cucaracha slide on October 16, 1913.

Figure 103. Suction dredge working the flooded toe of the Cucaracha Slide, as viewed on October 28, 1913. The dreding would continue for the next 10 months.

Life on the edge. In April 1909, David Gaillard was promoted to Lieutenant Colonel, 25 years after graduating from WestPoint near the top of his class. During the period lasting from April 12, 1907 to June 1, 1913, the total excavation carried out under Gaillard's watch in the Central Divison was 104,800,873 yards, a staggering figure even today. There is little doubt that he bore an increasing amount of stress battling the wildly unpredictable landslides, which were playing havoc with the entire project and delaying its completion, possibly for several years.

The Gaillard's son Pierre had remained in Washington, DC when they moved to Panama in 1907 to finish his junior year of high school. He then surpised everone by enrolling in engineering studies at the Massachusetts Institute of Technology at the age of 17 and bereft of a high school diploma. He studies electrical engineering and graduated from MIT in 1911. He then moved to Panama and lived with his parents (Figure 104) while commuting to the Gatun Locks each day working on the electrical machinery being installed there.

Gaillard's death. During the stressful summer of 1913, David Gaillard seemed to be breaking down under the strain of his arduous responsibilities. Everyone assumed he was having a nervous breakdown, even Goethals. Dr. W. E. Deeks, head of the Medical Clinic at Ancon Hospital feared it was more serious. In July, Deeks talked the senior Gaillard into taking a steamer back to the states, to have him diagnosed by some specialists. The Colonel, Mrs. Gaillard, their son Pierre, and the Chief Health Officer, Surgeon Colonel Charles Field Mason, traveled together by steamer back to New York, thence onto the Army Hospital in Washington. After a series of tests the medical team diagnosed his illness as an infiltrating brain tumor.

His wife and son sought the advice of the finest doctors, but there was few procedures that could be undertaken in those days. His considerable fame won him the opportunity to undergo exploratory brain surgery at the Peter Bent Brigham Hospital in Boston, close to where Colonel Goethals' younger son Tom, was attending Harvard Medical School at the time. Tom was part of the medical audience that witnessed the delicate operation peformed by Dr. Harvey Cushing.

Figure 104. Lt. Colonel and Katharine Ross Gaillard at their quarters in Culebra having afternoon tea. Mrs. Gaillard was an avid botantist who festooned their home with an attractive array of tropical plants during the six years they lived in the Canal Zone.

On August 17th, Gaillard was taken to Johns Hopkins Medical School in Baltimore, where he underwent follow-up treatments, without making progress. He died at Johns Hopikins in Baltimore on December 5, 1913, never seeing the canal to its completion and becoming one of its last victims. He was 54. Colonel Gaillard was buried with full military honors at Arlington National Cemetery. On the day of his internment Congress voted, and President Wilson approved, the conveyance of one's year's salary of Colonel Gaillard to his widow Katharine Gaillard. In 1915, the

Culebra Cut was renamed the *Gaillard Cut* in honor of David DuBose Gaillard's gallant efforts to see it completed. The men of the Third U.S. Engineers Regiment that he commanded during the Spanish American War prepared a biography of their beloved commander in 1916 (*David DuBose Gaillard: A Memorial.*), which contains a wealth of information on Gaillard's family tree, his professional career, and a series of written testimonaials from men that worked with him, mostly on the Panama Canal (US Army, 1916). Katharine Ross Gaillard passed away on December 30, 1937, and was intered at Arlington, next to her husband.

After his father's death, Pierre Gaillard settled in Washington, DC where he lived almost all of his life. He volunteered for duty in the First World War and was commissioned as a First Lieutenant, and discharged from active duty as a major in 1923. He remained in the Army Reserve and was recalled to active duty as a Lieuteannt Colonel in 1940, and was later promoted to Colonel. He served as one of the Army's experts on ordinance analysis during the Second World War. He passed away in 1982.

WILLIAM L. SIBERT

Brief sketch of military service. William Luther Sibert was born in Gadsden, Alabama on October 12, 1860. There being no high school in his community, his parents hired a tutor for him and he was accepted into the University of Alabama in 1878. After two years of study he took the competitive examination and received an appoinment to West Point in July 1880.

As a cadet "he was studious, reserved, tolerant, and loved by all." At the academy he was given the sobriqent "Goliath" because of his splendid physique and because he roomed with fellow Southerner David Gaillard, who was smaller of stature, the two being known as "David and Goliath." He graduated 7th of 37 cadets in the Class of 1884, and received one of three coveted Corps of Engineers billets (the other two billets were taken by Cassius E. Gillette and David D. Gaillard). Along with David Gaillard he was sent to Willets Point, NY to receive additional engineering traning from the Army's Engineering School of Applications, from which he graduated in 1887 (Figure 105).

Between July 1887 and the outbreak of the Spanish American War in the spring of 1898, Sibert was engaged almost exclusively on construction of river and harbor improvements. In September 1887, he married Mary Margaret Cummings of Brownsville, Texas, and the couple was blessed with five sons and a daughter. From 1887-92, he was given local charge of lock and dam improvements and construction on the Kentucky, Green,and Barren Rivers. He then served two years (1892-94) with Colonel Orlando M. Poe in the construction of the 800 foot long lock on the St. Marys Falls Canal between Lakes Superior and Huron at Sault Ste. Marie, Michigan, then the world's largest lock. He was also given charge of constructing the ship channel between the foot of Lake Huron and the head of Lake Erie. In 1894, he began a four year tour as District Engineer for the Little Rock District, supervising river and navigation improvements in that region, including improvments to the locks and dams on the White River. In March 1896, he was promoted to Captain.

Figure 105. Second Lieutenant of Engineers William L. Sibert of Alabama with is dog "Butch" while attending the Army's Engineering School of Applications at Willet's Point, New York (Clark, 1930).

In 1898-99, he commanded Company B, Engineer Battalion at Willets Point and was an Instructor in the Engineering School of Application. In July 1899, he was sent to the Philippines, where he served as Chief Engineer of the Eighth Army Corps, serving in the First and Second Divisions, and distingusihed himself in several actions, against insurgents in Northern Luzon, Southern Luzon, Cavite, Laguna, Batangas, and Tabayas Provinces. His courage and initiative caught the attention of his superiors and he was given command of the Engineering Battalion, then made Chief Engineer of the Eighth Army Corps, and later named Chief Engineer of the Department of the Pacific. Sibert also served as Chief Engineer and General manager of the Manila & Dagupan Railway.

In May 1900, he returned to the states and was appointed District Engineer of the Louisville District. Here he worked on the Louisville and Portland Canal and locks and dams along the Green, Barren, and Wabash Rivers. In December 1901, he was transferred to Pittsburgh to assume the duties as District Engineer, and was promoted to Major in April 1904. In Pittsburgh, he supervised a wide array of projects associated with the Corps of Engineers charge to develop and maintain a minimum 9-feet deep channel along the Ohio River. This work included the construction of five dams along the Ohio River, ten locks and dams along the

Monongahela River, and three along the Allegheny River. He was cited for his emergency work during flooding of the Allgheny River in January 1907 for *"prompt and energetic action in demolishing a part of Dam No. 3 at Springdale, PA, [using dynamite] which was undoubtedly the means of saving considerable property."*

Appointment to the Isthmian Canal Commission. As described in the prevous sections on Goethals and Gaillard, Sibert was recommended to Goethals by his old classmate David Gaillard. In retrospect, it was an excellent choice, as Sibert (Figure 106) had much more construction experience with locks and dams than just

Figure 106. William L. Sibert as he appeared at the conclusion of the canal project. He went on to hold many important posts after the canl was completed, including Director of the Chemical Warfare Service and Chairman of the Colorado River Board that reviewed the Boulder Canyon Project.

about any other Corps of Engineers officer who would have been eligible for the position in Panama (of commensurate rank and seniority). His appointment, was however, difficult, because he was being supervised by Goethals, who had much less experience and had been serving as a staff officer in Washington. He perceived Goethels as having been selected for the starring role in Panama not because of his engineering prowess or experience, but because of his close association with War Secretary Taft.

Sibert had little time to prepare before embarking for Panama on March 10, 1907, leaving his wife and six children to catch up with him later (Figure 107). He was immediately appointed to the Isthmian Canal Commission and given charge of all lock and dam construction on the canal project, an enormous responsibility.

Figure 107. The William L. Sibert family in Pittsburgh, as they appeared shortly before departing for Panama in the spring oof 1907. They included Major Siebert, his wife Mary, their sons William, Franklin, Harold, Edwin, and Martin, and their daughter Mary. Edwin and Franklin both became major generals in the Army (Clark, 1930).

Disagreements over the locks on the Pacific side. Sibert began working on the designs for the Gatun Dam and spillway structure, as well as the locks. The original scheme proposed in the Miniroty Report of the International Board of Consulting Engineers in 1906, was a continuous flight of three locks at Sosa Hill, built adjacent to the Sosa-Corozal earth embankment dam, three miles south of Miraflores. Colonel Goethals' trusted colleague Sydney B. Williamson was in charge of constructing the Sosa-LaBoca Locks on the Pacific side.

As described previously, excavation work for the locks and dam at that location had begun in July 1907, and after five months of dredging, a fill trestle was placed along the axis of the proposed embankment and fill placement began. The blue clay lying beneath the proposed embankment experienced a bearing capacity failure, and it was decided to abandon the site and move upstream, to Miraflores.

Crews then began drilling additional borings at Miraflores in December 1907, to ascertain the foundation conditions for a flight of locks. Sibert favored a continuous flight of three locks at Miraflores, similar to the arrangement being proposed for Gatun. He argued that these would be less expensive to construct and less expensive to operate in the long-term, because adjoining walls would be shared (for construction purposes) and adjoining machinery of their operation (for maintenance).

In the late spring of 1908, Sibert and his family took their annual months leave, to visit the United States. While he was gone Goethals and Williamson discussed the foundation conditions being exposed at Miraflores. Gun shy after what had occurred during the previous year trying to place fill on soft blue clay at the Sosa-LaBoca embakment, Williamson felt that two locks could be founded on sound material at Miraflores, but the third would require much deeper excavations, to -40 feet below sea level, which would oblige him to construct expensive cofferdams. He recommended to Goethals that the third lock be constructed two miles north, at Pedro Miguel, where the excavtions could easily reach bedrock.

When Sibert returned from his vacation he learned that the decision had been made to separate the Pacific locks. He felt that Goethals should have consulted him on the matter because he was in responsible change of the design of all locks and dams. Furious, he directed additional borings at Miraflores, to support his case for constructing a continuous flight of three locks, and that it would have saved $4,000,000 (Sibert and Stevens, 1916).

Sibert was correct in asserting greater construction and operatinsand maintenance costs associated with splitting up the Pacific Locks. They are more expensive to operate and maintain than the Gatun Locks, and their aggregate cost of construction was considerably higher (this is partially because their cost included the Miraflores Dam and gated spillway structure, while the Gatun Dam and spillway were recorded as separate costs).

Williamson provided his side of the story many years later (Williamson, 1934). His rationale for limiting himself to two locks were the unknowns associated with constructing the requisite cofferdams and mucking out the soft materials well below sea level. He was correct in pointing out that it is not possible to acurately gage how much these activities would have cost, because it would have depended on how well those measures worked once employed. But, the fact that he was the man in responsible charge of actualy supervising their construction, argues for the decsion that Goethals made. Richard Whitehead, "*one of Goethals' staunchest allies among the young engineers,*" told David McCullough that Goethals "*wasn't quite human enough for everyone*" (McCullough, 1977; p.573).

Reassignment. On June 30, 1908, Colonel Goethals re-organized the canal project according to geographic sectors and divisions of responsibility, including where the Division Engineers and their assistants would live. The Atlantic Division was placed under Major Sibert, where he would be responsible for constructing the seven mile long approach channel to the Gatun Locks on the Atlantic side, three massive locks at Gatun, as well as Gatun Dam, spillway, outfall channel, and powerhouse. Over the succeeding months the design of locks and dams would be reorganized as well, under Lieutenant Colonel Harry F. Hodges (described previously).

Major Sibert would have responsible charge of constructing the Gatun dam and spillway, eight miles of channel excavation, and the Gatun Locks along the right abutment of the dam. In carrying out these responsibilities, Sibert was ably assisted by some of the best and brightest military and civilian engineers. These included: Major Chester Harding, Assistant Division Engineer; Major James P. Jervey supervising masonry work of the triple locks; Major Edwin Jadwin in charge of dredging and construction of breakwaters; Major George M. Hoffman Assistant Engineer for work on Gatun Dam; Caleb Saville, a Harvard graduate who had worked on the North Dike of the Wachusetts Dam, he was given charge of assessing the hydrology of the Chagres watershed and assisted in the design and construction of the Gatun Spillway; Ben Johnson, a West Point graduate in 1889, who as a civilian engineer, had charge of constructing the locks; and, Captain H.W. Stickle was given charge of the rock quarry operations at Portobello, under Major Jadwin. All but Stickle are shown in Figure 108. A picture of Sibert with his office staff is presented in Figure 109.

Figure 108. Sibert's principal assistants in the Atlantic Division included, from top left, clockwise: Major Chester Harding, Major James P. Jervey, Major Edwin Jadwin, Caleb M. Saville, Ben Johnson, and Major George H. Hoffman (images from Jackson & Son, 1911).

Figure 109. Lieutenant Colonel Sibert (seated, seventh from right) and the office staff of the Atlantic Division at Gatun, as seen in 1911. He never wore his Army uniform the entire time he lived in Panama.

Restraining the Charges River. The kingpin structure of the American canal scheme was an earthen dam of unprecedented magnitude and scope at Gatun, to catch the aggregate flow of the unpredictable Chagres River and all its principal tributaries. Upon this structure, and this structure alone, the entire plan rested because it created the man-made lake rising 85 feet above sea level, across which ships would pass from ocean to ocean (Figure 110).

Everything about the Gatun Dam was enormous. Its proposed dimensions were without precedent: 6,400 feet long and a maximum width that was later enlarged to 2,300 feet. With a height of 110 feet above sea level it could store sufficient water to create a reservoir covering up to 164 square miles. In its center would sit the most critical structure; a mass concrete spillway crossed the Chagres River to create a 164-square-mile lake as part of the canal.

The biggest problem with the dam site was the underlying geology, which included deepley incised channels, shown in Figure 111. The massive embankment would be placd over these channels, which were up to 258 feet below sea level. The "good news" was that the materials filling these channels were of relatively low permeability, being sandy silts and clay. There were more pervious materials, such as sands and gravel, but they were at considerable depth.

The most pondered design considerations revolved around how best to handle subsurface seepage, the low bearing capacity of the surficial deposits, and how best to handle potetial differential settlement, across the filled channels. The concrete spillway would wisely be founded on the bedrock rise between the two channels, as would be the hydroelectric power plant and the outfall channel (Figure 110).

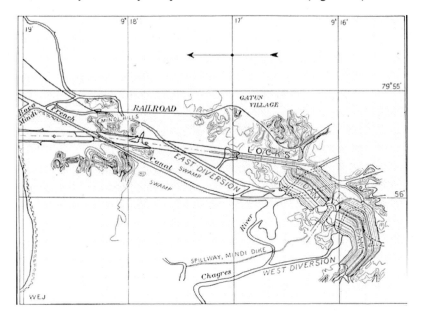

Figure 110. Map showing the positions of the approach channel to the Gatun Locks, the Gatun Dam and Spillway, the old French canal excavations, and the channels of the Chagres River. Note how the upper lock rests on the dam's right abutment (Goethals, 1911).

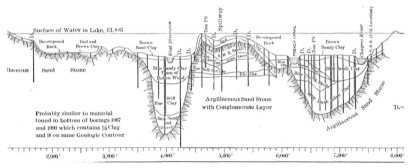

Figure 111. Detail of subsurface conditions exposed along the centerline of Gatun Dam, where two ancienyt channels of the Chagres River excavated deep into the surrounding bedrock (Wegmann, 1911).

The engineers who prepared the Minority Report of the International Board in February 1906, were of the opinion that a sheetpile seepage cutoff, like that used on the North Dike of the Wachusetts Dam, was not necessary at Gatun because the great majority of the valley fill was low permability clays. The differential settlement and bearing capacity isues went hand-in-hand with one another. Through 20 years before soil mechanics was even invented, the minority board members recommended that very gradual side slopes be utlized, so as to spread the dam's bearing load over as large of an area as possible.

Work began on constructing the townsite of Gatun two days after the Senate vote in late June 1906. Shortly thereafter laborers began clearing and grubbing the dam site, which ran 1-1/2 miles between two low hills. This tedious clearing and grubbing took the better part of an entire year. Construction of the embankment was initiated the following year (1908), a few months before the reorganization of the project into three geographic divisions. This filling began with the construction of upstream and downstream "containment dikes," common in those days for hydraulic fill embankments. These were placed from wooden trestles carrying side-dumping railcars. The dikes were originally spaced about 1,600 feet apart, so the dam would be 15 times as wide as it was high (and adding 100 feet for the crest width).

The crest elevation was to be 115 feet, or 30 feet above the normal operating pool, providing a record volume of flood storage, not exceeded until the completion of Hoover Dam in 1935.

Four dredges, ten steam shovels, and ten muck trains were assigned to the placement of fill for the dam. The fill material came from a number of sources: the excavations for the Gatun Locks, dredge spoils from the three mile long channel between Gatun and Limon Bay, and from muck trains emanating from the Culebra Cut and Mindi Hill, up to 23 miles to the south. It was a long trip, but about a dozen trainloads per day of fill were diverted from the waste dumps at Tabernilla and taken onto the dam.

Filling progressed smoothly until November 20, 1908, when the rock containinment dike along the toe of the upstream side of the dam began settling, and dropped as much as 60 feet over a zone about 300 feet wide, destroying the fill-laying timber trestle. Sibert's crews had been dumping rock in that area over the previous 10 days, to build up the containment dike. This had been during a period of intense rainfall, which they believed played a role at the time (it likely did not; they just surcharged the saturated clay foundation much too quickly to allow for adequate dissipation of the pore water pressure developed in the clay).

The negative publicity attached to such a failure naturally caught the media's attention, and was splashed across many newspapers back in the states, because the Chagres River burst through the gap created by the sudden settlement. This hastened the formation of a Special Board of Consultants to investigate the matter, appointed by War Secretary Taft, who was also the President-elect. Taft accompanied the special board to Panama in January 1909, and they made a thorough study of the situation, described later. Their recommendation was to broaden the rock dike to spread its load and thereby reduce the bearing pressures beneath it, which had exceeded the capacity of the underlyinhg clay. Although this recommendation was sound, the actual reason for the bearing failre was the build up of hydrostatic stress, or pore water pressure, within the caly, because the load of rock was placed upon it too quickly, not allowing sufficient time for the pore water pressures within the clay to dissipate, which takes many weeks, or even months in low permeability clay.

In 1910, a second series of slides occurred in the main embankment between the locks and the spillway. This time it was the dam's crest that gave way, slumping 10 to 15 verical feet over a distance of approximately 1,000 feet. This failure likely extended into the hydraulic fill core of the dam, shown in Figure 112. The upstream face inclination was then lowered from 4:1 to to 7.67:1 (horizontal to vertical), as shown in the upper portion of Figure 112.

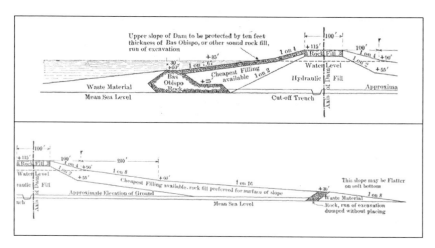

Figure 112. Upper view is the maximum upstream cross section through Gatun Dam as it was actualy built. Note 15 feet of freeboard. Lower view is the maximum dpownstream section. Years later a golf course was built on the 16:1 downstream face, which slopes just 3.6 degrees (Wegmann, 1911).

In the end, Sibert's engineers decided to construct an even more conservative cross section than that recommended by the Special Board of Consultants, because they had the fill to waste from all of the excavations for the canal, plus the landslides. What began as a 3:1 (horizontal to vertical) slope of 18.1 degrees was winowed down to a 11:1 slope of just 5.2 degrees, which was later dropped to 16:1 or just 3.6 degrees (Figure 110). This flatening of the side slopes had the added benefit of lengthening "seepage paths" through and around the dam and its foundation, lowering the potenial for destablizing hydraulic uplift or high seepage pressures.

The west diversion channel (Figure 113) was closed in late April 1912, before the main embankment had been completed by constructing a diversion dam across the channel, using fill dumped from a trestle. There had been some concern about whether the deeply incused channel would be able to sustain the surcharge of the diversion embanment, given how quickly it was placed. Water slowly began to build up behind the new dam. By the time the embankment was topped out it contained 22,000,000 cubic yards of fill materials, a world record until the completion of Fort Peck Dam in 1940.

Gatun Spillway. Tha Gatun Spillway was an arched concrete gravity dam 100 feet high with a crest length of 800 feet. It was designed according to the suggestions made by Isham Randolph and Frederic P. Stearns of the International Board of Consulting Engineers in February 1906. The design of the 30-foot wide Stoney Gates with sills submerged 16 feet below the desired operating pool were within 12 inches of that employed on the Chicago Sanitary and Ship Canal, overseen by Randolph.

This design allowed the gates to be lifted and the water to spill from a lesser crest elevation, but with increased hydraulic head, making them very hydraulically efficient. The only problem was velocity and impact force, and the drop (43 feet) was much higher than in Chicago. With two additional feet of head (elevation 87 feet), the spillway could safely pass about 182,000 cubic feet per second.

Caleb Saville was assigned to solve this problem because he had worked on the Wachusetts Dam, which was the prototype consulted for comparison with the conceptual designs for Bohio and Gatun dams. Saville made extensive surveys of the available literature and ran the calculations. He ended up designing reinforced concrete impact blocks nine feet high at the base of the spillway skirt, to hasten energy dissipation. Theorems were not yet inexistence to deal with dynamic forces, fatigue failure, or resonant frequency of vibration. Saville decided to place steel impact jackets around the upstream faces of the energy dissipator blocks (shown in Figure 114), and these have served their intended purpose remarkably well, comnsidering their design was without precedence.

In early 1909 the first concrete was poured for the spillway, and this work was completed 19 months later, after which the Stoney Gates were set into place, which took another year and a half. During the next year the discharge of the Chagres was allowed to build up behind the new dam and spillway. The rising waters of Gatun Reservoir eventually allowed testing of the 16 Stoney Gates along the spillway, during the second week of June 1913 (Figure 115). Much to the engineer's delight, they performed flawlessly. Saville had been transferred to Culebra in July 1908, but was able to return to observe the tests. On June 27th, when the reservoir elevation had risen to 48 feet the last of the Stoney Gates was closed, allowing the lake to now rise to its operating pool elevation of 85 feet.

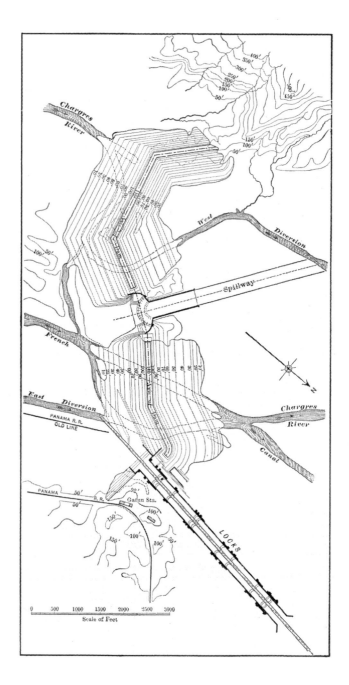

Figure 113. Plan view of the Gatun Dam as it was constructed between 1907-11. The gravity arch concrete spillway was constructed on the middle on a bedrock knob, while Gatun Locks rests on the dam's right abutment (Wegmann, 1911).

Figure 114. Panorama of the concrete spillway structure at Gatun Dam, situated between the massive earthen embankments. The large blocks of concrete around the base of the spillway skirt dissipate the energy of the flowing water to reduce its velocity in the outfall channel.

Figure 115. Aerial oblique view of the Gatun Spillway in operation some years after the project's completion. The concrete training walls of the outfall channel serve to protect the dam's embankments. The large structure at left center is the hydroelectric power station, which provides electricty to run the canal's locks (USACE).

Gatun Locks. From 1907-09, the Panama Railroad was relocated out of the Chagres River Valley and the excavations for the locks themseves, about 6,600 feet long, 600 feet wide, and 50 feet deep (on average), requiring about seven million cubic yards of excavation. The deepest excavtion would be for the lowest lock basin, extending to a depth of 66 feet below sea level. Since most of the excavtion involved relatively unconsolidated materials, much of this work was under taken using dredges, with the spoils being pumped through pipes to form the silty clay core of the two embankments comprising the Gatun Dam. Where more firm materials were encountered (in the uppermost lock chamber), steam shovels were used.

Army engineer Major James P. Jervey supervised the concrete work during construction of the triple locks. Before he resigned in September 1907, Frank Maltby had designed four pairs of steel tower cableways to handle all of the concrete, steel pipes, and forms to be placed at the Gatun Locks. These towers were fabricated by the Lidgerwood Manufacturing Company, who hired Maltby to supervise their erection on site, which he did in the summer of 1908, returning to Panama with his new bride (Maltby, 1945). Each tower was 85 feet high, capable of lifting two 6-ton buckets (2.2 cubic yards each) of concrete (Figure 116). The lock walls were poured in lifts using 36 foot long box forms, made of steel plate with timber rakers, as shown in Figure 116.

Figure 116. Traveling tower cableways used to place concrete for the Gatun Locks, shown just as the first bucket of concrete was placed in late August 1909.

The first concrete for the locks was poured on August 24, 1909, and the last concrete was placed on May 31, 1913, almost four years later. In part this was because of the very small batch sizes used in those days, but the concrete was of very high quality, even without mechanical vibration. Jervey would eventually supervise the placement of more than two million cubuc yards of concrete.

The walls of each lock were 1100 feet long and 81 feet high, which is higher than a six story building (a comparison they loved to cite at the time of construction). The locks at Gatun were equipped in the same manner as those on the Pacific side, described previously, except that they emplyed a curved transition with training walls that extended into Gatun Reservoir at the upper end.

The first vessel to pass through the Gatun Locks was one of the ICC tugboats, named, appropriately, the *Gatun,* on September 26, 1913 (Figure 117). This trial lockage took 1 hour and 51 minutes, and there were some teething problems with the passage. The lake had not yet reached its operating level, so the upper chamber took much longer to fill. But the head differential between the filled upper lock and the empty middle lock was 56 feet, so when the culvert valves were opened the water shot up like a group of 62 canons! The locks were also manually manipulated, because the electrical controls had yet to be installed and tested. The canal would have opened in late 1913 but for landslides in the Culebra Cut which had dammed the channel.

Figure 117. The ICC Tugboat Gatun entering the first chamber of the Gatun Locks on September 26, 1913 on a test run. The canal would not open for another 10-1/2 months while the landslides at Culebra were battled by the dredges.

Postscript on Sibert's career. Sibert was promoted to Lieutenant Colonel of Engineers in September 1909, which increased his salary to $6,750 per year, still a bargain for the government, considering the responsibilities he was shouldering. On April 1, 1914, he was detached by a special act of Congress to serve as Chairman of a Board of Engineers on flood prevention problems in China for the International Red Cross, and he spent June through October of that year in China, while the Panama Canal was being christened. Upon his return he was made a Division Engineer in Cincinnati. On March 12, 1915, he received the thanks of Congress for services rendered in constructing the Panama Canal, and was promoted to Brigadier General in the Line of the Army, effective March 4, 1915. In April, he was posted to San Francisco to assume command of the Pacific Coast Artillery District. The following month, on May 16, 1915, his wife Mary died. Later that year (1915) he published a book on the Panama Canal titled *The Construction of the Panama Canal,* which was co-authored by John Frank Stevens.

Shortly after War was declared on Germany in April 1917, Sibert was promoted to Major General and assumed command of the Army's First Division, the first unit sent overseas as part of the American Expeditionary Force in France. Just before departing for France in June he married Juliette Roberts of Pittsburgh, but she died 15 months later in the flu epidemic. General Sibert received some notoriety because all five of his sons served in the war: Lieutenant Colonel Franklin C. Sibert, Infantry; Major William O. Sibert, Chemical Warfare Service; Major Harold W. Sibert, Corps of Engineers; Lieutenant Edwin L. Sibert, Field Artillery; and Corporal Martin D. Sibert, Infantry. Edwin and Franklin became major generals during the Second World War.

In May 1918, Sibert was named the first Chief of the Chemical Warefare Service, which he retained until March 1920, when he took command of the Army's 5th Division and Camp Gordon, Georgia. He was retired on April 3, 1920, and awarded the Distinguished Service Medal.

In 1923, he accepted the position of Manager of the Alabama State Docks Commission, which constructed a marine terminal in Mobile. In May 1928, he was named Chairman of the Colorado River Board, appointed under an Act of Congress to investigate and report on the Boulder Canyon Project, the largest line item appropriation ever approved by Congress up until that time. The Commission was comprised of Chairman Sibert, Warren J. Mead, Charles Berkey, Daniel W. Mead, Robert Ridgway, with Reclamation Commissioner Elwood Mead as technical advisor. They spent six months reviewing all of the technical aspects of the project, issuing their report to President Coolidge in early December 1928, which fixed the location, height, flood storage, and spillway capacity of Hoover Dam, and satisfied Congress sufficiently to gain passage of the Boulder Canyon Act within two weeks of the report's release.

In 1930, his biography appeared, written by Colonel Edward B. Clark. In 1922, Sibert married Evelyn Clyne Bairnsfather of Edinburg, Scotland to whom he remained married until his death on October 16, 1935 in Bowling Green, Kentucky.

He was buried at Arlington National Cemetery. In mid-1937 the Canal Zone issued a black 14-cent stamp of General Sibert to honor his contributions to the canal project.

SYDNEY B. WILLIAMSON

Brief sketch of professional experience. Sydney Bacon Williamson (Figure 118) was born on April 15, 1865, in Lexington, Virgina. His father, General Thomas Williamson, was Professor of Engineering for 47 years at the Virginia Military Institute (VMI). At age 15, he entered VMI and graduated in the Class of 1884, which he served as Cadet Adjutant. He then took a position teaching at King's Mountain Academy in York, South Carolina, until February 1886, when he set off for St. Paul, Minnesota where he found employtment with the Chicago, Burlington & Quincy Railroad as a resident engineer to compute earth excavation quantities. He then accepted a position with the St. Paul & Duluth Railway to relocate their line between Thompson and Duluth, which was performed in bitter winter conditions. He then worked on realignment surveys between St. Paul and White Bear Lake, where he met James Jerome Hill, owner of the line.

In 1889, he was hired by the Northern Pacific Railway Co. as an assistant to the chief engineer. He was sent to Billings, Montana to blaze, survey, and manage the construction of a new rail line to Cooke City from Rocky Point, through Laurel, Montana, while surveying an alternate route to Cooke City through Stillwater Canyon. After building this line he retruned to St. Paul and was given the assignment of making a reconnaissance through northern Minnesota, over which the tracks of the Great Northern line later passed.

He then received an attractive offer from the firm Williamson & Earl in Montgomery, Alabama, as a junior partner with his half-brother W. G. Williamson. They designed and supervised the construction of sewerage systems throughout the South, including Columbus, Georgia and Montgomery, Alabama. In May 1890, he married Helen C. Davis of St. Paul and they had a son Lee Hoomes and a daughter Julia Lewis.

When work slackened in 1891, he accepted a position a U.S. Assistant Engineer on the Tennessee River, where he began making surveys of all the river improvments then contemplated between Milton's Bluff and Decatur, Alabama. The young lieutenant in charge of of the district was George W. Goethals. Recognizing Williamson's surveying abilities, Goethals had him detailed to the ditrict office in Florence, Alabama, to begin plans and estimates for the lock nearby at Riverton, intended to bypass Colbert Shoals, Alabama. Its extreme low water lift of 26.42 feet was the greatest attempted up to that time (Goethals had a difficult time getting the Chief Engineers office to approve the plans for a single lock lift even for two lock lifts of 13 feet each). The single lock lift was finally approved and Goethals and Williamson were obliged to overcome many construction obstacles before finally completing the project, which held the lift record of 26+ feet until exceeded by the locks of the Panama Canal 17 years later.

Unfortunately, the low bidder for the Riverton Lock got into a difficult situation that was beyond his experience and problem solving abilities. When he began excavating he soon encountred pervious channel sands, and water began pouring into the excavation. He decided to install a sheetpile cofferdam wall around the lock excavation, which took almost a month. Overconfident, he renewed the excavation and after a few days so much water was welling up inside the cofferdam that Williamson realized the scheme wasn't going to work. Within a short time the sheetpiles began kick-out and the failure quickly zippered around the lock excavation until just about all the sheets were toppled over. Ruined, the contractor walked off the job.

Figure 118. Civil engineer Sydney B. Williamson was placed in charge of the Pacific Division of the canal. He was an 1884 graduate of the Virginia Military Institute who had worked previously with George Goethals on Corps of Engineers projects (VMI Archives).

Goethals realized the gravity of the situation, which could have become a noticeable blemish on his record. Surprisingly, Williamson was cautiously optimistic. During the course of events he made careful oberservations and noted what he saw. If the lock excavtion were undertaken more slowly, in smaller sections and with less severe cut slopes, he felt that it might be completed. But, with all the past problems, he couldn't find any contactors wiling to "*work in that hole.*" He finally found someone with some Negro laborers who were desperate for work. The colored men said they would go into the excavation, but only if Williamson was in the hole with them, so he plunged himself into the work, like a section gang foreman, and that's where he was when Goethals, searching frantically for him, at last found him.

By "being in the hole" Williamson soon discovered that he could pretty accurately gage the hydrostatic (seepage) uplft pressure by the way "*the ground felt.*" The quick sand seemd to lose its strength very rapidly when it was disturbed, but of some means of drainage could be effected overnight, it would "*set up*" and "*regain much of its intrinsic strength.*" By employing sufficient drainage around and within the Riverton lock excavation, Williamson was able to get the masonry floors and sidewalls sufficiently established to pass off the project. After that experience, George Goethals trusted Sydney Williamson's judgment, and the two men would collaborate for many more years.

In 1896, when the Riverton Lock was 75% complete, Williamson was relieved of his duties at Riverton and sent to Muscle Shoals, where he was placed in charge of operation and maintenance of the Muscle Shoals Canal. While living next to Lock No.6 he carried out a series of field tests to verify the actual stresses developed against the lock gates in comparison with those deduced from design formulas.

When the Spanish American War broke out in April 1898 he obtained a leave of absence and accepted a commission in the Third Volunteer Engineers as a Captain, under Colonel David D. Gaillard. But he was soon sent for by another Colonel named Goethals, who was on the staff of General John R. Brooke in Puerto Rico, as the Chief Engineer. Williamson joined the command at Arroyo, where he was engaged in the construction of a dock for lighters carrying supplies for the American troops. After the armistice, he worked on making the San Juan Highway passable by repairing arch bridges that had been blown up by the Spanish. After being discharged he resumed his duties on the Tennessee River, working out of the Corps' Chattanooga office.

In August 1901, Goethals had him transferred to the Newport District to work on some projects then under construction, inlcudiung sea coast defenses of Narragansett Bay and modernization of existing forts. Williamson was also given charge of building emplacements for Forts Greble and Getty. One of the biggest problems with the area was the Army and Navy's inability to keep their powder magazines dry. Williamson collaborated in developing a sysetm of protected vents and drainage tiles to improve air circulation and drain off excessive moisture without sacrificing protection.

Keenly interested in applications of reinforced concrete, in 1903 Williamson took a job with the Expanded Metal Engineering Co. of New York. This position brought him into contact with the Corrogated Bar Co. of St. Louis, for whom he opened branch offices in Baltimore, at the Jamestown Exposition, and in Norfolk. He actively worked on preparing competing bids using reinforced concrete for jobs that had specified steel supports.

Move to Panama. In April 1907, Williamson (Figure 118) was invited by Goethals to "make haste for Panama at the earliest opportunity to accept an important position." He arrived at Cristobal on May 17th, but had no idea what sort of position he would be assigned. Goethals initially had him take a few weeks to see the entire project and "poke around a bit,"and get a feeling for things. In July, he was assigned as Division Engineer of La Boca Lock Construction Division, and this was where he was to remain for the duration of the project. In December 1907, his duties were expanded to include the "Pacific Locks and Dams Construction Division." His personal quarters were on the hill in Culebra, with Goethals, Gailard, and the other high ranking engineers (Figure 119).

Figure 119. President William Howard Taft (on left) visiting the Canal Zone in 1910. Sydney Williamson is sitting next to him in the dark coat and leather leggings, while George Goethals is standing at right (VMI Archives).

The Miraflores Locks controversy. It was in this capacity that he was supervising the placement of fill from rail tresles into La Boca Bay to construct the Sosa-Corozal dam for the locks on the Pacific side of the canal. Some 2,000,000 yards of fill had been placed when a bearing capacity and slope failure occurred in July 1907, which brought that scheme to a halt. From this point forward Williamson became much more careful about foundation exploration. He decided to take direct charge of the foundation evaluations at Miraflores and at Pedro Miguel. His program included test pits, wash borings, and diamond drill cores. He then made a careful review of the boring logs, searching them for signs of weak strata that could pose problems for construction, such as soft unconsolidated clay or uncemented "running sands."

After studying the boring logs made at the alternate site at Miraflores, and those 2.2 miles north, at Pedro Miguel, he became convinced that the scheme for split locks should be followed because the lowest chamber of three locks at Miraflores would have to extend to -40 feet below sea level to gain a solid foundation. Being in the channel of the Rio Grande River, he was also concerned about seepage-induced uplift impacting sheetpile or earthen cofferdams. This was the very problem that had plagued Williamson and Goethals at the Riverton Lock, years before, where the channel sands had liquefied because of high emergent seepage pressures in the floor of the excavation.

Goethals undoubtedly knew of Major Sibert's strong preference for constructing the Pacific Locks as one integral structure, like those on the opposite coast at Gatun. Sibert's arguments regarding economy of effort and materials were valid, as were his arguments that a single group of locks would cost less to operate and maintain. The variable was how much trouble they *might* have in advancing the lock excavation to a depth of -40 feet in river sediments so close to the coast.

While Sibert was away on his first annual vacation in the spring of 1908, Goethals chose to side with Williamson on the issue, which makes sense, given that Williamson was the person who would have responsible charge of the construction, and Goethals valued his engineering judgment, especially on the issue of cofferdams.

Reorganization of mid-1908. On June 30, 1908, Goethals announced that he was reorganizing the canal project into geographic divisions: the Pacific, Central, and Atlantic. Similar to the Corps of Engineers system delineating districts based on watershed boundaries, it seemed similar to a military structure, where general planning came from the headquarters, but the details were left to those placed in responsible charge, who knew they would be judged on the basis of "performance."

Williamson always wondered if Goethals geograhic reorganization was somehow related to the ongoing battle with Sibert over the Miraflores Locks. When Sibert returned he was very aggistated and out of sorts, feeling that his good opinion and vested authority as the Chief of Lock and Dam Construction had been ignored. After being informed of the decision made in his absence, Sibert worked doggedly to ascertain the actual foundation conditions at the Miraflores site, dubious of Williamson's fears of foundation problems.

Regardless of the spate between Goethals, Williamson and Sibert, the decision to divide the canal job up into geographic districts appears to have been a wise one. There is no way that Sibert, in 1908, could have kept his eye on all of the locks and dams being constructed simulatenously each day, across the 50 mile wide Isthmus. And, that sort of authority often precludes making the on-the-spot decsions that typify heavy construction, where time is always of the essence.

Sydney Williamson was given control of the Pacific Division, stretching from the Pacific Ocean approach channels, the Naos Island Breakwater, the three-mile long channel connecting Miraflores to the ocean, and the second set of locks at Pedro Miguel (Figure 120). Included in this were the dam, spillways, Stoney gates, and hydroelectric plant at Miraflores.

He was the only civilian engineer entrusted with such great responsibility. Oddly enough, none of the Corps of Engineers officers working for the ICC were assigned as his subordinates, as they were with Gaillard, Sibert, and Hodges (even though Williamson had spent many years of his career working for the Corps as a civilian engineer). What he did have was a free hand, and the confidence of the Chief Engineer and Chairman of the Isthmian Canal Commission. If Goethals left, he would leave as well. They worked together.

Figure 120. 70-ton steam shovel working the forebay of the Pedro Miguel Locks.

Innovations. Sydney Williamson was first and foremost a construction manager, but he was always toying with ideas to solve problems, and wouldn't hesitate to try something new, if he thoughtit could work. He probably had more experience with reinforced concrete dsign and construction than anyone else working on the canal job, because that technology was still in its developmental stage when the canal started in 1904.

One of his most-employed innovations was improving the wooden cantilever forms for placing concrete in the lock structures, which came from one of his subordinates, A.P. Craray, who had used it on a previous job. Williamson originated a scheme for casting hollow concrete cylinders, so they could be sunk in open caisson fashion in superposed sections, one upon another, like a vertical pipeline. These were used to construct the terminal piers in Balboa and for the North Approach Wall of the Miraflores Locks.

He demonstrated the concept worked by constructing the first dock, then left it as an example for those following him. After he departed the Isthmus this system was employed, following the general layout of the piers and wharves that Williamson had laid out.

In another area he decided to excavate a large volume of "blue gum mud" sitting in the canal prism (the channel area slatted for removal). It was difficult and expensive to handle becausae it stuck to the equipment, so he tackled it using hydraulic monitors and dredged the dislodged materials, pumped them to a nearby swamp, and filled it in with the dredge spoils, making new ground. As with most construction men, he was always sensitive to providing sufficient slopes to promote drainage, never leaving swales of pockets of low lying ground.

He was also very experienced in constructing hydraulic fill dams, using floating barge dredges that he would throw together, based on the size of the job. He used hydraulic monitors and lockbar pipe to pump the spoils uphill on occasion, filling the area where the puddled core of the dam was needed.

On other occasions he guided the design of some reinforced concrete barges, upon which he mounted his dredge pumps. He also designed the balanced berm chamber cranes (Figures 121, 122, and 123) used to mix and place the concrete for the locks. His foresight also alowed for the crane's subsequent dismantling and re-assembly at the site of the Balboa coaling station.

Williamson was also adroit at searching out quarry sites for aggregate. After investigating four different sites in his district he settled on the quarry at Ancon Hill, which he placed under the charge of a Scottsman named James A. Loulan. Loulan installed a rock crusher that provided two sizes of crushed stone that were used asconcrete aggregate for the port terminals at Balboa and Cristobal, and most of the coastal fortifications that were built on the Pacific side during the first few years. Williamson liked working with the men outside, helping them to solve problems. He usually entrusted as much of the administrative work as he could to trusted subordinates.

Those who worked on the locks at Miraflores or at Pedro Miguel could expect a daily visit from Sydney Williamson. Those who worked on dredges or in municipal positions woulkd see him less often. Sunburnt and always smoking cigarettes he looked older than he actually was (Figure 124). He seldom failed to visit each site where something was happening, reagrdless of the weather. He loved to speak to the foremen, always ready to discuss what might be done to speed operations up a bit.

Another remakable facet of the work performed under Williamson was the fact that there was very little turnover of the engineering staff. Hardly anyone left

Figure 121. Chamber cranes erected at Pedro Miguel Locks to hoist formwork and concrete.

Figure 122. Mixing crane and storage trestle in forebay of the Pedro Miguel Locks.

Figure 123. Berm cranes and storage trestles at Miraflores Locks (Williamson, 1915).

Figure 124. War Secretary Taft and Sydney Williamson during one of Taft's inspection tours of the Isthmus (VMI Archives).

the Division after he assumed command of it. He never asked the men to do anything he wouldn't do himself, and they knew that. A very friendly rivalry developed between the Pacific and Atlantic Divisions because their work was so similar in nature. Each one was always striving to beat the other's records, and so forth. Sometimes the workers would show up early, poised to jump as soon as the morning whistle blew, determined to better a record. This rivalry kept everyone one their toes.

Concrete production. Throughout the canal job unit costs were tacked by an army of office engineers whose original charge was to "watch every penny." The comparisons were healthy in the sense that they made everyone aware that progress was always being tracked, and that improvements were always being sought. From September 1907, onward unit costs were routinely published in *The Canal Record*, the weekly newspaper. Everyone read these figures and contemplated how they might figure some sort of competitive edge to make whatever they were doing a bit more efficient.

The Pacific Division also established several records for concrete yardage and unit cost efficiency that were unsurpased during construction of the Canal. The highest concrete output at any of the locks in one month was at Miraflores (Figure 125) during April 1912, when 97,735 cubic yards were mixed, handled, and placed, at an average rate of 3,759 yards per 8-hour shift. The highest yardage produced at one mix plant during an 8-hour shift was 5,700 yards.

Figure 125. Miraflores Locks under construction, as seen November 10, 1912 (USACE).

The unit price comparisons between the Atlantic and Pacific Divisions was skewd by the higher cost of aggregate on the Atlantic side, which had to be transported 20 to 30 miles by barge from the quarry at New Portobello, down the coast. The aggregate used on the Pacific side came from Ancon Hill, which was centrally located to the various construction sites, the longest railroad haul being just seven miles.

Postcript on post-Panama Canal career. By the close of 1912 Williamson could see that the work in his Pacific Division was substantially completed (Figure 126), and the installaion of the locks handling machinery was being coordinated by Colonel Hodges' Maintenance Division. When an offer was made by J.G. White & Company, Ltd of London, Williamson resigned his position on December 12, 1912. White performed commercial construction work all over the world. Working in a London office was far removed from the actual construction sites, but the variety of their work was stimulating for a problem solver like Sydney.

Figure 126. Excavation of foundations for Miraflores dam and spillway, as seen on May 8, 1913, after Williamson had departed for his new position in London (USACE).

When the first World War erupted in August 1914 their work came to a halt. Williamson's sudden availability came to the attention of the U.S. Secretary of the Interior, Franklin K. Lane, who was searching for experienced and savvy Chief of Construction for the U.S. Reclamation Service. Williamson joined the Reclamation Service in December 1914, and began by making a survey of the organization's activities and reporting back to Lane. In June 1915, he established his office in

Denver, to be closer to the reclamation staff and their construction projects, which were out west. They had 36 projects under construction and all of the project managers and construction engineers reported to Williamson.

In February 1916, he left to accept an assignment with Guggenheim Brothers in New York City and mining operations down in Chile, retaining the title Consulting Engineer with the Reclamation Service. He recruited the men he needed from the ranks of those he had worked with in Panama.

When the United States was drawn into the World War, Williamson was 52 years old. Guggenheim released him to serve in the War, promising him his old job when he returned. He offered his services to the War Department and was promptly commissioned a Lieutenant Colonel of Engineers in June 1918, and assigned to the 55th Regiment of Engineers, reaching France on July 16th. When the commanding officer was relieved to go elsewhere, Williamson assumed command. In late September, he was named section engineer of the District of Paris, and on November 4th promoted to Colonel (Figure 127).

Figure 127. Engineer Colonel Sydney Williamson at the age of 54 while serving in France in 1918-19 (VMI Archives).

He was responsible now for directing the single largest project in the theater, a supply depot at Gievres, France, which was two miles wide by six miles long, filled with hundreds of structures, shops, roundhouses, ice plants, refrigerated warehouses, etc. In March 1919, he departed France and returned to the United States, taking up his previous position with Guggenheim.

In January 1924, he began working in private practice with his son, Lee Hoomes Williamson, with offices in Birmingham and Charlottesville. He also assisted George Goethals with some consulting work on the Lake Worth Inlet in Palm Beach when the general took ill in 1927, and saw the work through to completion the following year, after Goethal's death.

In 1929, President Hoover named Williamson as one of three civilian members of the Interoceanic Canal Board, appointed to investigate the cost of building and maintaining a second canal across Nicaragua, as well as constructing larger locks in Panama. The board submitted its report to Congress on November 30, 1931. Williamson continued to serve on the Board of Engineers for Rivers and Harbors, which was examining many of the flood control issues along the lower Mississippi River, includeing the Bonne Carre Floodway.

In 1935, his health began to fail and on April 30th, he was retired under the provisions of the Civil Service Retirement Act (he had 27 years of federal service out of his 41 years of gainful employment). He died on January 13, 1939, in his 74th year, leaving his wife Helen and their children Lee Hoomes and Julia. In April 1940, the Panama Canal Zone issued a 30 cent stamp with his image (shown in Figure 1).

THE GEOLOGIST - DONALD F. MACDONALD

Donald Francis MacDonald (Figure 128) was born in 1875 in Egerton in Pictou County in Nova Scotia. As a young man of 15 he ventured west to seek his fortune by heading for British Columbia to engage in the fur trade with the Hudson's Bay Company. He then tried working as a miner, which interested him in the study of geology. In the summer of 1902, he began working for the U.S. Geological Survey as a Field Assistant, continuing each summer until 1911. In 1905 he received his bachelors degree in mining engineering from the University of Washington. He continued to graduate school at George Washington University in Washington, DC, receiving a master's degree in geology in 1906. In 1907, he was appointed a fellow in the graduate school of geology at the University of Chicago, and from 1908-09 he had charge of the geology program at Tulane University in New Orleans. In 1908, he was promoted from Field Assistant to Junior Geologist in the USGS; in 1909, to Assistant Geologist; and in the later part of 1910, to Geologist in service of the Isthmian Canal Commission.

Donald MacDonald was dispatched to Isthmian Canal Commison by the U.S. Geological Survey in late 1910, after reveiving a request for "geological help" relative to the excavation of rock and soil. Since MacDonald had a background and training in mining engineering, he was given the nod to go. He arrived on the Isthmus in January 1911, and was billeted in Culebra, where he remained for three years, until December 1913. His offical job title was "geologist to the Isthmian Canal

Commission." During the three years that he lived in Panama he made daily forays into the Culebra Cut to familiarize himself with the geologic units expressed there, as shown in Figure 128.

Figure 128. Geologist Donald F. MacDonald (1875-1942) was a native of Nova Scotia who worked on the canal in Panama from 1910-13, and again, between 1940-42 (Jackson & Son, 1911).

Figures 129, 130, 131, and 132 present representative images found in MacDonald's USGS files. MacDonald was most puzzled by the landslides and "slip-outs that occurred when the ground mass appeared relatively dry. He mused that these were likely triggered by all of the dynamite blasting, which occurred several times each day, somewhere up or down the line. The block failure shown in Figure 130 was a revelation to MacDonald about the role of structural control on slope instability. Several faults intersected at this location, and one of those was semi-parallel to the cut face, creating a "plane of weakness" that allowed the ground mass to "heave outward," destroying all but one of the railroad tracks running along the floor of the channel prism.

MacDonald penned his first article about Panama in the May 8, 1912 issue of *Science*, titled "Heating of Local Areas of Ground in the Culebra Cut, Canal Zone." MacDonald had observed heat emanating from the formations, but only after blasting holes were drilled and rainfall penetrated the rock. He took samples in April 1911 and sent these to Washington for chemical and mineral analyses. The results showed that they contained 1.92% sulfuric acid and minute crystals of gypsum. The heat was likely generated by the rapid oxidation of pyrite. Bluish sulfurous smoke had been observed emanating from the cut, and deposits of sulfur not also noted.

Figure 129. The La Pita Slide along the canal excavation, photographed in May 1910. This is one of the earliest slides that the engineers attempted to evaluate and understand, hastening their request for geological advice. Geologic unit contacts are shown in black ink (MacDonald, 1913).

Figure 130. One of the early landslides at Cucaracha, on the south side of Gold Hill, which occurred on July 5, 1911, toppling a 95-ton Bucyrus shovel, a locomotive, and a string of muck cars.

Figure 131. Cut slope failure on the Obispo Division on August 21, 1912. Note ponded water. MacDonald discovered that this slide was structurally controlled by a fault crossing the canal excavation at a low angle.

Figure 132. Flowage off the lower slopes of the Cucaracha Landslide on February 2, 1913. This sudden change in the physical character of the movement surprised MacDonald.

MacDonald's next article was for the International Geological Congress in Canada in 1913, titled "Excavation Deformations." This article includes Figure 129 with the inked annotations and contacts (the La Pita Slide). He blamed this small block glide pull-out on seepage from the unlined drainage ditch running parallel to the excavated slope. One of the figures in that article is his conception of the block kinematics involved in rotational slumping, shown in Figure 133, below. This is a fairly realistic interpretation. The lower half of the same figure presents a similar section he includes in his 1915 report (MacDonald, 1915).

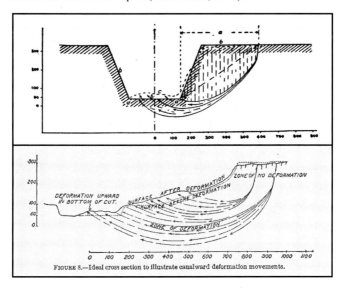

Figure 133. Examples of figures contained in MacDonald's 1913 article (upper) and his Bureau of Mines Bulletin 86, released in 1915 (lower). These illustrate his concepts of relative motion in shear.

In 1915, MacDonald wrote another article that appeared in the journal Science titled "Some Earthquake Phenomena Noted in Panama," which resulted from a series of earthquakes that were clustered around the Azuero Peninsula in October 1913.

MacDonald summarized his three years of studying landslides in Panama in an 88-page report titled: "*Some Engineering Problems of the Panama Canal in the Relation to Geology and Topography,*" released as U.S. Bureau of Mines Bulletin 86. This volume contains a remarkable collection of observations, test data, photographs, and engineering geologic mapping of the landslides that were plaguing the Culebra Cut in 1913. What's even more remarkable was that MacDonald was completely self-taught, he didn't have the benefit of any formal training in making assessments of slope stability, yet most of his concepts are well founded, based on observations. Figure 134 is a copy of MacDonald's geologic map of the West and East Culebra

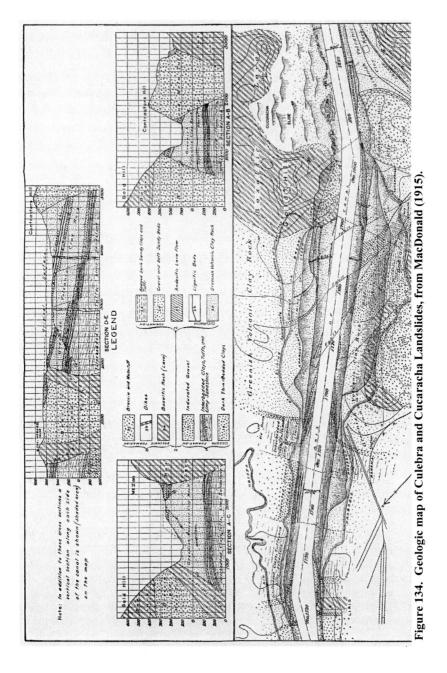

Figure 134. Geologic map of Culebra and Cucaracha Landslides, from MacDonald (1915).

Slides, as well as the Cucaracha Slideshow, showing the relationships between stratigraphy, faults, and physical features with the observed land slippage.

Postscript on MacDonald's career. MacDonald's 1915 report on Landslides along the Panama Canal holds a special place in the history of geological engineering and engineering geology in the United States because it is considered the first serious attempt to use the principles of geology to aid engineering efforts, known as "engineering geology." Its importance to the Panama Canal cannot be overstated, but it didn't garner nearly as much interest in the continental United States until the untimely failure of the St. Francis Dam near Los Angeles in March 1928, which killed about 500 people, and led to the incorporation of engineering geologic assessments in most large projects, especially dams (the U.S. Army Corps of Engineers hired their first geologist, Edwin B. Burwell, Jr. in 1938, following the failure of the upstream face of the Fort Peck Dam).

After the Panama Canal was closed a fourth time in two years by massive landslides in 1915, President Woodrow Wilson asked the National Academy of Sciences to appoint a committee of eminent scientists to study the problem and prepare a report of their findings. The members were appointed in November, and traveled to the Isthmus in December 1915. They issued a preliminary report in April 1916, but their final report was not published until 1924. One of those who testified before the panel and submitted much data was Donald F. MacDonald, whose assessment of the mineralogy of the Cucaracha Shale (along with Warren J. Mead) was included as an appendix to the 1924 report by the Academy.

Donald MacDonald returned to Panama many times. In July 1918, he published another article in Science titled "The Panama Slides That Were," which responded to some geology professor who had opined that the canal would be closed for several years. On November 7, 1918, he was married to Lucy Chapeze Hagen of Chapeze, Kentucky in Balboa, Canal Zone. He left the USGS to become a Professor of Geology at St. Francis Xavier University, where he remained until 1939, when he was coerced to return to Panama as the geologist for the Third Locks Project then just getting under way. He and Lucy moved to the Isthmus and he died of an apparent heart attack on Panama on June 1, 1942. He was 67 years old.

FREDERICK MEARS

Despite George Goethals' best efforts when he arrived on the Isthmus in March 1907, a number of key people John Frank Stevens had brought to the project decided to leave. These included dredging expert Frank Maltby, and several of the principal railroad engineers, including the two key figures operating the Panama Railroad, William G. Bierd and Ralph Budd.

Fortunately, in Frederick Mears (Figure 135), Goethals retained a young officer who became one of the most celebrated railway engineers in the world. Like John Frank Stevens, Mears was self-taught. Born in Fort Omaha, Nebraska in May

1878, as a young man he attended private schools in New York and San Francisco before enrolling in the Shattuck Military Academy in Faribault, Minnesota, a prestigious military prep school. His best friend and fellow cadet at Shattuck was John Frank Stevens' eldest son Donald, and during his tenure those years he was a frequent guest at the Stevens home in Minneapolis, located about 70 miles from the academy. For the remainder of his life, John Frank Stevens would refer to Mears as "his fourth son," and more than any other civil engineer, Mears followed Stevens' footsteps.

Figure 135. Left - First Lieutenant Frederick Mears as he appeared in 1906. He shouldered the responsibility of keeping the Panama Railroad working efficiently after Stevens' key railroad engineers departed. Right – Colonel Mears during the First World War (both images USACE).

After graduating from Shattuck in 1897, the elder Stevens hired him to work on the Great Northern Railway as part of a survey party on the Park Rapids Extension in Minnesota. He exhibited a remarkable grasp of the surveying craft and rose rapidly through the ranks, becoming resident engineer for the Kootenai Valley & Bedlington and Nelson line in Idaho and British Columbia.

This assignment was interrupted by the outbreak of the Spanish American War in April 1898, but he remained on the job to complete his assignment. Desperate to become an Army officer, he enlisting in the Army in October 1899, with the hope of taking the officer's examination as soon as practicable. He was dispatched to the Philippines in 1900, where he was assigned to the U.S. 3rd Infantry, which saw heavy fighting with Filipino insurgents. He was soon promoted to corporal, then sergeant, and in July 1901, received a commission in the Regular Army as a Second Lieutenant in the 5th Cavalry. Mears returned to the states in 1903, and the following year matriculated through the Army's Infantry and Cavalry School at Fort Leavenworth, followed by the Army Staff College, from which he graduated in 1905.

Mears was promoted to First Lieutenant in 1906, shortly after John Frank Stevens summoned him through Secretary of War William Howard Taft. In May 1906, Stevens appointed him track foreman in the Culebra Cut during the meticulous operation of expanding the engineering excavation plant. This work included the layout and construction of seven double-track benches sloping away from both sides of the Continental Divide, an unprecedented feat at that time.

The following September, he began surveys to relocate the Panama Railroad around and across portions of the proposed Gatun Lake, working as Ralph Budd's principal assistant. From May to October 1907, Mears was Resident Engineer in charge of constructing the relocated line of the Panama Railroad, and took a brief leave of absence to marry Jane P. Wainwright in Texas in April 1907, and settling in quarters in the Canal Zone. In October 1907, he succeeded Budd as Constructing Engineer of the relocated line, until its completion in December 1909. The 40 miles of relocated line cost approximately $9 million (Figure 136). When the new line was completed Colonel Goethals named him Chief Engineer of the Panama Railroad, an enormous responsibility on the canal project. During his last two years he served as General Superintendent of the Panama Railroad and Steam Ship Company.

Figure 136 The 40 miles of relocation for the Panama Railroad around the Gatun locks, dam and reservoir took more than two years to complete. This shows the elevated line crossing the Chagres River at Gamboa on the left, and piles being driven for a trestle fill across an arm of Gatun Reservoir.

When the canal was completed in August 1914, Goethals recommended Mears to President Woodrow Wilson when he was asked to assume the responsibility of constructing a railroad into the Alaska Territory. The only problem was Mears was still on the register as a cavalry officer and the Army was reluctant to dispatch him for what was essentially the highest visibility post that had been earmarked for an Army Engineer in 1914-15. So, a special Act of Congress was passed authorizing Mears to transfer to the Corps of Engineers. Mears established the town of Anchorage, Alaska, and blazed a railroad that extended 500 miles into the Alaskan interior, and was promoted to Captain in 1915.

When America declared war on Germany in April 1917 Mears was summoned to Washington by Major General William M. Black, the Corps' Chief Engineer (and the first Corps of Engineers officer to serve in Panama, back in 1903-05). Black promoted Mears to major in November 1917, and to temporary Colonel of Engineers in January 1918, when he was given command of the newly formed 31st Engineer Regiment. This regiment was the principal railway unit supplying the American Expeditionary Forces serving on the Western Front in France. He was then promoted to General Manager of American Army Transportation Corps in France, a force that numbered 50,000 troops.

In June 1919, Mears returned to the United States and was reduced in rank to Lieutenant Colonel of Engineers in the Regular Army. He returned to Alaska to resume his work on the Alaska Railroad, which was completed in 1923, at a cost of $56 million. That same year he retired from duty with the rank of colonel, and became Assistant Chief Engineer of the Great Northern Railway in 1925, where he supervised the design and construction of the Cascade Tunnel under Stevens Pass in Washington (Mears, 1932). The 7.9 mile long tunnel was completed in just 36 months and opened in January 1929, the only long rail tunnel of that era that was completed on time and within budget. He became Chief Engineer of the Great Northern in 1933, and served in this capacity until his death in January 1939, at the age of 60 (Hoffman et al., 2009).

SPECIAL BOARD OF CONSULTING ENGINEERS (1908-09)

In late November of 1908, the rock containinment dike along the upstream side of the Gatun Dam began settling rapidly, evenually dropping as much as 60 feet over a zone about 300 feet wide. Colonel Sibert's forces had been dumping rock in that area over the previous 10 days, seeking to enlarge and heighten the containment dike. Concurrently, it had been raining most of the month (28 inches in the previous 24 days at Gatun). This was termed "settlement," but was a bearing capacity failure of the soft blue sandy clay infilling old channels at this particular zone beneath the dam's sloping flanks. The bearing failure triggered what was actually a deep rotational slump landslide, which destroyed the railroad fill trestle, and allowed the waters of the Chagres River to flow unimpeeded through the new gap.

On November 24th, a New Orleans newspaper ran a front page story stating that the "Gatun Dam had failed," and photographs of the river running wildly across the wrecked tresle appeared ominous. The story was then covered in all the major newspapers the following day, across the United States. Roosevelt's political enemies were having a heyday.

Seeking to do as much "damage control" as possible, President Roosevelt asked President-elect Taft to go down to Panama to investigate the situation. A "Special Board of Consulting Engineers" was then appointed to accompany Taft. These included: Frederick P. Stearns, who had overseen the design of the Wachusetts Dam and had previously served on the International Board of Consulting Engineers in 1905-06; Isham Randolph, who had supervised the design and construction of the Chicago Sanitary & Ship Canal, and had also served on the International Board of Consulting Engineers in 1905-06; James Dix Schuyler, a consulting engineer from southern California who was one of the country's leading experts on hydraulic fill embankment dams; Allen Hazen, a consulting engineer in hydraulics and hydrology from New York City (and originator of the Hazen-Williams formula for hydraulic conductivity); John R. Freeman, an 1876 graduate of MIT who had invented interior fire sprinklers to protect textile mills, and owned his own consultancy in Boston, which specialzed in the develop of dams and water works; Arthur Powell Davis, a nephew of John Wesley Powell and one of the founders of the National Geographic Society in 1888, with extensive expertise in hydrography surveys for the U.S. Geological Survey and a number of foreign nations (including China); and, Henry A. Allen, a 1887 engineering graduate of the Naval Academy who had worked for Allis Chalmers before founding his own consulting firm, which had designed the pumping stations for the Chicago Sanitary District. All in all, it was a well-rounded and experienced group of engineers, with considerable technical expertise as well as real-world experience with earthen embankment dams.

In January 1909, Taft and his board of consultants (shown in Figure 137) traveled down to Panama to assess the situation with the Gatun Dam. The breach had long since been repaired, but they were shown photographs and presented with technical data documenting the bearing failure/landslide, and they correctly assessed that the soft clays comprising the foundation were of low permeability, and therefore, particularly susceptible to *"squeezing"* and *"plastic flow"* whenever they received concentrated surcharge loads, or those loads were brought on too quickly, of which the latter they believed to have been the cause.

They were impressed with the very conservative sections being employed for the dam's main embankment, which sloped downward at inclinations from 4:1 (near the dam's crest) to 16:1 (horizontal to vertical), which were without precedent at that time. They recommended that the dam's crest be lowered 30 feet, but that they could dispense with the driving of a sheetpile cutoff wall beneath the dam's central core because the underlying materials were of such low permeability. The Special Boards recommendations were summarized in the April 1, 1909 issue of *Engineering News*, with a comparison of three cross sections: one initially proposed in 1906; that

Figure 137. The Special Board of Consulting Engineers (SBCE) of the Isthmian Canal Commission pictured in early 1909. From left: William H. Taft, George W. Goethals, Frederick P. Stearns, James D. Schuyler, Allen Hazen, Isham Randolph, Henry A. Allen, John R. Freeman, and Arthur Powell Davis.

prepared by ICC Assistant Engineer Caleb Saville in August 1908, which was being followed when the slump occurred; and that of the Special Board of Consultants. The actual design section implemented was even more conservative than that proposed by the Special Board, as shown in the bottom of Figure 138.

CORPS OF ENGINEERS ROLE

For nearly 100 years, many people have believed that the Panama Canal was "built by the U.S. Army's Corps of Engineers." In actuality, there were only a handful of Army officers actually involved in constructing the canal. These included: Colonel George W. Goethals (West Point Class of 1880), Colonel Harry F. Hodges (West Point Class of 1881), Lieutenant Colonel David D. Gaillard (West Point Class of 1884), Lieutenant Colonel William L. Sibert (West Point Class of 1884), Lieutenant Colonel Carroll A. Devol (Pennsylvania Military College Class of 1879) who served as Chief Quartermaster; Major Tracey Campbell Dickson (West Point Class of 1888) as Inspector of Shops; Major Chester Harding (West Point Class of 1889) who would oversee construction in the Atlantic Division; Major Eugene T. Wilson (West Point Class of 1888) Supervisor of the Subsistence Department (billeting); Major Edgar Jadwin (West Point Class of 1892) as Resident Engineer of the Atlantic Division;

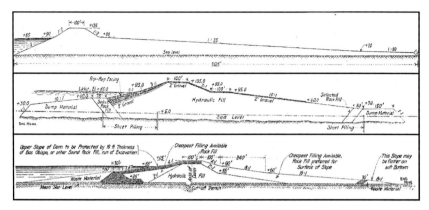

Figure 138. Comparisons of maximum sections through the proposed Gatun Dam. The upper section was that originally proposed in the Minority Report of the International Board in 1906; the middle section shows the section being used when the slump occurred in November 1908; and the lower diagram presents the section recommended by the Special Board of Consultants in 1909 (Engineering News, 1909).

Major James P. Jervey (West Point Class of 1892), Major George M. Hoffman (West Point Class of 1896), Captain Courtland Nixon (Princeton Class of 1895) Depot Quartermaster at Cristobal; Captain H.W. Stickle (West Point Class of 1899) as Assistant Resident Engineer, Atlantic Division; Captain Robert E. Wood (West Point Class of 1900) as Superintendent of the Department of Labor, Quarters and Supply at Cristobal; Captain F.C. Whitlock, Assistant Subsistence Officer; Lieutenant Frederick Mears, Superintendent of the Panama Railroad and Steamship Company, and Lieutenant George Rodman Goethals.

Of these, only eleven were Army Engineers: Goethals, Hodges, Gaillard, Sibert, Harding, Jadwin, Jervey, Hoffman, Stickle, Mears, and the younger Goethals. Although Mears was technically a cavalry officer, he was assigned to the Corps of Engineers shortly after the project's completion. Ben Johnson, Benjamin Harrod, Joseph Ripley, Frank Maltby, and Sydney Williamson had been trained as military engineers, or had previously worked as civilian engineers for the Corps of Engineers, in large part because it was the principal agency dealing with riverine navigation and port development during that era.

Post-war consultations. The [Panama Canal Zone Government was formed when the Isthmian Canal Commission was abolished in April 1914. It assumed responsibility for the management, operation, and maintenance of the Panama Canal. The new organization was staffed by about 500 Americans and 7,000 West Indians, held over from the original construction work. The company operated like a colonial enclave, where all goods were imported and sold by stores run by the company, such as a commissary, and so forth, similar to a military base. The U.S. Government provided its own Canal Zone Police, courts, and the United States District Court for the Canal Zone.

A Canal Zone Governor was appointed by the President, who was usually a Corps of Engineers Brigadier General, as well as his immediate assistant, a senior Colonel who served as the Zone's Engineer of Maintenance and Lieutenant Governor. Until 1951, the Lieutenant Governor always succeeded as Governor after a three-year tour. In 1951 the Panama Canal Company (PCCo) was formed to assume the role of the Zone Government.

Numerous studies were undertaken by the Army Corps of Engineers to either enlarge the canal's capacity or to construct a new sea level canal. These included the Third Locks Project of 1939-42; the Sea Level Canal Schemes of 1947-48; the Navigable Pass Plan of 1948; Canal widening projects between 1962-70; Pan-Atomic schemes to construct a sea level canal under the aegis of Project Plowshare by the Atomic Energy Commission, from 1961-77; and the Atlantic-Pacific Interoceanic Canal Study Commission convened in 1965-70 (Rogers, 2012).

In 1979 exclusive control of the Canal Zone by the United States ceased and the management of the canal was assumed by the Panama Canal Commission (PCC), which sought to integrate and transfer responsibility between the United States and Panama during the 30-year transition, completed on December 31, 1999.

The PCCo and PCC routinely called upon the Army Corps of Engineers to evaluate the Canal's slope stability problems, especially, after the Second World War (Canal Zone, 1947). More than 60 landslides, with volumes as great as 23 million cubic yards, occurred between 1912 and 1979. These slides required additional excavations of more than 59 million cubic yards to maintain the Canal during that same period. Most of this work was coordinated by the Geotechnical Engineering Group at the Corps' Waterways Experiment Station in Vicksburg, Mississippi, for the Panama Canal Commission.

In 1968 the Panama Canal Commission set up a Geotechnical Advisory Board to advise them on improving slope stability along the canal. They also established a Landslide Control Program focused on continual mitigation of landslide hazards. In October 1986 the Cucaracha Slide reactivated once again, spilling 526,000 cubic yards of debris into the canal, pinching the main channel to a width of just 115 feet! The Corps appointed a new Geotechnical Advisory Board, which meets in Panama about once per year. These responsibilities were gradually shifted to the Autoridad del Canal de Panama, or "ACP," which was established in June 1997, and assumed control of all canal operations on January 1, 2000.

LANDSLIDES CONTINUE

Two months after the Canal opened, the East Culebra Landslide reactivated, blocking the channel. This debris was cleared in about two months, but more landslides followed. Between January and September 1915 10,000,000 cubic yards of slide debris fell into the canal, filling a reach over one-half mile long, which required excavation of 10 million cubic yards of slide debris over nine months blocking the canal for seven months (except for small vessels). The material was excavated by floating dredges and shovels. (Figures 139 and 140). This closure triggered the appointment of a special commission drawn from the National Academy of Sciences, which included half a dozen eminent geologists. The panel studied the problem for years before issuing their report in 1924 (NAS, 1924).

Figure 139. Dipper and suction dredges working the toe of the East Culebra Slide, just north of the Continental Divide, on July 3, 1916

Figure 140. Zion Hill and West Culebra Slide in July 1916, taken from Contractor's Hill, looking north.

Landslides shut down the canal on 26 occasions between 1914 and 1986. Once the channel was re-opened tugs would assist smaller vessels, such as coastal freighters, pass through the canal in 1916-17. Much of the 1920s and 30s witnessed transiting vessels being taken through the Culebra Cut one at a time, which dredging continued (Figure 141)

Figure 141. Ships had to be tied up and taken through the Culebra Cut one at a time with tugs during much of the late 1910s, '20s, and '30s, when landslides plagued the canal, causing 26 closures.

CONCLUSIONS

In the final accounting, the Panama Canal project ended up costing just over $375 million, as well as the lives of about 6,000 workers, including 300 Americans. The French excavated 73,000,000 cubic yards of material between 1881-1903, of which the Americans utilized only 40% (29,200,000 yards). The Americans managed to dig 238,845,582 cubic yards of material between 1904-14, about 444% more than estimated in early 1906. Much of this excess volume was ascribable to landslides, and overexcavation of foundations (which also experienced slope failures, but on a smaller scale), and the larger size of the locks, requested by the Navy's General Board in 1907.

The Minority Report of February 1906 had estimated that 53,800,000 cubic yards of material would require excavation, with an overall cost of $139,705,200. This estimate was exclusive of the extensive activities undertaken to ensure sanitation (which accounted for almost 35% of the total cost), enlargement of the railroad, extensive harbor improvements, breakwaters, coastal fortifications ($15 million), or all of the problems with landslides. Another cost that was underestimated was aggregate on the Atlantic side, which had to be imparted from a considerable distance up and down the coast. The Minority Report was spot-on, however, in regards to their estimated time-to-completion, at nine to ten years.

The Canal was America's project of national pride. It was quickly appreciated as a priceless asset to project American influence on the Pacific Basin, where America had gained numerous far-flung possessions during its whaling days, Seward's purchase of Alaska, and the former Spanish possessions purchased by the United States after the Spanish American War.

There were only two capital ships ever built on the Pacific Coast prior to the Second World War (*USS Oregon and USS California*). The utility of the Canal as a strategic kingpin was not demonstrated until the July 1919 when 33 ships comprising the new U.S. Pacific Fleet were able to transit through the Panama Canal in a little more than 48 hours, a remarkable accomplishment.

The canal would prove to be of inestimable worth during the Second World War, when thousands of warships, landing craft, barges, and floating dry docks passed through its portals, most of the time employing one-way traffic in the parallel locks. The first warships too large to fit through the canal were the Midway Class aircraft carriers commissioned in late 1945 to mid-1947, too late to see action in the Second World War.

It remains today a solemn monument to the creativity and determination of American engineers at the turn of the 20th Century, who accomplished so much with so little experience going into the project; and to the Army doctors, whose efforts in combating infectious disease was probably America's greatest contribution to mankind.

ACKNOWLEDGEMENTS

The writer was fortunate to be stationed at Rodman Naval Station in the Canal Zone as a naval intelligence officer, where he was shown generous hospitality by engineers and geologists of the Panama Canal Commission, in particular, geotechnical engineer Luis Alfaro. The writer is also indebted to the staff of the old Panama Canal Commission Library and Technical Resources Center, who supplied access to thousands of photos; the staff at the Linda Hall Library in Kansas City, who provided access to their A.B. Nichols Collection of manuscripts, drawings, and photos; the archivists at the Virginia Military Institute for access to the Papers of Sydney Bacon Williamson; the U.S. Geological Survey Archives in Reston; and to the National Archives and Records Service. The historians at the United States Military Academy at West Point also provided valuable background materials on many of their graduates affiliated with the canal project. Grateful acknowledgement is also made to Professors Ralph B. Peck, and J. Michael Duncan, and Robert L.

Schuster, who provided a great deal of background information on the history of slope stability problems.

All of the photographs reproduced herein were taken from the National Archives and Records Service, except as noted at the end of each figure caption. The annotation "USACE" refers to the U.S. Army Corps of Engineers. Bernard Dennis, Chairman of the ASCE History & Heritage Committee, kindly led the peer review effort, providing valuable comments and suggestions, as well as format ideas.

REFERENCES

Abbot. Henry L. (1913). Hydrology of the Panama Canal. ASCE *Transactions*, 76: 986.
Bates, Lindon Wallace. (1906). *The Crisis at Panama*. L. W. Bates, New York.
Baugh, Odin. (2005). *John Frank Stevens: American Trailblazer*. Arthur H. Clark Co., Spokane.
Bennett, Ira E. (1915). *The History of the Panama Canal*, Historical Publishing Company, Washington, D.C.
Bishop, Joseph Bucklin. (1915). *The Panama gateway*. Scribner, New York.
Bishop, Farnham. (1916). *Panama, Past and Present*. The Century Co., New York.
Bishop, Joseph Bucklin, and Bishop, Farnham. (1930). *Goethals: Genius of the Panama Canal*. Harper & Brothers, New York.
Board of Consulting Engineers. (1906). Report of Board of Consulting Engineers for the Panama Canal, U.S. Government Printing Office, Washington, DC, 426 p.
Budd, Ralph. (1944). Memoir of John Frank Stevens. ASCE *Transactions*, 109:1440.
Burr, William H. (1876). Approximate Determination of Stresses in the Eye-Bar Head. ASCE *Transactions*, 6: 127.
Burr, William H. (1885). Niagara Cantilever Bridge. ASCE *Transactions*, 14: 596.
Burr, William H. (1887). Kentucky and Indiana Bridge. ASCE *Transactions*, 17: 184.
Burr, William H. (1903). Panama Canal. ASCE *Transactions*, 50: 198.
Canal Zone, Governor (1947). Report of the Governor of the Panama Canal, Isthmian Canal Studies, and Appendix 12: Slopes & Foundations.
Chambers, F.T. (1931). Memoir of Harry Harwood Rousseau. ASCE *Transactions*, 95:1597.
Civil Engineering. (1934). Carl Ewald Grunsky, Past-President, 1855-1934. *Civil Engineering* 4(7):373-74.
Complimentary Banquet given to Carl Ewald Grunsky by the Citizens of San Francisco on the eve of his departure to assume the duties of Isthmian Canal Commissioner (1904). Palace Hotel, March 15, 1904. Cubery & Co., San Francisco. Archived in the C. E. Grunsky Collection, University of California Water Resources Center Archives, Riverside.
Duval, Miles P. Jr. (1947). *And the Mountains Will Move*. Stanford University Press, Palo Alto.
Engineering News. (1909). Concerning the Gatun Dam and Earth Dams in General. *Engineering News*, 61(13) [April 1, 1909], p.354-58.

Foust, Clifford. (2013). *John Frank Stevens: Civil Engineer*. Indiana University Press, Bloomington.
Fteley, A., and Stearns, F. P. (1883). Description of Some Experiments on the Flow of Water Made during the Construction of Works for Conveying the Water of Sudbury River to Boston. ASCE *Transactions*, 12:1.
Gerber, E., Prout, H.G., and Schneider, C.C. (1905). Memoir of George Shattuck Morison. ASCE *Transactions*, 54 (B): 513.
Gibson, John M. (1950). *Physician to the World: The Life of General William C. Gorgas*. Duke University Press.
Goethals, George R. (1939). Memoir of Sydney Bacon Williamson. ASCE *Transactions*, 105:1940.
Goethals, George W. (1911). The Panama Canal. Address to the National Geographic Society, February 10, 1911, Washington, DC.
Goethals, George W. (1915). *Government of the Canal Zone*. Princeton University Press, Princeton.
Goldmark, Henry. (1928). Emergency Dam on Inner Harbor Navigation Canal at New Orleans, Louisiana. ASCE *Transactions*, 92: 1589.
Gorgas, Col. W.C. (1904). *A few general directions with regard to Destroying Mosquitoes, particularly the Yellow Fever Mosquito*. U.S. Gov't Printing Office, Washington, DC.
Gorgas, William C. (1915). *Sanitation in Panama*. Appleton, New York.
Gorgas, Marie C., and Hendrick, Burton J. (1924). *William Crawford Gorgas, His Life and Work*. Doubleday, Garden City.
Grunsky, Carl E. (1909a). The Sewer System of San Francisco, and a Solution of the Storm-Water Flow Problem. ASCE *Transactions*, 65: 294.
Grunsky, Carl E. (1909b). The Type of the Panama Canal. *Popular Science Monthly*, 74 (May):
Grunsky, Carl E. (1910). Discussion of Water Supply for the Lock Canal at Panama. ASCE *Transactions*, 67: 91.
Hains, Peter C. (1894). Reclamation of the Potomac Flats at Washington, D.C. ASCE *Transactions*, 31: 55 and 497.
Hains, Peter C. (1896). Foundations for heavy buildings. ASCE *Transactions*, 35: 469.
Hardy, Rufus. (1939). *The Panama Canal Twenty-Fifth Anniversary August 15, 1939*. Panama Canal Press, Mt. Hope, Canal Zone.
Haupt, Lewis M. (1898). Dredges and dredging. ASCE *Transactions*, 40: 340.
Haupt, Lewis M. (1905). Dredges: Their construction and performance. ASCE *Transactions*, 54 (C): 507.
Hoffman, J.T., Brodhead, M.J., Byerly, C.R., and Williams, G.F. (2009). *The Panama Canal: An Army's Enterprise*. Center of Military History, U.S. Army, Washington, DC.
Isthmian Canal Commission. (1908). *Minutes of Meetings of the Isthmian Canal Commission; March, 1904 to September 1905 Inclusive*. U.S. Government Printing Office, Washington, DC, 324 p.

Isthmian Canal Commission. (1914). *Minutes of Meetings of the Isthmian Canal Commission and of its Executive and Engineering Committees; April, 1905-March 1914.* U.S. Government Printing Office, Washington, DC, 349 p.

Jackson, F.E. & Son. (1911). *The Makers of the Panama Canal and Representative Men of the Panama Republic.* Chasmar-Winchell Press, New York, 261 p.

Karner, William J. (1921). *More Recollections.* T. Todd, Boston, 261 p.

Keller, Ulrich. (1983). *The Building of the Panama Canal in Historic Photographs.* Dover Publications, New York.

Lanyon, Richard. (2012). *Building the Canal To Save Chicago.* Xlibris Corporation, Evanston.

Le Prince, Joseph A. (1916). *Mosquito control in Panama: the eradication of malaria and yellow fever in Cuba and Panama.* G.P. Putnam's Sons, New York.

Lewiston Evening Journal. (1905). *Magoon Takes the Oath-New Governor of the Panama Canal Zone Sworn In; Bad Condition of Affairs.* Friday May 26, 1905, p. 2.

Lindsay, Forbes (1912). *Panama and the Canal To-day*, New Revised Ed., L.C. Page & Co., Boston.

Lutton, R.J., Banks, D.C., and Strohm, W.E., Jr. (1979). Slides in Gaillard Cut, Panama Canal Zone. Ch. 4 in B. Voight, ed., *Rockslides and Avalanches, 2.* Elsevier Scientific, New York, p. 151-224.

MacDonald, Donald F. (1915). Some Engineering Problems of the Panama Canal in the Relation to Geology and Topography. U.S. Bureau of Mines *Bulletin 86*, Washington, DC.

Maltby, Frank B. (1945). In At The Start At Panama. *Civil Engineering*, 15:6-9 (June-September 1945).

Marx, C. D., Dewell, H.D., Herrmann, W.L., Huber, W.L., Means, T.H., and Thurston, E.T. (1935). Memoir of Carl Ewald Grunsky. ASCE *Transactions* 100:1591-1595.

McCullough, David. (1977). *The Path Between the Seas.* Simon and Schuster, New York.

Mears, Frederick. (1932). Part II - Surveys, Construction Methods, and a Comparison of Routes; The Eight-Mile Cascade Tunnel, Great Northern Railway, A Symposium. ASCE *Transactions*, 96: 926-949.

Molitor, F.A., Noonan, E.J., and Safford, H.R. (1922). Memoir of John Findley Wallace. ASCE *Transactions*, 85:1635.

Morison, George S. (1902). The Bohio Dam. ASCE *Transactions*, 47: 235.

Morris, Edmund. (2001). *Theodore Rex.* Random House, New York.

National Academy of Sciences. (1924) Report of the Committee of the National Academy of Sciences on Panama Canal Slides. NAS Volume XVIII, Wash, DC.

Panama Canal Rule for Grading Lumber. (1915). *Proceedings* American Railway Engineering Association, 16:915-916.

Parker, Matthew. (2007). *Panama Fever: The Epic Story of the Building of the Panama Canal.* Doubleday, New York.

Rogers, J. David. (2012). 70 years of schemes to improve and enlarge the Panama Canal: in Loucks, E.D. ed., *Proceedings* 2012 ASCE-EWRI World Environmental & Water Resources Congress, Albuquerque, pp. 1013-1023.

Scientific American. (1884). The Interoceanic Ship Railway. *Scientific American*, 51:26 (Dec 27, 1884), p. 428-431.
Sibert, W.L., and Stevens, J.F. (1915). *The Construction of the Panama Canal*. S. Appleton & Co., New York.
Stearns, Frederic P. (1902). Discussion on the Bohio Dam. ASCE Transactions, 47: 259.
Stevens, John F. (1928). The Panama Canal: Address at the Annual Convention at Denver, Colorado. ASCE *Transactions*, 91: 946
U.S. Army, Third Volunteer Engineers. (1916). *David DuBose Gaillard: A Memorial*. Volunteer Engineers, St. Louis.
Vollmar, J.E., Jr. (2003). "The Most Gigantic Railroad." *Invention and Technology*, 18:4, p. 6
Walker, J.G., Pasco, S., Noble, A., Morison G.S., Hains, P.C., Ernst, O.H., Burr, W.H., Haupt, L.M., Johnson, E.R., and Staunton, S.A. (1902). *Report of the Isthmian Canal Commission 1899-1901*. U.S. Gov't Printing Office, Wash, DC, 263 p.
Wallace, John F. (1889). The Sibley Bridge. ASCE *Transactions*, 21: 97.
Wallace, John F. (1891). The Red Rock Cantilever Bridge. ASCE *Transactions*, 25: 722.
Wallace, John F. (1894). The form of railway excavations and embankments. ASCE *Transactions*, 32: 263.
Wallace, John F. (1897). The Lakefront Improvements of the Illinois Central Railroad in Chicago. ASCE *Transactions*, 38: 315.
Ward, C.D. (1904). The Gatun Dam, ASCE *Transactions*, 53:36-44.
Wegmann, Edward. (1911). *The Design and Construction of Dams*. John Wiley & Sons, New York.
Wegmann, Edward. (1916). Memoir of Charles Dod Ward, ASCE *Transactions*, 80:2228-31.
Wiggins, Sarah W. (2005). *Love And Duty: Amelia And Josiah Gorgas And Their Family*. University of Alabama Press, Tuscaloosa.
Williamson, S. B. (1915). Methods of Construction of the Locks, Dams, and Regulating Works of the Pacific Division of the Panama Canal. Paper No. 12, Transactions of the International Engineering Congress, San Francisco.
Williamson, Sydney B. (1930). The Civil Engineers of the Panama Canal. [16 pages handwritten notes]. Sydney B. Williamson Papers 1909-39, Virginia Military Institute Archives.
Williamson, Sydney B. (1934). Autobiography of Sydney B. Williamson [54 pages handwritten notes]. Sydney B. Williamson Papers 1909-39, Virginia Military Institute Archives.

Remembering Joseph Pennell and the Panama Canal

Augustine J. Fredrich, F.ASCE, D.WRE

Professor Emeritus of Engineering, University of Southern Indiana; 10 Old Delmonte Drive, Little Rock, AR, 72212; (501) 219-4280; email: ajfredrich@gmail.com

ABSTRACT

Joseph Pennell, one of the greatest lithographers and print-makers in the United States in the first quarter of the 20th century, saw in the work of the engineers of his time the pursuit of both utility and beauty. He also understood that the process of creating structures results in a series of fleeting scenes, each with a beauty and vitality that is best captured on the construction site. He called his efforts to capture these scenes "The Wonder of Work," and those efforts took him to the sites of some of the most spectacular projects in what some call "The Golden Age of American Engineering." Pennell traveled to Panama in 1912 to create images of the Panama Canal under construction—the greatest engineering work of his time. The 28 lithographs he produced there are some of the greatest images ever created of a project under construction.

INTRODUCTION

Joseph Pennell was born in Philadelphia on July 4, 1857, the only child of Larkin and Rebecca Pennell—both descendants of Quakers who had accompanied William Penn to America in the seventeenth century, He had the conventional education of a Quaker child of that time and place. Although Quaker homes, schools and meeting houses were generally devoid of any art, he developed an early interest in art that set him apart from his peers.

In his autobiography Pennell described himself as "a solitary little Quaker" who felt out of place most of his childhood. His years at the Germantown Friends Select School, he said many years later, were "the worst of my life." After completing high school he worked as a clerk for a coal company and attended evening classes at the Philadelphia School of Industrial Art.

Pennell attended the Philadelphia Academy of the Fine Arts sporadically from 1878 until 1882. His talent as an illustrator blossomed there, and the opportunities for competent illustrators were substantial at a time when newspapers and magazines were rapidly increasing in popularity despite the fact that the technological capability to reproduce photographs was in its infancy. While still a student, he began accepting commissions to illustrate articles for magazines and other publications. After trips on assignment to New Orleans in 1881, England in 1882, and Italy in 1883, Pennell became convinced that he could support himself as an artist.

In June 1884, Pennell married Elizabeth Robbins, a young writer who had already had articles published in *Atlantic* and *Century* magazines. Shortly thereafter they decided to move to London to pursue opportunities to write and illustrate magazine articles and books on European travel. For the next thirty years the Pennells traveled across Europe and recorded what they saw and experienced in dozens of articles and books and hundreds of etchings and drawings.

It had been obvious from the first that Pennell had a special talent for producing images of structures and their environs. His earliest published work in the United States included many scenes with a building or bridge as the focal point. He recognized the inherent beauty that often exists in the built environment and was convinced that he could capture it and convey it to others. In Europe he honed his artistic skills with hundreds of pen-and-ink drawings and etchings of cathedrals, old bridges, castles and other similar picturesque structures. Brief trips back to the United States in 1893, 1904, 1908, 1909, 1910, 1912 and 1915 gave him opportunities to depict American engineering triumphs that were spectacular in their own right.

During the years the Pennells lived in London they became central figures in a colony of expatriate American artists there. The most important artist in their circle was James Abbott McNeil Whistler, the son of one famous American civil engineer and the brother of another. Whistler himself had studied engineering at West Point and had worked for a brief time as a draftsman and illustrator in an engineering office. The Pennells and Whistler became close friends, and by the time of Whistler's death in 1903, Pennell's work and his interest in printmaking showed unmistakable marks of Whistler's influence. Their relationship was so close that Whistler named Pennell as the executor of his estate.

"GREAT ENGINEERING IS GREAT ART"

"I understand nothing of engineering," Pennell once wrote, " but I know that engineers are the greatest architects and the most pictorial builders since the Greeks, and this is why they are carrying on the tradition." He saw in the work of the engineers of his time the pursuit of both utility and beauty, where others saw only the utility.

A number of times, in speaking to different audiences, Pennell pointed out that in his opinion "Great engineering is great art," although not always in those exact words. However, regardless of the audience or the specific words he chose, the intent of the statement was the same: well-executed engineering projects not only achieve the intent of their designers and builders but also lend themselves to the talents of artists. Pennell proved the truth of his assertion by producing hundreds of images of man-made structures in Europe and the United States.

Pennell's interest in creating images of bridges and buildings was not unusual for an artist. Many renowned artists have been captivated by the beauty of man-made structures. What set Pennell apart from his fellow artists was his interest in depicting the process of creating the structures. He understood that the process of creating structures results in a series of fleeting scenes, each with a beauty and vitality that is best captured on the construction site, many of which are far more interesting than the completed structure, and all of which are lost forever once the structure is completed. "It is not rendering the subject as it is," he said, "but giving the sensation it makes on you, and if that sensation is strong enough, others will feel it." This sensation is what Pennell called "The Wonder of Work." He first used that term in 1909 to describe what had become and was to continue to be his most abiding interest during the latter part of his career as an artist. The construction process, together with other similar kinds of human endeavor beginning to emerge as a result of the Industrial Revolution, formed Pennell's "Wonder of Work" theme.

PENNELL AT PANAMA

By 1912, Pennell's interest in "The Wonder of Work" had become compelling, and he knew he wanted to visit the site of the greatest human undertaking of his time—the Panama Canal. After several pleas, each more insistent than those preceding it, he wrote to the editor of *Century* magazine in this unusual, but appropriate, format:

> "What
> I want
> Is
> To Go
> To
> Panama
> NOW
> and do the picturesque side of a great
> engineering feat before it is finished—
> and ruined from my point of view."

That message persuaded *Century* magazine to under-write Pennell's trip to Panama in 1912 to create images of the canal while the canal construction work was in full swing. The 28 lithographs he produced there, on sites from one end of the future canal to the other, are the most well known of the thousands of drawings, etchings and lithographs he created.

Figure 1. Joseph Pennell

THE PANAMA CANAL IMAGES

Joseph Pennell's eye for the striking construction scene is evident in his Panama Canal images. 'Working from nature,' as he called it, he would visit a construction site, find the location that gave the most interesting perspective on the work underway, and quickly produce the image that captured the most dramatic moment. During Pennell's career, photographic technology advanced to the state where the camera could be used to record construction scenes, but Pennell maintained that cameras could not capture the vigor, the scope, or the drama that a good artist could work into an image by manipulating light, shadow, scale and perspective. His work on the Panama Canal lithographs demonstrates his ability to use his skill as an artist to do just what he maintained cameras of his era could not do.

The 28 images Pennell created during his time in Panama covered the full gamut of scenes resulting from work on the canal, including depictions of the housing of both the native and American workers, the natural scenery in the area the canal traversed, and Panama City itself. But the large majority of his lithographs were created to show the work underway on the various elements of the canal. His fascination with the work is clear in the images themselves and in the notes he made as he worked, both of which are included in his book, "Joseph Pennell's Pictures of the Panama Canal," published in 1912 after his return from Panama. The scope and scale of the work are dramatized by scenes from the construction of the locks at Gatun, Miraflores and Pedro Miguel and from excavation of the massive cut at Culebra, Paraiso, and Bas Obispo. Inclusion of people and machinery in the images contribute more to one's appreciation for the scale of the enterprise than to recognition of anyone or any specific machine. And although his images focus more on the work than on the people carrying it out, his respect for the workers is clear in his notes. In his introduction to the images, Pennell wrote of his arrival at the canal construction scene, "Soon people in authority came up – I supposed to stop me – instead it was only to show pleasure that I found their work worth drawing. These men were all Americans, all so proud of their part in the Canal, and so strong and healthy – most of them trained and educated, I knew as soon as they opened their mouths. These engineers and workmen are the sort of Americans worth knowing. And all this is the work of my countrymen and they are so proud of their work."

Figure 2. Gatun Dinner Time

A few of the images that Pennell created and his reflections recorded at the time of their creation give some insight into what he accomplished and how he felt about what he had seen. About the image above, (Figure 2), for example, Pennell wrote: "At Gatun, the first time I stopped, I saw the workmen, in decorative fashion, coming to the surface for dinner. This lithograph was made from a temporary bridge spanning the locks and looking toward Colon. The great machines on each side of the locks are for mixing and carrying to their place, in huge buckets, the cement and concrete, of which the locks are built."

About the image of work in the Culebra Cut (Figure 3) Pennell wrote, "This shows the cut and gives from above some idea of the different levels on which the work is carried out. It is on some of these levels that slides have occurred and wrecked the work. The slides move slowly, not like avalanches, but have caused endless complications; but Colonel Gaillard, the engineer in charge, believes he will triumph over all his difficulties."

Figure 3. In the cut at Las Cascadas

Pennell did not go about his work without interacting with the workers building the canal. He respected them and their opinions as much as he appreciated their accomplishments. When he visited the work underway at Miraflores lock (Figure 4) he wrote, "And the interest of these Americans in my work and in their work was something I had never seen before. A man in huge boots, overalls and ragged shirt, an apology for a hat, his sleeves up to his shoulders, proved himself in a minute a graduate of a great school of engineering, and proved as well his understanding of the importance of the work I was trying to do, and his regret that most painters could not see the spleendid motives all about; and the greatest compliment I ever received came from one of these men, who told me my drawings 'would work.'"

Figure 4. The walls of Miraflores Lock

And finally, in a reflection that demonstrates that Pennell understood both the beauty of the work created by the engineers who built the canal and the fact that his images would be the only testimony to the care with which they created the project while knowing that much of its real beauty would never be seen again once the project was completed, he wrote of the approaches to the lock at Gatun (Figure 5), "These huge arches, only made as arches to save concrete and to break the waves of the lake, are mightier than any Roman aqueduct, and more pictorial, yet soon they will be hidden almost to the top by the waters of the lake."

Figure 5. Approaches to Gatun Lock

Pennell knew that he had completed the signature work of his career when he finished his time in Panama. In the conclusion of his introduction to the published version of his images he wrote, "I did not bother myself about lengths and breadths and heights and depths. I went to see and draw the Canal, and during all the time I was there I was afforded every facility for seeing the construction of the Panama Canal, and from my point of view it is the most wonderful thing in the world; and I have tried to express this in my drawings at the moment before it was opened -- for when it is opened, and the water turned in, half the amazing masses of masonry will be beneath the waters on one side and filled in with earth on the other, and the picturesqueness will have vanished. But I saw it at the right time, and have tried to show what I saw. And it is American – the work of my countrymen."

FINAL YEARS

By 1917, World War I had made it impossible for the Pennells to work in Europe. Also, life for Americans in England was becoming increasingly difficult. So, in July of 1917 they placed many of their possessions in storage in London and returned to the United States. Pennell considered World War I the greatest tragedy of his life, but he produced images of war-related industry for use by both the British and American governments in promoting war bond sales. His images of bridge and skyscraper construction after the war document an era of American construction that punctuated the United States' emergence as a world economic and industrial power.

Joseph Pennell was not popular with his contemporaries. He was opinionated and outspoken, and his views often ran contrary to the views of those more powerful and influential than he was. One of his friends wrote: "I have often told him, he was the most quarrelsome Quaker that ever was . . ." That, together with his insistence on choosing "unartistic" subjects such as bridge and skyscraper construction, caused many contemporary artists and critics to ignore his work for most of his career. Despite the fact that he refused to cultivate "the right people," the quality of his work eventually gained for him the success and recognition that he deserved as both an artist and a recorder of human achievement.

On April 23, 1926, Joseph Pennell died of congestive heart failure. His will specified that his collection of his own prints and the collection he had amassed of prints by his friend James Whistler were to be left to the Library of Congress. Unfortunately, many of his Whistler prints (and many of his own early prints as well), which had been left in storage in London, were ruined because of careless handling by warehouse personnel during the last months of World War I. Despite those losses, the Pennell collection at the Library of Congress contains hundreds of images depicting many outstanding achievements of the American civil engineering profession during one of its most productive eras.

(*Note: The images in this paper are reproductions made from negatives provided by the Office of History of the U.S. Army Corps of Engineers, except for Figure 3 which is provided courtesy of the author*)

REFERENCES

Bryant, Edward (1980). Pennell's New York Etchings, Dover, New York.
Mashek, Joseph (1971). "The Panama Canal and Some Other Works of Work." Artforum, 9, 38-41.
Palumbo, Anne Cannon (1982). Joseph Pennell and the Landscape of Change. Dissertation, University of Maryland, College Park.
Palumbo, Anne Cannon (1986). "The Cathedral and the Factory: The Transformation of Work in the Art of Joseph Pennell." Journal of the Society for Industrial Archeology, 2, 39-50.
Pennell, Elizabeth R. (1929). The Life and Letters of Joseph Pennell, (2 Volumes), Little, Brown and Company, Boston.
Pennell, Joseph (1912). Joseph Pennell's Pictures of the Panama Canal, J. B. Lippincott Company, Philadephia.
Pennell, Joseph (1916). "My Views of the 'Wonder of Work'." The American Architect, 109, 363-366.

Pennell, Joseph (1919). "The Need of a National Scheme in American Art, Architecture and Engineering." Journal of the Engineers Club of Philadelphia, 36, 278-281.

Pennell, Joseph (1925). The Adventures of an Illustrator. Little, Brown and Company, Boston.

Akira Aoyama's Achievements on Panama Canal Project
JSCE International Activities Center USA Group[1]

[1]International Activities Center, Japan Society of Civil Engineers, Yotsuya 1-chome, Shinjuku-ku, Tokyo JAPAN 160-0004; PH +81(3) 3355-3452; FAX +81(3) 5379-2769; email: iad@jsce.or.jp; URL: http://www.jsce-int.org

ABSTRACT

Japan Society of Civil Engineers (JSCE) discusses Akira Aoyama's achievements and contributions to the Panama Canal Project a hundred years ago. Akira started his professional career as a land surveyor on the project. His knowledge, effort, and dedication to his work earned the trust, respect and cooperation from his colleagues. His achievements have significantly contributed to mankind and civil engineering progress. JSCE also introduces the ASCE-JSCE cooperation and collaboration in this paper. JSCE believes that the cooperation between the two societies will not only strengthen their tie, but also produce more positive outcomes for their members, progress of civil engineering, and even society.

ABOUT JSCE

Introduction. JSCE was established in 1914, representing approximately 36,000 members. Since its establishment, JSCE has contributed to civil engineering progress, infrastructure development the well-being of society and has worked with 300 committees and 9 International Sections to pursue its goal. JSCE's mission is to contribute to the creation of a sustainable society through civil engineering practices. In order to achieve the mission, JSCE focuses on the following goals:

1) Enhance professional knowledge, skills and practices;
2) Strengthen the contribution of civil engineering to society; and
3) Promote communication, cooperation and collaboration among the members.

The members commit themselves to improve their professional practices, tackle various challenges and fulfill their responsibilities to society. Their commitment to civil engineering instills a strong sense of pride, honor and integrity among themselves.

In 2014, JSCE marks its 100th anniversary and carries out centennial projects on the theme "social contribution," "international contribution," and "citizens' interaction" for achieving a resilient society, and building the society in which next generations will enjoy. JSCE will host an International Symposium that will offer a comprehensive review of the past achievements and future civil engineering.

JSCE's 100th Anniversary and international activities. JSCE celebrates its 100th anniversary in November 2014 (Table 1). In the past 100 years, the Society has seen dramatic changes in society as well as a civil engineer's responsibility to society in those 100 years. While facing several social issues such as frequent natural disasters, natural resource shortages, energy related problems and environmental problems, we civil engineers realize that we have to develop antennas to catch diversified needs and to continue our best efforts to respond to those needs in order to contribute to society, holding a broad and long-term perspective.

Table 1. JSCE Timeline.

1879	The Japan Federation of Engineering Societies (JFES) is established by 23 graduates of the Imperial College of Engineering, including 3 civil engineers. With disciplinary development of engineering, many professional societies have stemmed from the JFES.
1914	Japan Society of Civil Engineers is founded.
1937	The Beliefs and Principles of Practice for Civil Engineers is created.
1999	Code of Ethics for Civil Engineers is drawn
2014	JSCE celebrates the 100th anniversary

Looking back on our past achievements and international technical exchange, we would like to examine future civil engineering and its responsibility at an International Symposium under the theme Contribution of Infrastructure to Life of Affluence to be held during the 100th Anniversary Celebration.

AKIRA AOYAMA'S CONTRIBUTION TO THE PANAMA CANAL

Akira Aoyama's Profile (1878-1963) - Table 2. Soon after graduating from the College of Engineering, Tokyo Imperial University in 1903, Akira Aoyama went to the United States and started a land surveyor career at a railway company in New York. He engaged in the Panama Canal Project as a leader of survey team and one of the core engineers for seven and a half years from 1904 through 1911.

He stayed there longer than most of American engineers.

Table 2. Akira Aoyama Biographical Sketch.

1878	Born in Iwata, Shizuoka, Japan
1899	Entered Tokyo Imperial University and majored in Civil Engineering
1903	Graduated at the top of his class (Summa Cum Laude) from the University, and left Yokohama in a passenger liner s/s Ryojun Maru for Seattle-Vancouver where he learned to make his both ends meet before proceeding to New York to see Prof. William H. Burr of the Civil Engineering Faculty of Columbia University as well as a member of the Isthmian Canal Commission (ICC) appointed by Theodore Roosevelt, 26th President of the United States. In New York, while waiting for his travel to Panama, he was introduced to a railway company by Prof. Burr, where he had to work for three months without pay. His experience helped him in the way of maneuverability of surveying technique in Isthmus of Panama.
1904	The United States commenced the construction of Panama Canal. Akira Aoyama was employed by ICC and started as a rodman on the Panama Canal Project
1905	Promoted to Assistant Surveyor
1907	As a surveyor, engaged in port facility construction on the Atlantic Division
1910	Promoted to Design Engineer, and was in charge of designing the central approach wall on the lower side of Gatun Locks. Then, promoted to Deputy Chief Engineer at Gatun Locks, Atlantic Division
1911	Resigned from ICC and returned to Japan
1912	Joined the Ministry of Interior, and took in charge of Arakawa Flood Discharging Channel construction
1914	Panama Canal opened to traffic
1915	Got married
1924	Arakawa Flood Discharging Channel was completed
1927	Engaged in the Shinano River Ohkouzu Diversion Channel project as a director of Regional Bureau
1934	Appointed to the 5th Vice Minister, Ministry of Interior
1935	Selected as the 23rd President of Japan Society of Civil Engineers
1936	Retired from the Ministry of Interior
1963	Passed away at the age of 84

Upon returning home, Akira Aoyama joined the Ministry of Interior and engaged in the construction of Arakawa Flood Discharging Channel, which completed twenty years later. Meanwhile, he conducted the Shinano River Ohkouzu Diversion Channel repair project (Figure 1) as Director of Regional Development Bureau. The project faced several difficulties and repeated delays, and he completed the project in 1931.

Figure 1. Diversion Weir Completed at the Shinano River Ohkouzu Diversion Channel in 1931

Akira was appointed to the 5th Vice Minister, Ministry of Interior, in 1934 and was elected as the 23rd President of Japan Society of Civil Engineers in the following year.

Akira Aoyama and Construction of Panama Canal. Behind his motivation in participating in the Panama Canal Project, we must note that he had two influential figures who inspired him to become a Christian and to be of service to his country and to God through all his life as a civil engineer: Becoming a Christian at the time meant something unusual and uncommon in society. That was the reason why he lost no time in making a decision to depart for Panama right after his graduation from the University. He was sincere, humble and dedicated to his work, and never behaved arrogant or boosted about his achievements and contribution to the Panama Canal Project and his success in the Arakawa River Improvement project.

Since he had a dream to participate in the Panama Canal Project which he believed all through his college years would mark a milestone in history and provide significant benefits to humanity, he devoted himself to the project faced by both unbearable tropical climates and a high risk of contracting local diseases of yellow fever and malaria. In fact, he got twice malaria, and also escaped death from hairbreadth in the Bohio jungle while surveying.

Akira the only Japanese civil engineer on the project (Figure 2) started his career as a field survey team member who had to make a daily report to a high-ranking engineer outside of his own ethnicity. He measured the area planned to be the canal route and surveyed the geology there, wearing a mosquito net in a tropical jungle day after day. He patiently and diligently did his assignments and gained skills and knowledge working his way up from the low-rank position. His diligence and efforts earned him continued promotions. He was promoted to a level-man, transit man, draftsman, design engineer, and then to deputy chief engineer at Gatun Locks, Atlantic Division. His responsibilities included the design and construction of the approach wall at Gatun Locks (Figure 3). He conducted the survey of the Gatun Dam and drafted a spillway of the Gatun Dam.

Figure 2. Drafting Force Division Engineering Office of Atlantic II/7 (Akira is in the middle)

During World War II around 1943, when the Imperial Japanese Navy asked him for his opinion about plan to destroy the canal locks in Panama, he answered "I know how to build the Panama Canal, but I don't know how to destroy it," then refused to cooperate with that plan.

Akira Aoyama's Achievement in Japan after Panama. After returning to Japan, Akira was employed by the Ministry of Interior as an engineer and dealt with several flood prevention works, utilizing the knowledge and skills that he had gained at the Panama Canal Project.

Figure 3. Central Approach Wall under Construction

Among the assignments given by the ministry, he faced the challenge of the Arakawa River improvement project (Figure 4). The Arakawa River was a violent river due to repeated floods causing damages to the areas. He took up the challenge; he lived near the river and conducted surveys, designed and supervised the construction of flood control channel. Moreover, he introduced innovative ideas, devices, and technologies such as steel-reinforced concrete system which he had acquired in Panama, and succeeded in digging the river bed down to 20 meters. Akira's excellent measurement of river flow and flood control method for the Arakawa Flood Discharging Channel were proved solid and fight off against the catastrophic Great Kanto Earthquake (M.7.9) of 1923.

CURRENT COOPERATION AND RELATIONSHIP BETWEEN ASCE AND JSCE

JSCE Event in Tokyo, 2014. JSCE celebrates its 100th anniversary in November 2014, holding three-day-long commemorative events, including the 100th Anniversary Commemorative Ceremony and Banquet, International Forum, and 3rd Roundtable Meeting on Disaster Management from November 19 to 21. Over the past 100 years, Japanese civil engineers have improved their practices to ensure the safety and wellbeing of people in cooperation with their colleagues around the world.

Figure 4a. Iwabuchi Flood Surge Gate on Arakawa Flood Discharging Channel

Figure 4b. Iwabuchi Flood Surge Gate on Arakawa Flood Discharging Channel

As the turn of the 21st century, civil engineers have played an increasingly important role in addressing globalization challenges, including natural disasters, food and water shortages, energy scarcity, poverty, and traffic problems. The civil engineers are called on to foster collaborations among themselves across borders so as to more effectively, respond to those challenges and to achieve societal resilience and sustainable development.

The International Symposium—International Forum on "Contributions of Infrastructure to Life of Affluence" will take place on November 20, 2014 (Figure 5). The symposium will discuss civil engineering contributions to achieving an affluent society, reviewing the past infrastructure development and international collaborations in the civil engineering field. Also, the 3rd Roundtable Meeting (changed from Asian Board Meeting) will be held on November 19 and 20.

The 3rd Roundtable Meeting will discuss civil engineering contributions to natural disaster prevention and mitigation with experts from many countries. Through the discussions, we will examine and identify how civil engineering will be in the coming century.

These international meetings will be held at the JSCE HQ and the JP Tower Hall and Conference near Tokyo Station.

Figure 5. JSCE International Symposium Announcement

ASCE and JSCE Cooperation. JSCE and ASCE reached an agreement on mutual cooperation in celebrating JSCE's 100th Anniversary and ASCE's Celebration of Panama Canal 100th Anniversary at the Global Engineering Conference during ASCE's 143rd Annual Civil Engineering Conference held in Charlotte, N.C. in 2013.

JSCE is planning to discuss Akira Aoyama's achievements and contribution to the Panama Canal Project at ASCE's Celebration. We hope to share what difficulties and efforts he made to achieve in the project with the audience in Panama. Moreover, we will remember that the knowledge, skills and technologies that we have obtained are developed from his and his colleagues' achievements.

Last we would like to thank ASCE for giving us the opportunity to celebrate these two significant events together. We believe that the cooperation between the two societies will not only strengthen their tie, but also produce more positive outcomes for their members, progress of civil engineering, and even society.

*Figures 2 & 3 were taken by Akira Aoyama, and captions were given by him as well.

REFERENCE

Takahashi, Y. (1962). "Visiting Akira Aoyama, one of the distinguished members in JSCE." *JSCE Magazine, vol.47 January 1962*, JSCE, 36-39. (in Japanese)

Gatun Dam History and Developments

Luis D. Alfaro[1], Manuel H. Barrelier[2] and Maximiliano De Puy[3]

[1]Vice-president of Engineering, Panama Canal Authority, Edif. 721 Corozal Oeste, Ciudad de Panamá, Panamá, Tel:(507)276-1585, email: lalfaro@pancanal.com.
[2]Geotechnical Engineer, Panama Canal Authority, Edif. 721 Corozal Oeste, Ciudad de Panamá, Panamá, Tel:(507)276-1657, email: mbarrelier@pancanal.com.
[3]Manager of the Geotechnical Branch, Panama Canal Authority, Edif. 721 Corozal Oeste, Ciudad de Panamá, Panamá, Tel:(507)276-1737, email: mdepuy@pancanal.com

ABSTRACT

The paper describes the main decisions leading to the construction of Gatun Dam, cited as the world's largest when completed. Subsequently, a brief summary of the dam's construction is presented, as extracted from documents of the era. Then, a description of the dam's post-construction performance is given. Its uneventful history contrasts sharply with its controversial origins.

Developments in engineering in the 20th Century have highlighted issues unknown to the designers and builders of the dam. These have motivated a detailed risk analysis of the structure, given that hydrologic and seismic demands, now known, are greater than previously believed. This evaluation process generated plans for remediation so it can reliably carry out its function, as the Panama Canal begins its second century of operation. The case also highlights, in a single structure, how much engineering practices have developed in this period.

DECISIONS LEADING TO THE NEED FOR GATUN DAM

The recent success the French had with the Suez Canal in the latter half of the19th Century, led by Ferdinand de Lesseps, motivated them to build a sea-level Canal in Panama. McCullough (1977) and Parker (2009) provide detailed accounts of these events. Two issues, unknown at the time, proved insurmountable to this courageous effort: tropical disease, and the high complexity of the geologic materials of volcanic and marine origin, weathered to varying degrees. In addition, the meteorological and hydrological conditions prevalent in Panama were substantially underestimated.

Ferdinand de Lesseps was the great promoter of the French effort. His resolve led him to disregard Baron Adolphe Godin de Lepinay's warning, who recommended a Lock Canal during the Canal Congress de Lesseps held in Paris in 1879. Godin de Lepinay's plan was remarkably similar to what was built several decades later. The Lock Canal reduced excavations, and provided much better regulation of the tempestuous Chagres River.

An epidemic of Yellow Fever, and other tropical disease, posed a serious challenge to the initial Panama Canal builders. Additionally, the lack of understanding of the behavior of weathered tuffaceous rocks, when excavated, challenged the resources of the private enterprise attempting to build the Canal. Both conditions eventually led the venture to financial failure.

However, in its final phases, the French, led by Phillipe Bunau-Varilla, did modify their design of the Canal, converting it to a Lock Canal to reduce the amount of excavation required. In 1887, eminent French engineer Alexandre Gustave Eiffel was engaged to design and build the locks for the Panama Canal. The Lock Canal needed a number of dams to impound water for operations, from the Chagres River and other smaller rivers in the general area. The plan was to build a main dam and Atlantic locks in Bohio, some 15 km south of the Gatun Dam site. Figure 1 presents a longitudinal profile of the Atlantic side of the Canal, with Bohio shown on the far right, (Wyse, 1890-91). The effort failed in 1891, due to lack of funds.

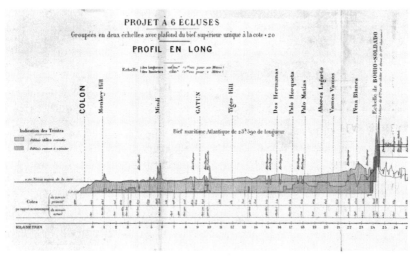

Figure 1. Location selected by the French to build the proposed Bohio Dam (Wyse, 1890-91)

When the U.S. took over the construction of the Canal in 1904, again plans shifted to a sea-level Canal. However, the issue on the type of canal to be built was heatedly debated. Sibert and Stevens (1915) summarize the conclusion as follows: *"It is the consensus of opinion among the best-informed men, for the reasons set forth in the report of the Chief Engineer of the Commission, under date of January 26, 1906, already quoted, that the selection of the 85-feet lock type for the Canal was eminently wise, and that such wisdom will become more and more apparent as time goes on."* With time, this proved to be absolutely correct.

An unprecedented storm occurred on December 2-4, 1906 in the Canal's watershed (ACP, 1876-2014). It was the basis for designing the main spillway structure for the project, and likely reinforced the decision to build a Lock Canal.

DESIGN AND CONSTRUCTION OF GATUN DAM

The U.S. plan included an earth dam on the site shown in Figure 2. This dam was the subject of much debate.

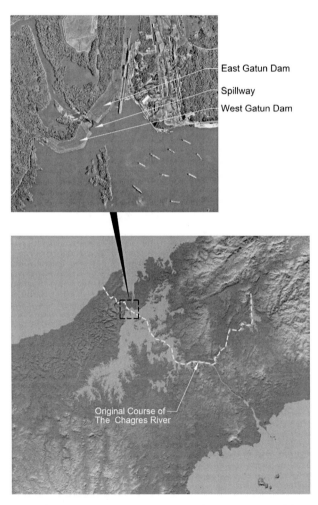

Figure 2. Location of Gatun Dam. The dotted line shows the original course of the Chagres River.

Gatun Dam takes its name from the Gatun River, a name that appears in Spanish maps dating as far back as 1750 (Isthmian Canal Commission, 1906). It was conceived as a hydraulic fill structure, confined by dry fill dikes. A very important feature is that parts of the dam rest over deposits of very soft marine sediments, up to 80 m thick, that filled two old river valleys created by the Chagres River in its geological past. The two valleys were separated by a small hill of sedimentary rock that separates the dam into east and west segments. This hill was used to build the spillway. This can be appreciated in Figure 2. The design of the dam was very innovative at the time. Hydraulic fill was gradually built up, confining it with dry fill perimeter dikes. Long debates ensued between proponents of the design, and those who considered it unsafe. Notice that this was done before the formal advent of Soil Mechanics, at a time when the concepts of effective stresses, pore pressures, shear strength of soft saturated soils and liquefaction, were yet to be defined.

Bennett (1915) points out: "*Colonel Goethals, as soon as he took charge, determined to ascertain just what the situation was beneath the surface of the valley. He called to his aid Caleb M. Saville, one of the foremost earth-dam experts of the world, who had been in charge of the work on the Wachusetts Dam in Massachusetts. They honeycombed the site of Gatun Dam with borings, and sank test pits here and there, so that they could go down and see with their own eyes the various strata of the proposed foundation of the dam.*"

Figure 3 shows a photograph of a 1/12 scale physical model made by Mr. Saville of the Gatun Dam (Isthmian Canal Commission, 1908). After the model tests were performed, Mr. Saville indicated:

"*1. That suitable material is available and near at hand for the construction of the Gatun dam by hydraulic process.*

2. That the foundations are suitable for such a structure as the proposed Gatun dam if they are properly treated.

3. That it is practically possible to construct a stable and water-tight earth dam at Gatun of the materials available.

4. That the hydraulic method of construction, as proposed for this work, is feasible if proper conditions are observed."

Figure 4 shows a photograph from the archives of the Panama Canal, of the construction of the hydraulic fill portion of Gatun Dam, from August, 1910, when construction was well underway.

Figure 5 presents a transverse geological cross section of Gatun Dam, along the West valley of the Chagres River.

Figure 3. Model Dam made in 1908 to test the technical viability of Gatun Dam (Isthmian Canal Commission, 1908)

Figure 4. Photograph of construction of the hydraulic fill portion of Gatun Dam (Panama Canal archives)

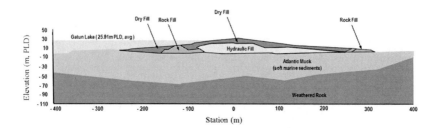

Figure 5. Cross Section of Gatun Dam through the West Chagres River gorge

The dam was built between 1907 and 1913. Its volume was 16.8 Mm3 (about 50 % is hydraulic fill). It was 800 m thick at the base, 91 m thick at the water line, crest width was 30 m, crest elevation was 32 m (105 feet) above mean sea level, and the maximum height of the embankment was 32 m. The upstream slopes were in the order of 1 on 7.7, while the downstream slopes were about 1 on 11.6.

Landslides and bearing capacity failures occurred during construction of the dam, but a delicate balance was maintained and construction of the dam was not nearly as difficult as the excavation of the channels through the continental divide (the Culebra Cut, now called Gaillard Cut in honor of David Dubose Gaillard who was in charge of its construction).

Bennett (1915) points out that *"When the stone in one of the toes of the dam sank in 1908, a sensational dispatch was sent to a New Orleans newspaper stating that the dam had given way. This report so stirred the nation – for the people had not stopped to think that there was as yet, no dam at Gatun to give way – that President Roosevelt asked President-elect Taft to go to Panama to investigate the situation. The net result of this investigation was that the engineers who accompanied Mr. Taft – Messrs. F.P. Stearns, Arthur Davis, Henry A. Allen, James D. Schuyler, John R. Freeman and Allen Hazen – decided that the dam was being built with a greater margin of safety than even the utmost precaution required. They reported that the crest of the dam should be cut down thirty feet, and that there was no necessity for driving interlocking sheet piling across the valley as a precaution against imaginary underground rivers."*

Figure 6 presents a photograph from the Panama Canal Authority's archives, of the group of consultants with President-elect Taft and Col. George Goethals.

Figure 6. Board of Consulting Engineers convened with the purpose of reporting to President Roosevelt their opinion as to the proposed Gatun Dam. Front row: President-elect William Taft, Frederick P. Stearns (Hydraulic Engineer from Boston), Allen Hazen (Hydraulic Engineer from New York), Henry A. Allen (Civil Engineer) and Arthur P. Davis (Chief Engineer of the Reclamation Services). Back Row: Col. George Goethals (Chief Engineer of the Panama Canal), James Dix Schuyler (Civil Engineer from Los Angeles), Isham Randolph (Civil engineer from Chicago) and John L. Freeman (Civil Engineer from Providence, R.I.). The photograph was taken in early February, 1909 at the Canal's offices in Culebra, Panama.

Haskin (1914) summarizes the issues surrounding Gatun Dam by stating: *"Around no other structure in the history of engineering did the fires of controversy rage so furiously and so persistently as they raged for several years around Gatun Dam."*

In late 1908, President Roosevelt sent President-elect Taft to Panama with a group of distinguished consulting engineers (The New York Times, 1908). In the same publication, John F. Stevens, then former Chief Engineer of the Panama Canal, stated that: *"the dam is being built actually much wider and higher than safety requires, but merely as a concession to prejudice, and that if the canal were being built by private interests, a much less massive structure would have been considered entirely secure."*

Figure 7 shows a 1/32 scale hydraulic model built in 1910 to finalize the hydraulic design of the spillway (Isthmian Canal Commission, 1910). The data used to design the spillway was based on about 20 years of records collected intermittently between 1876 and 1905 (ACP, 1876-2014).

Figure 7. Photograph of hydraulic model of the spillway
(Isthmian Canal Commission, 1910)

The successful building of this large earth dam, under so difficult geologic conditions, was one of the major technical accomplishments of Canal designers and builders.

Post-construction settlements make the elevation of its lowest points on the crest about 30.0 m (98.5 feet), indicating a settlement in the order of 2.0 m (6.5 feet). Given that maximum lake elevation has been 87.5' (26.67 m) above mean sea level, freeboard of the dam is still conservative (3.3 m).

Additional site investigation at the dam site, and of the dam itself, has been carried out since the 1990's. Figure 8 presents a summary of material parameters that characterize the hydraulic fill, specifically, Atterberg Limits and Standard Penetration Test values. These enable a reasonable understanding of the nature of this material.

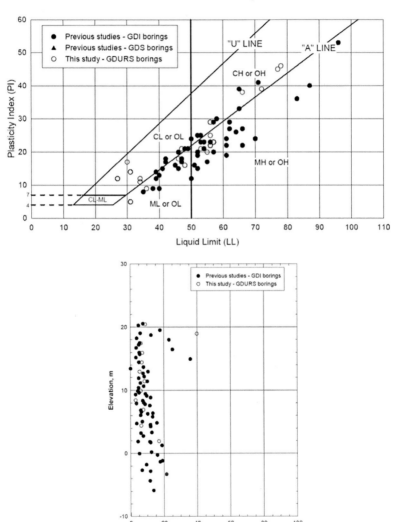

(a) Plasticity Chart and Normalized Blowcount for the Gatun East Dam
(URS Corporation, 2012)

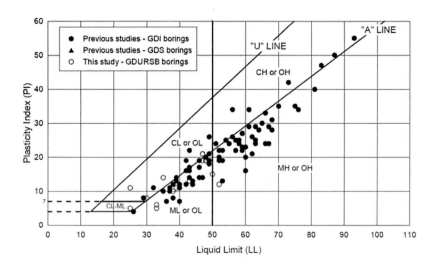

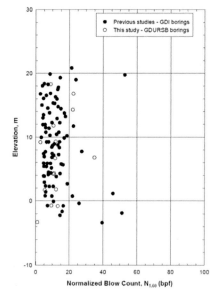

(b) Plasticity Chart and Normalized Blowcount for the Gatun West Dam (URS Corporation, 2012)

Figure 8. Summary of material parameters of the hydraulic fill in Gatun Dam

PERFORMANCE OF THE DAM (1914-2014)

The dam has performed well during its 100 year history. Figure 9 presents a recent aerial photograph of the Dam. The seepage through the embankments has been monitored in 15 weirs, without incidents of concern.

Figure 9. Aerial photo of Gatun Dam (circa 2009)

Springs on the downstream slope were detected in 1988. An inverted filter was placed in this area shortly afterwards, to minimize the risk of internal erosion.

Old drain pipes, left in place during construction to dewater the hydraulic fill, were detected in 1992. They were then filled with porous (gap-graded) concrete that provided structural support while continuing to provide drainage.

An effort was carried out to evaluate the risk of post-liquefaction failure of Gatun Dam in the late 1990's. Results were marginal, which motivated a recent independent peer review carried out by URS Corporation (2012).

The largest storm in the Canal's history occurred on 7-9 of December of 2010. Unprecedented spilling operations led to erosion on the upstream embankments in the vicinity of the spillway intake. Erosion protection was promptly built, to mitigate this problem (Figure 10).

Surface monuments indicate that displacements of the dam have been slight and considered acceptable (ACP, 1980-2014). The same source also indicates that pore water pressures have been consistent with the unconfined flow regime established in the dam.

Figure 10. Erosion protection on the upstream face of the dam, near the spillway intake built after the December 2010 storm caused significant erosion

PRESENT RISK ANALYSIS OF GATUN DAM

Given the relative quiescence of the dam described above, there were no great concerns regarding its reliability up to the mid-1970´s. However, two issues arose in the following decades that made further analyses a necessity.

Hydrologic. The failure of Teton Dam in Idaho in 1976, led the U.S. Congress to emit the Dam Safety Act of 1977 that required the inspection of all Federal Dams. As the Panama Canal Commission was a Federal Agency of the U.S. Government at the time, a detailed inspection of the Canal´s dams was ordered.

A technical team of the U.S. Corps of Engineers came to Panama in 1978-1979 to perform these inspections. U.S. Corps of Engineers (1979) is the report that describes their findings.

As part of this effort, a "Probable Maximum Flood" (PMF) was defined for the Canal watershed. It is a six-day event with a peak inflow of 23,400 cubic meters per second. An analysis reveals that the existing Gatun spillway has roughly half the capacity to route the PMF in order to restrict damages to acceptable levels. This scenario is aggravated by operational practices that have resulted from modernization schemes (increasing maximum Gatun lake level by the end of the rainy season).

After the December 7-9, 2010 storm in the Canal´s watershed (the greatest on record, and close to half the PMF), the Panama Canal Authority made the decision to build a second spillway in Gatun lake. Design of this structure is currently underway.

Seismic. In 1991, Canal engineers became aware of a large historical earthquake that occurred on 7 September 1882 on the Atlantic coast, as the French Canal effort was about to begin. Knowing now that hydraulic fill dams are notoriously vulnerable during earthquakes, efforts began immediately to: (a) improve characterization of the dam, and (b) characterize the seismicity of the area.

Site investigation efforts were carried out with borings to collect undisturbed samples and to perform Standard Penetration Tests. Additionally, Cone Penetration Tests were performed. Piezometers, surface movement monuments, and traveler pipes were installed in clusters along six cross sections of the dam: two on the east dam and four on the west dam. Figure 5 presents one of the cross sections through the dam, with a geologic interpretation that includes the information collected with these recent exploration campaigns.

Liquefaction susceptibility maps and liquefaction triggering maps were prepared to identify zones vulnerable to liquefaction. Then stability analyses and deformation analyses were performed with post-liquefaction material parameters. Multiple localized pockets of the heterogeneous mass of the hydraulic fill are most likely to undergo liquefaction during an earthquake. However, the very flat slopes of the embankments are an asset. Stability and deformations were characterized as marginal. It is uncertain that failure mechanisms that could compromise the integrity of the dam would occur, but given the uncertainties in the distribution and characterization of the various materials involved, it cannot be ruled out.

An additional complication would ensue if transverse cracking of the dam occurs during earthquake induced deformations in the zones where the deep gorges transition to their surrounding areas. The varying stiffness of the foundations in these zones could aggravate this potential cracking.

In 2011, a work order was given to URS Corporation to make an independent analysis of the seismic stability of Gatun Dam. The results confirm previous findings: liquefaction is likely in certain regions of the hydraulic fill, but perhaps not enough to generate a critical failure mechanism in the structure.

Roughly a third of Panama's GNP is generated by the Canal and the cluster of industries that depend on the Canal (Colon Free Zone, Ports, Trans Isthmian Railroad, suppliers to these industries). The consequences of a failure of this dam are so large for Panama's economy, and in general for the shipping industry that uses the Canal, that a conservative approach for mitigating these risks is well warranted.

PLANS FOR REFURBISHING GATUN DAM

A workshop was held in early February 2014 to address these matters. Included were key Canal technical personnel, the Canal's Geotechnical Advisory Board, the Canal's Physical Risks consultants, and URS Corporation personnel who performed the 2012 analysis of the seismic stability of Gatun Dam. The specific objective of this workshop was to reach a consensus on the best way to reinforce the dam so that its risk level is commensurate with other structures of similar importance worldwide.

The group's conclusion was that risk mitigation measures should be implemented. The work will involve the construction of filters and a berm on the downstream slope of the dam to increase the dam's reliability under potential post-liquefaction failure mechanisms. Soil improvement of the Atlantic Muck at the downstream toe will also be required to control settlements under the weight of the new berm. Figure 11 shows a conceptual design of this work. The filter is especially important to provide protection against piping, as the dam has no filters.

These relatively simple solutions can provide an appropriate level of conservatism to such a vital structure. In addition, the berm will enable the construction of a road to facilitate communication with the West side of the Canal.

These enhancements, will condition the century-old earth structure, so it can continue working reliably to serve the Panama Canal as an important component of the world shipping infrastructure, and the backbone of Panama's economy.

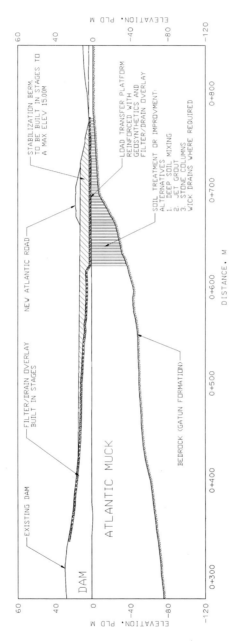

Figure 11. Proposed remediation work on the downstream slopes of Gatun Dam

REFERENCES

Abbott, Willis J. (1913), "Panama Canal in Picture and Prose," New York, Published In English and Spanish by Syndicate Publishing Company.

Autoridad del Canal de Panama (ACP), (1876-2014) Internal documents from the Meteorological and Hydrological Branch.

Autoridad del Canal de Panama (ACP),(1980-2014) Internal documents reporting results of field instrumentation in Gatun Dam.

Autoridad del Canal de Panama (ACP), (1993-2014) ACP's Geotechnical Advisory Board reports for meetings in which the Gatun Dam issue was addressed (Feb. 1993, Jan. 1995,Jul.1998, Sep. 1999, Nov. 2001, Jan. 2013 and Feb. 2014).

Bassell, Burr (1904), "Earth Dams. A Study," The Engineering News Publishing Company, New York.

Bennett, Ira E. (1915), "History of the Panama Canal Its Construction And Builders," Historical Publishing Company.

Burr, William H. (1902),"Ancient and Modern Engineering and the Isthmian Canal," First Edition, New York: John Wiley & Sons, London: Chapman & Hall, Limited.

Friar, W.K. (1972), "Porqué se llaman....?" The Panama Canal Review en Español, Spring issue.

Haskin, Frederic J. (1914), "The Panama Canal," Doubleday, Page & Company.

Isthmian Canal Commission (1906), Annual Report for the year ending December 1, 1906, Washington: Government Printing Office.

Isthmian Canal Commission (1907), Annual Report for the fiscal year ended June 30, 1907, Washington: Government Printing Office.

Isthmian Canal Commission (1908), Annual Report for the fiscal year ended June 30, 1908, Appendix E: Report of C.M. Saville, Assistant Engineer, on Gatun Dam Investigations. Washington: Government Printing Office.

Isthmian Canal Commission (1910), Annual Report for the fiscal year ended June 30, 1910, Washington: Government Printing Office.

McCullough, David (1977), "The Path between the Seas", Simon Schuster.

McDonnell, John B. (2011), "The Story of the Panama Canal" A.J. Publications.

Panama Canal A.B. Nichols Collection – Linda Hall Library.
http://lhldigital.lindahall.org/cdm/search/collection/panama

Parker, Matthew (2009) "Panama Fever," Anchor Publishing.

Sibert, William L. and Stevens, John F. (1915), "The Construction of the Panama Canal,", D. Appleton and Company.

Taft, William Howard (1908), "Present Day Problems: A Collection of Addresses Delivered on Various Occasions" New York: Dodd, Mead, LC Call Number: JK246 .T2

The New York Times (1908), "Engineers Chosen for Canal Inquiry – Davis, Freeman, Hazen, Randolph, Schuyler, and Stearns will accompany Taft – Two Favored Lock Plans – President Wants an Unbiased Report on Which to Act Before He Leaves Office," published December 30, 1908.

URS Corporation (2012) Report on "Geotechnical Evaluation of Gatun Dam" ACP Contract No. CMC-172538, Task Order No. 8.

Transactions of the International Engineering Congress- 1915, (1916) Neal Publishing, San Francisco, California.
U. S. Corps of Engineers (1979) "1977-1979 Inspection of Panama Canal Flood Control Facilities", (1979), prepared by U.S. Army Engineer District, Mobile Corps of Engineers, Mobile, Alabama.
Wheaton, James K., (2011), "The Panama Canal," Golgotha Press.
Wyse, Lucien N.B. (1890-91), "Canal Interoceanique de Panama. Plan General Project a 6 Escluses," Imp. Erhard Fres, Paris.

The 1915 Panama-Pacific International Exposition in San Francisco and Panama Canal Model, Conference and Proceedings

Jerry R. Rogers, Ph.D., P.E., D.WRE, Distinguished M. ASCE,
12127 Old Oaks Drive, Houston, TX 77024 (rogers.jerry@att.net) and
Luis D. Alfaro, Ph.D., Vice- President, Division of Engineering, Panama Canal Authority, Panama City, Panama (LAlfaro@pancanal.com)

ABSTRACT

The Panama–Pacific International Exposition (PPIE) was a world's fair in San Francisco between February 20 and December 4, 1915 that attracted 18,876,438 attendees. Its purpose was to celebrate the completion of the Panama Canal, and it was an opportunity to showcase the San Francisco recovery from the 1906 earthquake. Panama Canal papers were published in the *Transactions of the International Engineering Congress*, September 20-25, 1915. The canal transactions had two volumes by Neal Publishing of San Francisco in 1916: volume 1 presented papers 1-13, 527 pp. and volume 2 contained papers 14-25, 483 pp., totaling 1010 pp. All papers were by Panama Canal engineers, employees, or consultants. This paper summarizes several papers and covers highlights of the PPIE: Planning, Construction, Exhibits, Buildings, Landscaping, Commemorations, etc. Completed on time and on budget with large attendance, some people said the 1915 world's fair was the most successful ever held.

INTRODUCTION

The world's fair was constructed on a 635 acre (2.6 km^2) site in San Francisco, along the northern shore now known as the Marina District. Taking twelve months to build, a main attraction at the exposition was a large scale, topographic model of the Panama Canal covering five acres (depicting 5,000 square miles of the canal). A moving platform (1,440 feet long for 1,200 people in 144 cars endlessly connected) with seats priced at 50 cents each carried people around the exhibit with each seat having a duplex telephone receiver to transmit Panama Canal information for the 23-minute ride. The miniature Miraflores and Pedro Miguel Locks worked with water flowing, and ships traveled back and forth, controlled by magnets. Among the official PPIE souvenirs sold were paperweights filled with soil from the Culebra (Gaillard) Cut, plaques made from cocobolo wood used for railroad ties across the Isthmus of Panama, and each with a souvenir letter from the Governor of the Canal Zone: G.W. Goethals, former Chairman and Chief Engineer of the Isthmian Canal Commission.

TRANSACTIONS OF THE INTERNATIONAL ENGINEERING CONGRESS

Almost a thousand conventions and congresses met in San Francisco for the PPIE fair, including international engineering sessions September 20-25, 1915 by the

American Society of Civil Engineers, American Institute of Mining Engineers, the American Society of Mechanical Engineers, American Institute of Electrical Engineers, and the Society of Naval Architects and Marine Engineers. From the Panama Canal Commission Library, the following lists of papers in two volumes were published in 1916, totaling 1010 pages of Panama Canal details. Figure 1 shows the Table of Contents of the Panama Canal papers in the *Transactions of the International Engineering Congress*, September 20-25, 1915.

Paper No. 1: "INTRODUCTION" (pp. 1-30) and Paper No. 10: "THE DRY EXCAVATION OF THE PANAMA CANAL" (pp. 235- 286) were both presented in 1915 by Major General George W. Goethals, Member- ASCE, Governor of the Canal Zone, formerly Chairman and Chief Engineer of the Isthmian Canal Commission, Balboa Heights, Canal Zone, Panama. Also likely presented was Paper No. 11: "CONSTRUCTION OF GATUN LOCKS, DAM AND SPILLWAYS" (pp. 287-424) by Brig. General William Luther Sibert, U.S. Army, Member-ASCE, former Member Isthmian Canal Commission and Division Engineer, Atlantic Division, relocated in San Francisco.

In volume 2, Paper No. 19: "HYDRAULICS OF THE LOCKS OF THE PANAMA CANAL" pp. 165-234) was presented in October 1915 by Richard H. Whitehead, Associate Member-ASCE, Assistant Superintendent Pacific Locks, Panama Canal. No other papers in the *Transactions* were specifically noted as being presented at the 1915 San Francisco Conference, but several may have been. (On p. 167, Richard H. Whitehead, still living, made Congressional Record remarks on March 31, 1954 at the dedication of the Goethals Memorial at Balboa, Canal Zone.)

As a side note, in the first chapter of "The Battleship and the Canal," Lipsky noted that the *USS Oregon*, built in San Francisco, was ordered from the West Coast to Cuba early in the Spanish American War in 1898 (Lipsky, 2005). The battleship took 67 days to travel around the tip of South America. For military strategic reasons to move ships more quickly, the U.S. supported the construction of the Panama Canal. After Panama Canal completion, the canal trip took only 1-2 days. The *USS Oregon* and the Pacific Fleet steamed into San Francisco Bay during the 1915 PPIE.

GOETHALS 1915 "INTRODUCTION" IN THE *TRANSACTIONS*

In Paper No. 1: "INTRODUCTION" in volume 1, George W. Goethals provided a detailed early history of Isthmian Canal plans, surveys, studies, etc., by people and engineers from Spain, England, France, Netherlands, the United States, Central America, and other countries. Most of these early studies were favorable in recommending a canal via Panama or Nicaragua and some included some preliminary cost estimates. Goethals gave early credit to the Spanish explorer Vasco Nunez de Balboa, who, accompanied by engineer Alvaro de Saavedra Ceron, discovered the Pacific Ocean on September 25, 1513. Goethals noted: "Saavedra...prepared plans for a canal to be built along the route which he followed with Balboa in 1513. Death overtook Saavedra as he was about to lay his plans before the King of Spain."

CONTENTS

PAPERS

No.		Page
1	INTRODUCTION. By George W. Goethals.	1
2	COMMERCIAL AND TRADE ASPECTS OF THE PANAMA CANAL. By Emory R. Johnson.	31
3	OUTLINE OF CANAL ZONE GEOLOGY. By Donald F. MacDonald.	67
4	SANITATION IN THE PANAMA CANAL ZONE. By Chas. F. Mason.	85
5	PRELIMINARY MUNICIPAL ENGINEERING AT PANAMA. By Henry Weiss Durham.	117
6	MUNICIPAL ENGINEERING AND DOMESTIC WATER SUPPLY IN THE CANAL ZONE. By George M. Wells.	155
7	THE WORKING FORCE OF THE PANAMA CANAL. By R. E. Wood.	189
8	PURCHASE OF SUPPLIES FOR THE PANAMA CANAL. By T. C. Boggs.	205
9	THE CLIMATOLOGY AND HYDROLOGY OF THE PANAMA CANAL. By F. D. Wilson.	223
10	THE DRY EXCAVATION OF THE PANAMA CANAL. By George W. Goethals.	255
11	CONSTRUCTION OF GATUN LOCKS, DAM AND SPILLWAY. By W. L. Sibert.	297
12	METHODS OF CONSTRUCTION OF THE LOCKS, DAMS AND REGULATING WORKS OF THE PACIFIC DIVISION OF THE PANAMA CANAL. By S. B. Williamson.	425
13	DREDGING IN THE PANAMA CANAL.	

CONTENTS

PAPERS

No.		Page
14	GENERAL DESIGN OF THE LOCKS, DAMS AND REGULATING WORKS OF THE PANAMA CANAL. By H. F. Hodges.	1
15	THE DESIGN OF THE SPILLWAYS OF THE PANAMA CANAL. By Edward C. Sherman.	55
16	DESIGN OF THE LOCK WALLS AND VALVES OF THE PANAMA CANAL. By L. D. Cornish.	95
17	LOCK GATES, CHAIN FENDERS AND LOCK ENTRANCE CAISSON. By Henry Goldmark.	97
18	EMERGENCY DAMS ABOVE LOCKS OF THE PANAMA CANAL. By T. B. Monniche.	131
19	HYDRAULICS OF THE LOCKS OF THE PANAMA CANAL. By R. H. Whitehead.	165
20	ELECTRICAL AND MECHANICAL INSTALLATIONS OF THE PANAMA CANAL. By Edward Schildhauer.	215
21	THE RECONSTRUCTION OF THE PANAMA RAILROAD. By Frederick Mears.	291
22	PERMANENT SHOPS, PACIFIC TERMINALS, PANAMA CANAL. By A. L. Bell and H. D. Ehrman.	325
23	TERMINAL WORKS, DRY DOCKS AND WHARVES OF THE PANAMA CANAL. By H. H. Rousseau.	371
24	COALING PLANTS AND FLOATING CRANES OF THE PANAMA CANAL. By F. H. Cooke.	433
25	AIDS TO NAVIGATION FOR THE PANAMA CANAL.	

Figure 1. Table of Contents for the Two Volumes: Transactions of the International Engineering Congress, September 20-25, 1915 in San Francisco. Neal Publishing. San Francisco. 1916

After covering many other historical events for an Isthmian canal and railroad history, Goethals cited the French involvement: "Under the presidency of Ferdinand Marie de Lesseps, a 'Congres International d'Etudes de Canal Interoceanique' was invited to assemble in Paris in 1879, to consider and pass upon the whole question." Goethals summarized: This congress consisted of 135 delegates..., most of them favorably disposed toward de Lesseps, who had attained such success at the Suez Canal...de Lesseps...advocated a sea level canal.... Opposition to his views was manifested mainly by abstaining from voting...forty members absented themselves, ten refrained from voting, and only nineteen (of the 135) voted, sixteen of whom favored a sea level canal." Goethals referred to the 1879 concept (ultimately adopted): "Godin de Lepinay proposed a dam across the valley of the Charges at Gatun...with flights of ...locks on either side... This project would reduce very materially the amount of excavation...." Following some U.S. Isthmian canal studies in the 1880s and 1890s, the 1915-1916 Goethals paper primary excerpts include: "...the Act of March 3, 1899, authorized the U.S. President to make a full and complete investigation of the Isthmus of Panama, with a view to the construction of a canal, the Nicaragua route and the Panama route..." Nine appointed commission members submitted their report on November 16, 1901: "...the most practical and feasible route...for the United States...is that known as the Nicaragua route..." When the payment of $40M to France was deemed acceptable, the Isthmian Canal Commission changed to support the Panama route...with a summit level dam at Bohio.... Goethals summarized: "By an Act of Congress approved June 28, 1902, commonly known as the Spooner Act, the President was authorized to acquire...not to exceed $40M rights, ...property owned by the New Panama Canal Company, and to secure from the Republic of Columbia...a strip of land...subsequently failed of ratification by the legislative body of Columbia. The secession of the Province of Panama and the establishment of an independent Republic followed."

Goethals also wrote: "...shortly after the United States took possession on May 8, 1904, the question of a sea level versus a lock canal was agitated...the President convened a board of (thirteen) engineers to consider the entire subject.... The board submitted its report on February 5, 1906; a majority, eight in number, favored the sea level canal, and a minority, five in number (Alfred Noble, General Henry L. Abbot, Frederick P. Stearns, Joseph Ripley and Isham Randolph) favored the lock type of canal... The report of the board was submitted first to John F. Stevens, Chief Engineer of the Isthmian Commission, who had...expressed himself in favor of a high-level canal. He recommended the adoption of the type proposed by the minority of the board, modified by the withdrawal of the locks at Sosa Hill..." (It is interesting to note that Alfred Noble served as ASCE President in 1903, Frederick P. Stearns was ASCE President in 1906, and John F. Stevens became ASCE President in 1927.) Congress voted for the lock type design on June 29, 1906.

Goethals added: "By Executive Order, the dimensions of the locks were changed to a length in the clear of 1,000 feet and a width of 100 feet.... In a memorandum dated October 29, 1907, ...the Navy...recommended (the locks) increase to a clear width of 110 feet... and consequently, the width of 110 feet was adopted." Other design parameters and details were changed, some by Executive Order in 1908.

Goethals included cost details at the end of his paper, noting a 1906 change from a ten-hour work day to eight hours and higher cost of U. S. supplies/materials had increased the Panama Canal construction cost. Goethals ended his "INTRODUCTION" paper with: "...nine years would be required to complete the summit-level canal, or to January 1, 1915.... The result was that the work in all its parts was advanced to such a stage that the first ocean-steamer was passed through the canal on August 3, 1914, and, but for the unexpected slide which occurred north of Gold Hill on October 14, 1914, the canal would have been completed in its entirety within the estimated time."

A 1991 LESSON LEARNED FROM THE 1915 PANAMA CANAL CONGRESS

Goethals also wrote the tenth 1915 paper: "THE DRY EXCAVATION OF THE PANAMA CANAL," with landslide control statements, which was re-discovered by the canal Geotechnical Group in 1991. From the Goethals paper: "In the Culebra (Gaillard) Cut, the French had undertaken the construction of diversion channels, one on either side of the excavated area, to take care of the (drainage) water of the surrounding country.... With the canal adopted by the United States of a bottom width of 200 feet and a depth of 45 feet, a considerable change in the existing diversion channels was made necessary, and as carried out there was a channel on either side - the Obispo diversion on the east and the Camacho on the west. The Obispo diversion was constructed to carry the largest recorded flow.... Because of its proximity to the (Culebra/Gaillard) Cut, this diversion gave a great deal of trouble, and was undoubtedly responsible for some of the slides which developed.... In May, 1910, it caused the breaks which occurred at La Pita Point, and the subsequent break north of La Pita Point on August 20, 1912, at which time the Cut to the north was so completely flooded as to stop all excavation for three miles during the remainder of the wet season...it was necessary to build a concrete flume around La Pita Hill." This 1915 information led the 1991 canal Geotechnical Branch to examine the correlation between the original diversions and landslides, which proved to be quite strong. Immediately plans were revised for routine maintenance of the diversions, and a number of channels were built to reduce to a minimum the flow along these old diversions. Water was evacuated quickly from the diversion channels along perpendicular outlets to the Canal. This lowered the water level in the diversion channels, and had a positive impact by lowering the groundwater regime during rainy seasons. The Geotechnical Branch found this reduced the number of areas exhibiting incipient instabilities and/or fully-formed slides.

Figure 2 shows the rivers in the slide area, the position of the navigation channel through the Gaillard Cut and the three diversion channels constructed (highlighted). La Pita Point is on the East Bank between the Masambi River and Sardinilla River.

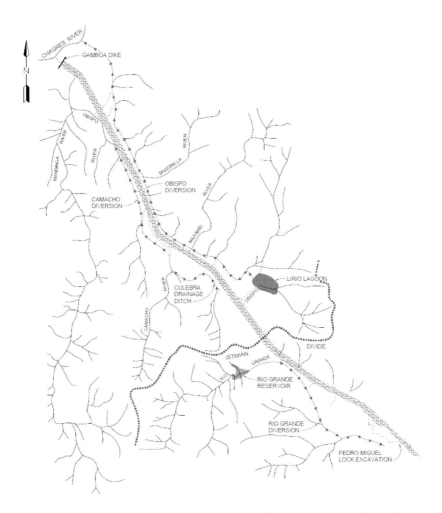

Figure 2: Gaillard Cut Region Showing Original Streams along the Canal Route and Diversions Required for Dry Excavation.
(Source: Lutton, Richard J., April 1975. Figure 1, Technical Report S-70-9).
(Re-drawn by the Panama Canal Authority in 2014)

PLANNING FOR THE 1915 WORLD'S FAIR IN SAN FRANCISCO

In 1904, Reuben Hale, a San Francisco department store owner, wrote a proposal to the Merchants Association that the city host an international exposition in

1915 to celebrate the scheduled opening of the Panama Canal (Ewald, 1991). After the Great San Francisco Earthquake and Fire on April 18, 1906, Hale met with other merchants in December 1906 to form the Pacific Exposition Company, renamed three years later as the Panama-Pacific Exposition Company. A meeting of more than 2,000 people held April 28, 1910 at the Merchants Exchange Building raised $4M in stock in two hours, including $25,000 from the Southern Pacific Corporation (Lipsky, 2005; Ewald, 1991). The State of California levied a $5M tax fund for the Exposition and San Francisco voted a $5M bond issue for the Exposition. On January 31, 1911, the U.S. Congress recommended San Francisco for the 1915 World's Fair and on February 15, 1911, President W.H. Taft signed a resolution designating San Francisco as the 1915 Exposition site. On February 2, 1912, Taft issued a proclamation inviting all nations to participate in the PPIE. President Taft came to San Francisco on October 14, 2012 for a ceremonial ground breaking, attended by more than 100,000 people. After considering multiple fair sites, the PPIE directors selected the Harbor View (Marina District) of 635 acres. The Exposition had to buy or lease 76 city blocks containing 200 parcels of land from 175 owners and tear down or move 200 buildings (Ewald, 1991).

SIGNIFICANT BUILDINGS IN THE 1915 WORLD'S FAIR

George W. Kelham was selected as PPIE Chief of Architecture, having taken part in the Paris Exhibition of 1900, and Kelham was the designer of the iconic Palace Hotel and Bohemia Club buildings in San Francisco. Kelham planned four Exposition zones, including:
1. Joy Zone (with 70 acres borrowed from Fort Mason to the east) consisting of seven main blocks in length, 100 feet wide, for 60 attractions, restaurants, thrill rides, adventures, and souvenirs (such as the popular Watch Palace). The well attended Panama Canal Exhibit was near the entrance to the world's fair along with models of the Grand Canyon and Yellowstone Park.
2. Great Showplaces of the World Zone with 70,000 exhibits in 220 acres,
3. Foreign and State Pavilions Zone, and
4. Livestock, Athletic Competitions, and Drill Grounds Zone (110 acres on some of the Presidio grounds).

In one core group, Kelham planned eleven main exhibition palaces: Palace of Machinery, the first building begun in January of 1913 (the largest wooden and steel building in the world, nearly 1,000 feet long, 367 feet wide and 136 ft. high, containing 250 exhibitions, displaying over 2,000 exhibits); Mines and Metallurgy; Transportation (including the displays of the Westinghouse Exhibit of a Pennsylvania RR electric locomotive and automobiles); Manufactures and Varied Industries; Agriculture and Food Products (such as Heinz and Sun-Maid Raisins); Education and Social Economy; Liberal Arts (with a ten ton, 20-inch Equatorial Telescope and Underwood Typewriter exhibit); Horticulture (with a larger glass dome of 152 feet in diameter than St. Peter's Basilica); and, the Fine Arts Palace (framed in steel for fireproofing, rebuilt in the 1960's for the Exploratorium, now re-located at Pier 15 on the Embarcadero at Green St.). To put the size of the Palace of Machinery in perspective, in January 1914, aviator stunt pilot Lincoln Beachey flew his plane through the building in the first indoor flight.

The great buildings were divided by one longitudinal and three lateral streets that intersected into magnificent courts: the Court of the Universe, the Court of Abundance, and the Court of the Four Seasons. The nearby Festival Hall, seated 3,500+ for more than 2,000 concerts. In the center was the Tower of Jewels (at 435 feet height) and Great Arch. The arch was 110 feet high and 60 feet wide and had six murals painted by William de Leftwich Dodge that allegorically presented the history of the Panama Canal (Lipsky, 2005). On the tower, there were 102,000 tustrian cut-glass jewels, each backed by reflecting mirror, made by the Novagem Company. W. D'Arcy Ryan, on loan from General Electric, selected PPIE innovative illumination by indirect lighting. Also, at night located on a platform in San Francisco Bay, the Scintillator sent 48 beams of light in seven colors above the PPIE, projected through steam created by a stationary locomotive (Ewald, 1991). Figure 3 shows the PPIE fairgrounds map.

West of the core, countries built 21 international pavilions (such as France, Italy, Portugal, Denmark, China, Siam, Japan (Golden Temple), Canada, Guatemala, Argentina, …) with 31 foreign countries participating. U.S. states built exhibition buildings (such as California (with exhibits from 58 counties), Pennsylvania (with the Liberty Bell obtained via 250,000 school children signatures), Texas, Ohio, Hawaii, Washington, Oregon (with Parthenon design and 48 Douglas fir columns, each 5 to 6.5 feet in diameter and 42 feet tall, the maximum length that would fit on a railroad flatcar), …). Expected to sell many railroad excursions to the PPIE, railroads had their own buildings: the Southern Pacific, the Grand Trunk System, the Canadian Pacific, and the Great Northern Railroad. "THE GLOBE" had dioramas of scenes along the route of the Western Pacific, Denver & Rio Grande, and Missouri Pacific. From San Francisco to St. Louis, the scenic trip at the PPIE fair took three minutes.
Figure 4 is a 1915 photo from the San Francisco History Center, San Francisco Public Library, showing the PPIE fountains in the south gardens and the Tower of Jewels facing the Avenue of Palms.

THE FEATURED 1915 PPIE WORLD'S FAIR MODEL OF THE PANAMA CANAL

Taking twelve months to build, the greatest educational attraction at the exposition was a large scale, topographic model of the Panama Canal covering five+ acres (depicting 5,000 square miles of the canal) near the Fillmore entrance to the exposition. Figure 5 is a photo of the 1915 Panama Canal Building in The Zone-PPIE.

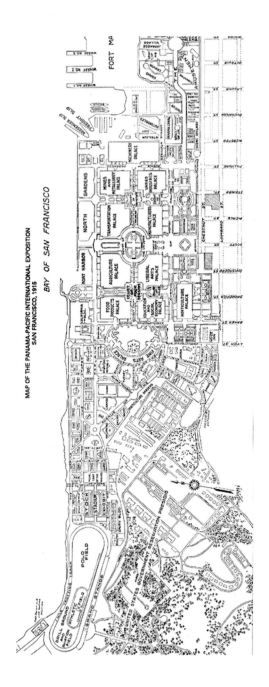

Figure 3. In the core, the Panama Canal building was on the right near the Machinery Palace and the Fort Mason entrance on the map of the Panama-Pacific International Exposition- San Francisco 1915 from the website of San Francisco Memories (http://sanfranciscomemories.com/ppie/map.html)

Figure 4. 1915 photo of the PPIE fountains in the south gardens and the Tower of Jewels facing the Avenue of Palms
(Photo ID No.: AAD-5040, PPIE Collection, San Francisco Public Library)

Figure 5. Photo of the 1915 Panama Canal Building in The Zone-PPIE. (Cardinell- Vincent Co. Photographers) (Photo ID No.: aaf-034, PPIE Collection: San Francisco History Center, San Francisco Public Library)

A novel moving platform (consisting of 144 cars, each ten feet long endlessly connected, totaling 1,440 feet - with seats priced at 50 cents each) carried 1,200 people around the exhibit with each seat having a duplex telephone receiver to transmit Panama Canal information by phonographs for the 23-minute ride. As each phonograph transmitted information to every third section of the platform, it took three sets or 45 phonograph records, and spectators heard 15 different records with 3,000 word talks (Panama Canal Exhibition Co., "The Panama Canal at San Francisco, 1915"). The miniature Miraflores and Pedro Miguel Locks worked with water flowing, and ships traveled back and forth, controlled by magnets, along with moving miniature electric Panama Railroad trains. After engineer L.E. Meyers visited the Panama Canal in 1911, the PPIE model was devised and patented over a two-year period by L.E. Meyers of Builders and Operators of Public Utilities, Chicago, IL. Figures 6 and 7 are photos of building the model of the Panama Canal for the PPIE.

The Panama Canal building and canal used more than 2,000,000 feet of lumber and 217 tons of cement and plaster. Panoramic painting was done on the walls of the nearby topography to depict an additional 4,000 square miles (Panama Canal Exhibition Co., "The Panama Canal at San Francisco," 1915). The Panama Canal model required 85 miles of copper wire, 104 motors, and seven different voltages from 2.4 to 10,000 volts, as well as alternating and direct current.

F. C. Boggs, Chief of the Washington Office of the Panama Canal, inspected and reported to The Panama Canal Exhibition Company on this Panama Canal reproduction on February 26, 1915: "This is to advise you that I have completed the checking and examination of your reproduction of the Panama Canal, as arranged with Colonel Geo. W. Goethals, and I find that it is so accurate that it will in half an hour import to anyone a more complete knowledge of the Canal than would a visit of several days to the waterway itself.... I congratulate you on the results you have achieved, and would recommend that everyone should see it."

Figure 8 is a photo of the Panama Canal Topographic Model at the 1915 San Francisco International Exposition.

Figure 9 is a photo of the Panama Canal Locks showing the details in the model at the 1915 PPIE.

LANDSCAPING AND SCULPTURE FOR THE 1915 PPIE WORLD'S FAIR

Landscape architect John McLaren, who had designed Golden Gate Park, selected elegant PPIE landscaping with many flowers, hedges and 30,000 sand cypress, acacia, spruce, eucalyptus and other trees. Chief of Sculpture, A. Stirling Calder, commissioned more than 1,500 sculptures in the fairgrounds.

ENGINEERING THE PANAMA CANAL 395

Figure 6

Figure 7

Figures 6 and 7. Photos showing the building the model of the Panama Canal. (Cardinell- Vincent Co. Photographers) (Photo ID Nos.: aaf-031 - September 25, 1914 and aaf-0032 - December 31, 1914, PPIE Collection: San Francisco History Center, San Francisco Public Library)

Figure 8. Photo of the Panama Canal Topographic Model (Charles Caldwell), 1868-1932, (UC-Berkeley, Bancroft Library), was published in the U.S. before 1923 and therefore is in the public domain in the U.S.

Figure 9. Photo of the Panama Canal Locks showing the details in the model (Photo ID No. aaf-0033, PPIE Collection: San Francisco History Center, San Francisco Public Library)

MAJOR PPIE WORLD'S FAIR ATTENDANCE DAYS

George Goethals had his own PPIE recognition on Goethals Day on September 7, 1915, and he returned for the Panama Canal International Engineering Congress, September 20-25, 1915. Figure 10 is a photo of Goethals being honored at this event.

Figure 10. Photo of Goethals being honored at "Goethals Day ceremony at the PPIE," "Fairs P.P.I.E. Days," ID: aad-6686 - C.C. Moore Photographer: was published in the US before 1923 and therefore is in the public domain in the US.

Some of the largest attendances were Opening Day on February 20, 1915 with a huge parade and 250,000 people; Dedication Day March 24 attended by U.S. Vice President Thomas Marshall; Independence Day July 4 with Speech by William Jennings Bryan; Roosevelt Day July 23 with Speech by Past President T. Roosevelt; Taft Day September 2 with Past President William Howard Taft; San Francisco Day including a Parade of Floats on November 2 with 348,472 (as the largest West Coast crowd ever to that time), and Closing Day December 4, 1915 brought the largest attendance (459,022 people) to the fair.

THE SAN FRANCISCO STREET RAILWAYS FOR THE FAIR

San Francisco's City Engineer M. M. O'Shaughnessy and consultant Bion J. Arnold improved the street railways to transport people to the world's fair by starting the Municipal Railway of San Francisco (Muni) on Dec. 28, 1912 to augment the private United Railways (URR) (Ule, 2005). By the fair opening, Muni had 200 street cars on ten lines and URR built a loop for the regular 19- Polk Street line. These street railways transported people to the PPIE, arriving on ferries at the Ferry Building, the Southern Pacific Railroad, or from other areas.

COMMEMORATIONS OF THE 1915 PPIE WORLD'S FAIR

Among the official PPIE souvenirs sold were paperweights filled with soil from the Culebra (Gaillard) Cut; plaques made from cocobolo wood used for railroad ties across the Isthmus of Panama; and each with a souvenir letter from the Governor of the Canal Zone George W. Goethals, former Chairman and Chief Engineer of the Isthmian Canal Commission. The PPIE promoted its world's fair extensively with press releases, preview booklets, post cards, coins, and a U.S. Postage Stamp. A "Memorial Certificate of Visitation" in Celebration of the Opening of the Panama Canal featured portraits of George Goethals, President Woodrow Wilson, and vignettes of the Palaces of Machinery and Fine Arts, cherubs, and signatures of government/fair officials (Lipsky, 2005).

OTHER WORLD'S FAIRS AND THE PANAMA CANAL FAIRS IN 1915-1916

As shown in Figure 11, a 1992 Smithsonian Exhibition covered World's Fairs from 1850 to 1940, with the 1915-1916 Panama Canal Fairs in San Francisco and a separate fair in San Diego Jan. 1- Dec. 31, 1916 as CASE 5. When San Diego did not obtain the official 1915 World's Fair bid, the city decided to hold its own San Diego Panama California International Exposition.

SUMMARY OF CONTRIBUTIONS OF KNOWLEDGE OF THE PANAMA CANAL BY THE 1915 PPIE FAIR, CONGRESS, MODEL AND PROCEEDINGS

Almost 19 million people visited the Panama-Pacific International Exposition world's fair in San Francisco between February 20 and December 4, 1915, including seeing and learning about the innovative Panama Canal topographic model of locks, dams, ships, ... The September 20-25, 1915 International Engineering Congress, led by George W. Goethals, Governor of the Canal Zone, had presentations on the history, design, construction, and cost of the Panama Canal. Over 1,000 pages of technical details by Panama Canal engineers, employees, or consultants were published in the *Transactions of the International Engineering Congress*, September 20-25, 1915, Neal Publishing, San Francisco, 1916. The civil engineering aspects and other information conveyed by the September 1915 Panama Canal Congress and 1916 Proceedings papers were most important educational presentations and a publication for people everywhere.

The 1915 PPIE world's fair was the first fair to receive a transcontinental telephone call; the first to exhibit a periscope in the U.S.; the first exposition to exhibit a million-volt electric transformer; the first to demonstrate steam pyrotechnics; the first fair to use indirect lighting; the first to use colored lighting; the first to use trackless streetcars; the first exposition to have a working automobile assembly plant on site (with 4,338 cars built and sold); and, the first to offer airplane rides to the public (Lipsky, 2005).

With all the exposition innovations and successful planning, careful construction on budget and on time, large attendance, and information on the Panama Canal, the Panama Pacific International Exposition was one of the best World's Fairs!

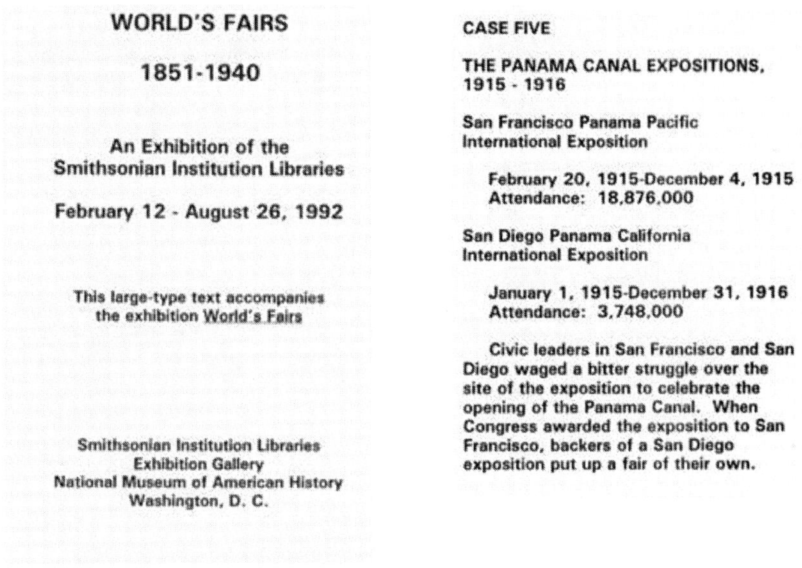

WORLD'S FAIRS

1851-1940

An Exhibition of the
Smithsonian Institution Libraries

February 12 - August 26, 1992

This large-type text accompanies
the exhibition World's Fairs

Smithsonian Institution Libraries
Exhibition Gallery
National Museum of American History
Washington, D. C.

CASE FIVE

**THE PANAMA CANAL EXPOSITIONS,
1915 - 1916**

San Francisco Panama Pacific
International Exposition

February 20, 1915-December 4, 1915
Attendance: 18,876,000

San Diego Panama California
International Exposition

January 1, 1915-December 31, 1916
Attendance: 3,748,000

Civic leaders in San Francisco and San Diego waged a bitter struggle over the site of the exposition to celebrate the opening of the Panama Canal. When Congress awarded the exposition to San Francisco, backers of a San Diego exposition put up a fair of their own.

Figure 11: Smithsonian Institution Exhibition February 12- August 26, 1992 on World's Fairs 1951- 1940 and the Panama Canal Expositions 1915-1916 in San Francisco and San Diego

REFERENCES

There have been many publications summarizing the 1915 Panama-Pacific International Exposition in San Francisco. Only a few selected PPIE publications, websites, and references are listed here.

Avery, Ralph Emery, (Regan) 1913. *Picturesque Panama and the Great Canal: The Eighth Wonder of the World. A Panorama of the Canal Zone, with the Complete Story of the Building and Operation of the Great Canal under the Supervision of Colonel George W. Goethals*, (Google ebook) (384 pp.).

Ewald, Donna and Peter Clute, 1991, *San Francisco Invites the World: The Panama-Pacific International Exposition of 1915*, Chronicle Books, San Francisco, 128 pp.

Lipsky, William, 2005, *San Francisco's Panama- Pacific International Exposition*, Arcadia Publishing, Charleston, 128 pp.

Lutton, Richard J., April 1975. Figure 1, Technical Report S-70-9 "Study of Clay Shale Slopes along the Panama Canal, Report 2: History, Geology, and Mechanics of Development of Slides in Gaillard Cut, Volume I: Text." Soils and Pavements Laboratory, U.S. Army Engineer Waterways Experiment Station, Vicksburg, MS. Re-drawn by Panama Canal Authority in 2014.

Map of the Panama Pacific International Exposition- San Francisco 1915, from San Francisco Memories website: http://sanfranciscomemories.com/ppie/map.html

Moore, Charles C., 1868-1932, "Albums of the Panama Pacific International Exposition" 190 Photos (Charles Caldwell), (UC-Berkeley, Bancroft Library) (in the public domain since published before 1923).

"Panama–Pacific International Exposition," Wikipedia website.

Panama Canal Exhibition Co., "The Panama Canal at San Francisco," 1915, 9pp.

San Francisco Public Library Website PPIE Photo Collection, San Francisco History Center (photo ID. No.: AAD-5040, aaf-031, aaf-032, aaf-033, aaf-034.

Transactions of the International Engineering Congress, September 20-25, 1915, San Francisco, Neal Publishing, San Francisco, 1916; vol. 1: papers 1-13, 527 pp.; vol. 2: papers 14-25, 483 pp. The Library of Congress has 12 volumes: [v. 1] The Panama canal 2 v. - -[v. 2] Waterways and irrigation- -[v. 3] Municipal engineering- -[v. 4] Railway engineering- -[v. 5] Materials of engineering construction- -[v. 6] Mechanical engineering- -[v. 7] Electrical engineering and hydroelectric power development- -[v.8] Mining engineering- -[v. 9] Metallurgy- -[v. 10] Naval architecture and marine engineering- -[v. 11]Miscellany- -[v. 12] Index volume, LC control no. 16024238, LC classification TA5 16 1915, Published San Francisco, Cal [1916].

Ule, Grant, January 10, 2005. "Fair, Please:" Streetcars to the 1915 Panama- Pacific Exposition, 6 pp., http://www.streetcar.org/blog/2005/01/fair-please-streetcars-to-the.html